Student Study C

Fundamentals of Mathematics
SIXTH EDITION

William M. Setek, Jr.
Monroe Community College

Prepared by
Michael A. Gallo
Florida Institute of Technology

Jane A. Edgar
Brevard Community College

Macmillan Publishing Company
New York

Collier Macmillan Canada, Inc.
Toronto

Maxwell Macmillan International Publishing Group
New York   Oxford   Singapore   Sydney

Copyright © 1992 by Macmillan Publishing Company, a division of Macmillan, Inc.
Printed in the United States of America

All rights reserved. No part of this book may be reproduced or
transmitted in any form or by any means, electronic or mechanical,
including photocopying, recording, or any information storage and
retrieval system, without permission in writing from the publisher.

Earlier editions copyright © 1989, 1986 by Macmillan Publishing Company and
copyright © 1983, 1979 and 1976 by William M. Setek, Jr.

Macmillan Publishing Company
866 Third Avenue, New York, New York 10022

Macmillan Publishing Company is part
of the Maxwell Communication Group of Companies.

Maxwell Macmillan Canada, Inc.
1200 Eglinton Avenue East
Suite 200
Don Mills, Ontario M3C 3N1

ISBN 0-02-3403438

Printing: 1 2 3 4 5 6 7 8   Year: 2 3 4 5 6 7 8 9 0 1

PREFACE

This study guide is a supplement to the textbook, *Fundamentals of Mathematics*, sixth edition, by Professor William M. Setek, Jr. The material in this publication represents a streamlined version of the material in the textbook, and provides added depth and insight to important topics and concepts.

The format of the study guide is consistently maintained throughout. Sections begin with a formal review of the corresponding material from the textbook. This review is then followed by a set of solved examples. Supplementary exercise sets are given at the end of the sections, and chapter tests are given at the end of the chapters. Answers to all supplementary exercises and chapter tests are provided at the back of the study guide. Three appendixes also are included. Appendix A contains a list of notation and symbols used in both the study guide and textbook; Appendix B contains eight tables of information used in various chapters of the study guide and textbook; and Appendix C provides a summary of important geometric facts and formulas.

In order to benefit from the material presented here, we suggest that the study guide be used *with* the textbook and not in place of it. To encourage such use, odd-numbered exercises from the textbook are used as examples. The examples are referenced to the textbook by the following heading:

Example x (page y, #z)

where x = the example number in the study guide;
y = the page number in the textbook where the example is located;
z = the actual textbook exercise number

Each example used in this guide represents a group of textbook exercises that are similar in nature and involve similar solutions. If you understand the method of solution of an example, you then should be able to transfer this understanding to help you solve the remaining exercises of that group. Complete, detailed solutions are given for all examples. (In this context the study guide serves as a solutions manual for selected odd-numbered exercises.)

The Florida CLAST Supplement

Supplementing the material in the textbook, a separate section relating to the Florida College Level Academic Skills Test (CLAST) is included in the study guide. This material is a thorough and comprehensive review of the mathematics subtest of the CLAST, and was written by Dr. Eunice Everett of Seminole Community College, Sanford, Florida. The CLAST material contains

- a general introduction to the mathematics subtest of the CLAST (Revision, Fall 1990);

- Suggestions for preparing and taking the mathematics subtest of the CLAST;

- a formal presentation of the current skills, by subject area; (The discussion of these skills consists of a listing of the skill itself; information about how the skill might be tested; and various examples, with solutions, that demonstrate the type of questions that could be asked relating to the skill.) and

- two sample CLASTs with answers for students who are preparing to take the mathematics subtest.

Acknowledgements

Many individuals provided support to this project. The following people are hereby acknowledged (and their names immortalized in print):

- Dee Dee Pannell, Director of Academic and Research Computing Services at Florida Institute of Technology — You are terrific Dee Dee; we could not have completed this project without your support.
- Eunice Everett, Professor of Mathematics at Seminole Community College
- William Setek, Professor of Mathematics at Monroe Community College
- Robert Pirtle, Executive Editor at Macmillan Publishing Company
- Florence Whittaker, who provided us with a list of errors from the first edition
- Colleagues at Florida Institute of Technology, in particular Dan Stack and Jim Wells

A special noteworthy recognition is extended to Julia, who became part of our family on January 2, 1992.

Note: Please report any errors found in this study guide, or suggestions for improvement to:

> Michael A. Gallo
> Academic and Research Computing Services
> Florida Institute of Technology
> 150 West University Boulevard
> Melbourne, FL 32901
> Internet: gallo@zeno.fit.edu

Contents

1 Sets 1

1.1	Introduction	1
1.2	Notation and Decsription	1
1.3	Subsets	6
1.4	Set Operations	9
1.5	Pictures of Sets (Venn Diagrams)	12
1.6	An Application of Sets and Venn Diagrams	18
1.7	Cartesian Products	23
1.8	Chapter 1 Test	26

2 Logic 27

2.1	Introduction	27
2.2	Statements and Symbols	27
2.3	Dominance of Connectives	32
2.4	Truth Tables	36
2.5	More Truth Tables—Conditional and Biconditional Statements	36
2.6	DeMorgan's Law and Equivalent Statements	44
2.7	The Conditional (Optional)	46
2.8	Valid Arguments	48
2.9	Picturing Statements with Venn Diagrams (Optional)	52
2.10	Valid Arguments and Venn Diagrams (Optional)	55
2.11	Switching Networks	61
2.12	Chapter 2 Test	65

3 Probability 67

3.1	Introduction	67
3.2	Definition of Probability	67
3.3	Sample Spaces	70
3.4	Tree Diagrams	70
3.5	Odds and Expectations	73
3.6	Compound Probability	77
3.7	Counting, Ordered Arrangements, and Permutations (Optional)	82
3.8	Combinations (Optional)	86
3.9	More Probability (Optional)	87
3.10	Chapter 3 Test	90

4 Statistics 91

4.1	Introduction	91
4.2	Measures of Central Tendency	91
4.3	Measures of Dispersion	93
4.4	Measures of Position (Percentiles)	96
4.5	Pictures of Data	98
4.6	The Normal Curve	106
4.7	Chapter 4 Test	111

5 An Introduction to the Metric System 113

5.1	Introduction	113
5.2	History of Systems of Measurement	113
5.3	Length and Area	116
5.4	Volume	119
5.5	Mass (Weight)	122
5.6	Temperature	124
5.7	Miscellany	126
5.8	Chapter 5 Test	127

6 Mathematical Systems 128

6.1	Introduction	128
6.2	Clock Arithmetic	128
6.3	More New Systems	133
6.4	Modular Systems	135
6.5	Mathematical Systems Without Numbers	138
6.6	Axiomatic Systems	140
6.7	Chapter 6 Test	142

7 Systems of Numeration 144

7.1	Introduction	144
7.2	Simple Grouping Systems	144
7.3	Multiplicative Grouping Systems	147
7.4	Place-Value Systems	150
7.5	Numeration in Bases Other Than 10	154
7.6	Base 5 Arithmetic	156
7.7	Binary Notation and Other Bases	159
7.8	Chapter 7 Test	162

8 Sets of Numbers and Their Structure 163

8.1	Introduction	163
8.2	Natural Numbers—Primes and Composites	163
8.3	Greatest Common Divisor and Least Common Multiple	167
8.4	Integers	171
8.5	Rational Numbers	175
8.6	Rational Numbers and Decimals	180
8.7	Irrational Numbers and the Set of Real Numbers	183
8.8	Scientific Notation (Optional)	186
8.9	Chapter 8 Test	188

9 An Introduction to Algebra 190

9.1	Introduction	190
9.2	Open Sentences and Their Graphs	190
9.3	Algebraic Notation	192
9.4	More Open Sentences	195
9.5	Problem Solving	198
9.6	Linear Equations in Two Variables	203
9.7	Graphing Equations	204
9.8	The Slope of a Line (Optional)	207
9.9	The Equation of a Straight Line (Optional)	210
9.10	Graphing $y = ax^2 + bx + c$	215
9.11	Inequalities in Two Variables	218
9.12	Linear Programming	220
9.13	Quadratic Equations (Optional)	223
9.14	Chapter 9 Test	226

10 An Introduction to Geometry 228

10.1	Introduction	228
10.2	Points and Lines	228
10.3	Planes	234
10.4	Angles	237
10.5	Polygons	244
10.6	Perimeter and Area	250
10.7	Solids	255
10.8	Congruent and Similar Triangles	260
10.9	Networks	264
10.10	Chapter 10 Test	266

11 Consumer Mathematics 269

11.1	Introduction	269
11.2	Ratio and Proportion	269
11.3	Percents, Decimals, and Fractions	272
11.4	Markups and Markdowns	276
11.5	Simple Interest	281
11.6	Compound Interest	283
11.7	Effective Rate of Interest	286
11.8	Life Insurance	287
11.9	Installment Plans and Mortgages	290
11.10	Chapter 11 Test	293

12 An Introduction to Computers 295

12.1	Introduction	295
12.2	History of Computers	295
12.3	How a Computer System Works	295
12.4	Using BASIC	297
12.5	More BASIC Statements	300
12.6	Chapter 12 Test	303

Florida CLAST Review 305

1	Introduction	307
2	Suggestions for Preparing and Taking the CLAST	310
3	Arithmetic	313
4	Geometry	332
5	Algebra	359
6	Probability and Statistics	385
7	Logic and Sets	405
8	Practice CLAST: Form A	423
9	Practice CLAST: Form B	432
10	Practice CLAST Answers	441

Appendix		443
A	Key to Notation and Symbols	444
B	Tables	445
C	Geometry Summary	450
Answers		454

1. SETS

1.1 INTRODUCTION

In this chapter, we discuss the concept of sets. Sets, or more specifically, **set theory**, is a branch of mathematics that enables us to describe mathematics in a clear and accurate manner. It is important that students of mathematics be familiar with the notation and language of sets.

1.2 NOTATION and DESCRIPTION

1.2.1 Definition of a Set

A set can be thought of as a collection of objects. Every object that belongs to a set is called a **member** or an **element** of the set. We shall define a set to be a well-defined collection of objects such that we will always be able to determine if an object is or is not a member of a set. For example, the collection of all teachers in your school forms a set, and a mathematics teacher from the school would be a member of this set. The secretary of a department within the school, however, would not be a member of this set.

1.2.2 The Roster Form Representation of a Set

There are many ways in which we can specify (i.e., describe) a set. One way is to provide a word description of a set. For example,

the set whose elements are the months of the year that begin with the letter J

is a word description of a particular set. Sometimes it can be cumbersome to write a word description of a set. In such cases a more practical way to specify a set would be to list all of its elements. Such a method is referred to as the **roster form**, and conforms to a particular design. To specify a set by the roster form we enclose the name(s) of the set's element(s) between braces, { }, and separate each element's name by a comma. Using this notation for the set whose elements are the months of the year that begin with the letter J, we have

{January, June, July}

When specifying a set by roster form, the order in which the elements are listed is not important. As a result all of the following are roster form representations of the set of the months of the year that begin with the letter J.

{January, June, July} {January, July, June}
{June, January, July} {June, July, January}
{July, January, June} {July, June, January}

1.2.3 Symbolic Representation of Sets and Members

Sets are usually denoted (or named) by capital letters such as A, B, or C. Therefore, using the example above if we write

$A = \{$January, June, July$\}$

then set A is the set whose elements are the months January, June, and July.

The symbols, $\in$ and $\notin$, are used to indicate whether or not an element is a member of a set. The symbol $\in$ is read *is an element of* or *is a member of*, and the symbol, $\notin$, is read *is not an element of* or *is not a member of*.

2 STUDENT STUDY GUIDE

Using the example $A = \{\text{January, June, July}\}$, we would write

$$\text{June} \in A \quad (\textit{June is an element of set A})$$

$$\text{or March} \notin A \quad (\textit{March is not an element of set A})$$

1.2.4 The Use of an Ellipsis

It is sometimes impractical to list all of the elements of a set. For example, in mathematics the set of natural numbers from 1 to 500 consists of five hundred elements. To specify this set by roster, although possible, would be most inconvenient. To remedy this problem we utilize an abbreviated roster form to list these elements. This form employs the use of three consecutive dots ... called an **ellipsis** to indicate that a previously established pattern is to continue. For example, if B were the set of natural numbers from 1 to 500, then we would represent B as

$$B = \{1, 2, 3, \ldots, 498, 499, 500\}$$

The use of an ellipsis indicates that there are elements in the set that follow the indicated, established pattern, but for convenience have not been written. In the illustration above, $19 \in B$, $256 \in B$, and $402 \in B$. However, $0 \notin B$, $0.555 \notin B$, and $502 \notin B$.

1.2.5 Set-Builder Notation

An alternative method to the roster form for specifying a set is the **rule method**, or **set-builder notation**. Rather than list all of the elements of a set, we specify a set by writing within braces a formal description that describes or identifies the members of a set. As part of this formal description we use a variable (such as x) to represent any one of the members of the set. Set-builder notation uses the following convention:

$$\text{Name of Set} = \{x \mid x \text{ is a } \textit{word description of the set}\}$$

As a result, the following statement is used to represent the set of months of the year that begin with the letter J:

$$A = \{x \mid x \text{ is a month of the year that begins with the letter J}\}$$

This statement is read

Set A is equal to all x such that x is a month of the year that begins with the letter J

Note that the vertical bar within the braces is read *such that*, and the phrase used to describe the set denotes what the variable x represents.

To distinguish among the three methods for specifying a set, consider the following illustration. (All are equivalent.)

- **Word Description**

  The set whose elements are the days of the week

- **Roster Form**

  $A = \{\textit{Sunday, Monday, Tuesday, Wednesday, Thursday, Friday, Saturday}\}$

- **Set-Builder Notation**

  $A = \{x \mid x \text{ is a day of the week}\}$

1.2.6 Finite Sets

A **finite set** is a set that contains a definite number of elements. The following are examples of finite sets.

$$A = \{1,2,3,4\}$$
$$B = \{1,2,3,...,48,49,50\}$$
$$C = \{1,3,5,...,99\}$$

Notice that in each set listed above it is possible to count the number of elements in the set. There also is an end to the listing of the elements of the set. These are the characteristics of a finite set.

1.2.7 Infinite Sets

Unlike finite sets, **infinite sets** have an unlimited number of elements. The following are examples of infinite sets.

$$A = \{1,3,5,7,9,...\}$$
$$B = \{2,4,6,8,...\}$$
$$C = \{1, \frac{1}{2}, \frac{1}{4}, \frac{1}{8}, \frac{1}{16},...\}$$

Notice that in each set listed above an ellipsis is used to indicate that the established pattern will continue indefinitely. Listing of the elements of an infinite set is an endless process. Consequently it is not possible to count the total number of elements of the set.

1.2.8 Empty Sets

It is also possible for a set to contain zero elements (i.e., no elements). Such a set is referred to as an **empty set** or **null set**. The empty set is designated by placing nothing between braces, { }, or by the symbol $\emptyset$. We must exercise caution not to write $\{\emptyset\}$ to indicate the empty set. This notation does not represent the empty set because there is an object contained within the braces. The symbol $\{\emptyset\}$ represents the set with the element $\emptyset$.

1.2.9 Equal Sets and Equivalent Sets

If two sets contain exactly the same elements we say that the two sets are equal. For example, if $A = \{1,3,5\}$ and $B = \{3,1,5\}$, then set A is equal to set B, denoted $A = B$. If, however, $Q = \{2,4,6\}$ and $R = \{2,4\}$, then set Q and set R are not equal to each other, denoted $Q \neq R$.

Consider now the two sets $A = \{1,3,5\}$ and $B = \{a,b,c\}$. Clearly, $A \neq B$. Notice, though, that both A and B have exactly the same *number* of elements; each set contains three elements. Two sets such as A and B that contain exactly the same number of elements are said to be **equivalent sets**. Observe that if two sets are equal, then they are also equivalent. However, if two sets are equivalent, they are not necessarily equal.

Example 1 (page 7, #1a-f)

Are the following statements true or false?

Solution

a. $2 \in \{1,2,3,4\}$

True; 2 is an element of the set that contains the elements 1, 2, 3, and 4.

b. $8 \in \{2,4,6,...\}$

True; although 8 is not specifically written as a member, the pattern established indicates that 8 is indeed a member.

Continued on next page ...

c. {m,o,r,e} = {r,o,m,e}

True; since both sets contain the same elements, they are considered to be equal, regardless of the order in which they are listed.

d. {1,2,3} is equivalent to {4,5,6}

True; although these two sets are not equal (their members are not the same), they are equivalent since they both contain the same number of elements, namely three.

e. $12 \in \{x \mid x \text{ is a counting number}\}$

True; the word description of this set indicates that its members are counting numbers; twelve is a counting number.

f. $0 \in \{\ \}$ False; the empty set contains no elements. Thus the number zero cannot be an element of this set.

Example 2 (page 7, #3a-f)

Are the following statements true or false?

Solution

a. $\{\emptyset\}$ is a finite set.

True; we can both count the number of elements in this set (one) and there is an end to the listing.

b. $\{\emptyset\}$ is an empty set.

False; an empty set does not contain any elements; this set has one element.

c. {0} is an empty set.

False; same reason as (b) above.

d. The set of students enrolled in this course is a finite set.

True; same reason as (a) above.

e. The set of big cars manufactured in the United States is a well-defined set.

False; the word *big* does not have the same meaning to everyone.

f. Equivalent sets are equal sets.

False; an equivalent set merely needs to have the same number of elements. An equal set, however, must have exactly the same members.

Example 3 (page 7, #5a)

List the elements of the given set in roster form.

Solution

a. The set of Great Lakes

There are 5 Great Lakes: {Huron, Ontario, Michigan, Erie, Superior}

Example 4 (page 8, #7a,d,e)

List the elements of each set in roster form.

Solution

a. $\{x \mid x$ is a letter in the English alphabet$\}$

The word description of this set indicates that its members are the 26 letters of the English alphabet. Thus the roster form representation is

$$\{a,b,c,...,z\} \text{ or } \{A,B,C,...,Z\}$$

Notice the use of the ellipsis to abbreviate the listing. Notice further that this set is not well-defined since there are two possible answers: the set with lowercase letters and the set with uppercase letters.

d. $\{x \mid x + 3 = 5\}$

The phrase used to describe this set is the mathematical equation, $x + 3 = 5$. Solving this equation for x we get an answer of 2. Thus the roster form representation is $\{2\}$.

e. $\{x \mid x + 1 = x\}$

The mathematical equation, $x + 1 = x$, has no solution. That is, there are no values that can be substituted for x to make the equation a true sentence. Thus the answer is the empty set, $\varnothing$.

Example 5 (page 8, #9b)

Write the given set in set-builder notation.

Solution

b. $\{a,e,i,o,u\}$

The elements listed represent the vowels in the English alphabet. Therefore the set-builder notation for this set is

$$\{x \mid x \text{ is a vowel in the English alphabet}\}$$

SUPPLEMENTARY EXERCISE 1.2

In 1 through 8 list the elements of each set in roster form.

1. The set of letters in the work *musk*.
2. The set of days in the week that begin with the letter *T*.
3. The set whose elements are the digits of the number *two thousand, three hundred fifty-eight*.
4. The set whose elements are the odd numbers that can divide evenly into the number 18.
5. $\{x \mid x$ is a month of the year with exactly 30 days$\}$
6. $\{x \mid x$ is the name of all the *bases* on a baseball diamond$\}$
7. $\{x \mid x$ is a (single) digit in the decimal number system$\}$
8. $\{x \mid x + 0 = 2\}$

Continued on next page ...

In 9 through 11 specify the given set using set-builder notation.

9. {5,10,15,20,...,90,95,100}
10. {Sunday, Saturday}
11. {*reading, 'riting, 'rithmetic*}

In 12 through 25 answer true or false.

12. $6 \in \{1,2,3,...,99,100\}$
13. $6 \in \{1,1,2,3,5,...\}$
14. $! \in \{x \mid x$ is a punctuation mark$\}$
15. $3 \notin \{345\}$
16. $L \notin \{a,b,c,...,x,y,z\}$
17. $21 \in \{x \mid x$ is divisible by 3$\}$
18. The set of female U.S. presidents is well-defined.
19. The set of two-digit numbers greater that 100 is finite.
20. The set of the number of miles of highway in the United States is infinite.
21. {M,I,C,R,O,S} = {S,O,R,C,I,M}
22. {125} ≠ {5,2,1}
23. {1,3,9} is equivalent to {139}
24. {∅} is equivalent to {0}
25. {0,2,4,6,...,96,98,100} is equivalent to {1,3,5,...,97,99}

1.3 SUBSETS

In our work with sets we frequently encounter two sets that are neither equal nor equivalent, but nevertheless contain common elements. For example, in the two sets $A = \{1,3,5\}$ and $B = \{1,2,3,4,5\}$ note that every element in set A, namely 1,3, and 5, is also contained in set B. Notice further that the two sets are not equal. In cases such as this we say that set A is a **proper subset** of set B, and denote this as $A \subset B$.

Definition of Proper Subset

> *If every element in set A is also an element in set B,*
> and $A \neq B$, then A is a proper subset of B.

If we remove the restriction, $A \neq B$, from the definition of a proper subset, we have the definition of a **subset**.

Definition of Subset

> *If every element in set A is contained in set B*, then A is a subset of B, denoted $A \subseteq B$.

This second definition implies that every set is a subset of itself. This is why it is necessary to make a distinction between a subset and a proper subset. The notation used to represent subsets ($\subseteq$) and proper subsets ($\subset$) should serve as an aid in distinguishing between these two concepts. If one set is a subset of another set, then the elements of the first set must be contained in the second set, or the two sets must be equal. Note that the symbol $\subseteq$ suggests $\subset$ or =. $A \subseteq B$ means A is contained in B ($A \subset B$), or A is equal to B ($A = B$). On the other hand, a subset of a given set that is not the set itself, is a proper subset. The absence of the underline in the proper subset symbol ($\subset$) can be thought of as an indicator of this characteristic. Thus $A \subset B$ means the elements of A are contained in B and the two sets are not equal.

1.3.1 Notes on Subsets

1. When using the symbol for subsets ($\subseteq$ or $\subset$), the item on either side of the symbol must be a set. This should not be confused with the symbol used to show set membership ($\in$).

2. The number of subsets contained within a given set is equal to the power 2^n, where n is equal to the number of elements in the set.

3. The empty set is a subset of every set.

1.3.2 Universal and Complement Sets

Whenever we consider problems dealing with sets we frequently employ the use of a **universal set**. This universal set, denoted by U, is a general set and contains all of the elements being discussed in a particular problem. As a result, every set that is being considered in a problem is a subset of the universal set. The selection of a universal set is usually determined by the nature of the discussion and can change from problem to problem.

If we are given a universal set and a subset of the universal set, then the subset will have a **complement set**. This complement set consists of elements that are members of the universal set excluding those elements that are in the given subset. For example, if $U = \{1,2,3,4,5,6,7,8,9,10\}$ and $A = \{3,4,5,9\}$, then the complement of A, denoted A', contains the elements 1,2,6,7,8, and 10. A visual method that is used to obtain the complement of a given set is to cross out the elements of the given set from the universal set. The complement set then consists of the remaining elements in the universal set. This technique is illustrated below.

$$\text{If } U = \{1,2,3,4,5,6,7,8,9,10\} \text{ and } A = \{3,4,5,9\}$$

$$\text{then } A' = \{1,2,\cancel{3},\cancel{4},\cancel{5},6,7,8,\cancel{9},10\}$$

$$\text{or } A' = \{1,2,6,7,8,10\}$$

With practice, though, a set's complement can be determined mentally.

Example 1 (page 12, #1a,b,c,e)

Tell whether each statement is true or false.

Solution

a. $\{1,2\} \subset \{1,2,3,4,5\}$ *proper subset*

   True; the symbol $\subset$ means that every element in the first set must be contained in the second set, and the two sets cannot be equal. This is exactly the situation in this problem.

b. $0 \subset \emptyset$

   False; the symbol $\subset$ indicates a specific relationship between two sets. Zero is not a set.

c. $\{2,4,6\} \subset \{6,4,2\}$ *subset $\subseteq$*

   False; although every element of the first set is also a member of the second set, the two sets are also equal. Thus the first set is a subset of the second set, not a proper subset. (This statement would be true if the symbol $\subseteq$ were used.)

e. $\{0\} \subset \emptyset$

   False; all of the elements of the first set, namely the element 0, are not contained in the second set. (The empty set has no elements.)

8 STUDENT STUDY GUIDE

Example 2 (page 12, #3c-f)

Tell whether each statement is true or false.

Solution

c. $\{2\} \in \{2,4,6\}$

False; although the element 2 of the first set is contained in the second set, the symbol $\in$ does not represent this relationship.

d. $\{d\} \subset \{a,b,c,...,z\}$

True; although the element d is not shown in the second set it is considered to be a part of the set by the use of the ellipsis. Thus the single element of the first set is contained in the second set, and the two sets are not equal.

e. $\{2\} \subseteq \{2,4,6\}$

True; the symbol $\subseteq$ means the elements of the first set, namely 2, must be contained in (or equal to the elements of) the second set. This is indeed the case.

f. $\{B\} \subset \{a,b,c\}$

False; the uppercase letter B, which is the element of the first set, is not contained in the second set.

Example 3 (page 12, #5a,b,d)

List all possible subsets of each set.

(**Note:** When asked to list all possible sets of a given set, first determine the number of subsets the given set has by evaluating the power, 2^n, (where n is the number of elements in the given set), and then arrange the listing in the order of zero elements at a time (i.e., the empty set), one element at a time, two elements at a time, three elements at a time, ..., and so forth.)

Solution

a. $\{10,4\}$

Since this set contains two elements, there will be 2^2 or 4 subsets. They are:

- Zero at a time: $\{\ \}$
- One at a time: $\{10\}$, $\{4\}$
- Two at a time: $\{10, 4\}$

b. $\{m,a,t,h\}$

This set consists of four elements. Therefore there will be 2^4 or 16 subsets. They are:

- Zero at a time: $\{\ \}$
- One at a time: $\{m\}$, $\{a\}$, $\{t\}$, $\{h\}$
- Two at a time: $\{m,a\}$, $\{m,t\}$, $\{m,h\}$, $\{a,t\}$, $\{a,h\}$, $\{t,h\}$
- Three at a time: $\{m,a,t\}$, $\{m,a,h\}$, $\{m,t,h\}$, $\{a,t,h\}$
- Four at a time: $\{m,a,t,h\}$

d. $\emptyset$

Since the empty set contains zero elements it will have 2^0 or 1 subset, namely itself.

Example 4 (page 12, #7a,e)

If $U = \{m,e,t,r,i,c\}$, evaluate the given sets.

Solution

a. $\{m,e,t\}'$

The complement of a set is the set of all elements found in the universal set that are not contained in the given set. Thus the complement of the set $\{m,e,t\}$ is $\{r,i,c\}$.

e. $\{\ \}'$

The complement of the empty set is the universal set. Thus the answer here is $\{m,e,t,r,i,c\}$.

SUPPLEMENTARY EXERCISE SET 1.3

Given $A = \{1,3,5\}$, $B = \{2,3,4\}$, and $B = \{1,2,3,4,5\}$. Answer true or false in problems 1 through 10.

1. $A \subset B$
2. $A \subseteq A$
3. $B \not\subset A$
4. $A \subset B$
5. $B = B$
6. $\emptyset \subseteq B$
7. $B \subset B$
8. $A \not\subset B$
9. $A \subset \emptyset$
10. $B \subset B$

In 11 through 20, answer true or false.

11. $\emptyset \subset \{a\}$
12. $\{a\} \subseteq \{a\}$
13. $1 = \{1\}$
14. $\{0\} \subseteq \{\emptyset\}$
15. $8 \notin \{1,2,3\}$
16. $\{8\} \in \{1,2,3,...,10\}$
17. $\{5\} \subseteq \{x \mid x \text{ is a prime number}\}$
18. $3 \subset \{x \mid x \text{ is a single digit number}\}$
19. $X \subseteq X$
20. $4 \in (1,2,3,4)$

Given $U = \{1,2,3,4,5,6,7,8,9,10\}$, find the complement of the sets in problems 21 through 24.

21. $\{1,3,5,7,9\}$
22. $\{2,4,6,8,10\}$
23. $\{4,5,6,...,9\}$
24. $\{1,2,3,...,7\}$
25. List all subsets of $\{2,8,9\}$

1.4 SET OPERATIONS

1.4.1 Intersection

Given two sets A and B, the **intersection** of A and B, denoted $A \cap B$, is a new set that contains all the elements common to both A and B. If $A = \{1,2,3,4,5\}$ and $B = \{2,4,6,8,10\}$, then $A \cap B = \{2,4\}$ since the elements 2 and 4 are the only elements contained in both A and B. If two sets were to have no common elements, that is their intersection is the empty set, then we say that the two sets are **disjoint**.

1.4.2 Union

Given two sets A and B, the **union** of A and B, denoted $A \cup B$, is a new set that contains all the elements in either A or B, or both A and B. If $A = \{1,2,3,4,5\}$ and $B = \{2,4,6,8,10\}$, then $A \cup B = \{1,2,3,4,5,6,8,10\}$.

1.4.3 Notes on Set Operations

1. The operations of union and intersection involve two sets and hence are referred to as *binary operations*.
2. The operation of complement (section 1.3) involves only one set. Thus it is referred to as a *unary operation*.
3. Frequently, an expression involving sets may employ several operations. (See example 2, which follows this discussion.) In such instances we must follow a prescribed order of operations. This order is given here.

Working from left to right:

1. Evaluate expressions within parentheses.
2. Evaluate all complements.
3. Evaluate all unions and intersections (as they occur from left to right).
   In the event of nested grouping symbols, begin with the innermost and work towards the outermost.

As an illustration of this order of operations consider the following three evaluations.

- Evaluate $A \cap (B \cup C)$

    [1] First evaluate $B \cup C$

    [2] Now evaluate the result of [1] with set A

- Evaluate $A' \cap B' \cup C$

    [1] First evaluate A'

    [2] Next evaluate B'

    [3] Now find the intersection of A' and B'

    [4] Finally, find the union of set C with the result of [3]

- Evaluate $A \cup (B \cap C')'$

    [1] First evaluate C'

    [2] Next find the intersection of set B with the result of [1]

    [3] Now find the complement of the result of [2]

    [4] Finally, find the union of set A with the result from [3]

Example 1 (page 15, #1a,e,f)

For each pair of sets find $A \cap B$ and $A \cup B$.

Solution

a. $A = \{2,4,6,8\}$ and $B = \{1,3,4,6,7\}$

   <u>Intersection</u>

   The elements common to both A and B are 4 and 6. Thus $A \cap B = \{4,6\}$.

Continued on next page ...

Union

The union of the two sets will contain a listing of all the elements from both sets with those elements that are common to both sets being listed once. Thus, $A \cup B = \{1,2,3,4,6,7,8\}$.

e. $A = \{5,10,15,...\}$ and $B = \{10,20,30,...\}$

Intersection

First note that B is a proper subset of A. Therefore those elements that are common to both sets are all contained in B. As a result, $A \cap B = \{10,20,30,...\}$ or more simply, B.

Union

Since $B \subset A$, all the elements in A or B, or both A and B, are contained in set A.
Hence, $A \cup B = \{5,10,15,...\}$ or more simply, A.

f. $A = \{1,3,5,7,...\}$ and $B = \{2,4,6,8,...\}$

Intersection

Set A consists of the odd counting numbers and set B consists of the even counting numbers. Clearly, they are *disjoint* sets and $A \cap B = \emptyset$.

Union

The union of sets A and B will be the set of counting numbers. Thus, $A \cup B = \{1,2,3,...\}$.

Example 2 (page 16, #3c,e)

Given the sets $U = \{1,2,3,4,5,6,7\}$, $A = \{2,4,6,7\}$ and $B = \{1,3,5,6,7\}$, evaluate the given sets.

Solution

c. $A' \cap B'$

- Find A' and B'

$$A = \{2,4,6,7\} \rightarrow A' = \{1,3,5\}$$
$$B = \{1,3,5,6,7\} \rightarrow B' = \{2,4\}$$

- Now find the intersection of A' and B'

$$A' \cap B' = \{\ \}$$

The answer is the empty set, $\emptyset$.

e. $(A \cap B)'$

- First evaluate $A \cap B$

If $A = \{2,4,6,7\}$ and $B = \{1,3,5,6,7\}$, then $A \cap B = \{6,7\}$

- Now find the complement of $A \cap B$

$$A \cap B = \{6,7\} \rightarrow (A \cap B)' = \{1,2,3,4,5\}$$

The answer is the set $\{1,2,3,4,5\}$

Example 3 (page 16, #5e,h)

Given $U = \{1,2,3,...,10\}$, $A = \{2,3,4,5\}$, $B = \{4,5,6,7,8\}$, and $C = \{4,6,7,8,9\}$, evaluate the given sets.

Solution

e. $(A \cap B) \cap C$

- First evaluate $A \cap B$

  If $A = \{2,3,4,5\}$ and $B = \{4,5,6,7,8\}$ then $A \cap B = \{4,5\}$

- Now find the intersection of $(A \cap B)$ with set C

  $\{4,5\} \cap \{4,6,7,8,9\} = \{4\}$

The answer is $\{4\}$.

h. $A' \cap (B' \cap C')$

- First evaluate A', B', and C'

  $A = \{2,3,4,5\} \rightarrow A' = \{1,6,7,8,9,10\}$

  $B = \{4,5,6,7,8\} \rightarrow B' = \{1,2,3,9,10\}$

  $C = \{4,6,7,8,9\} \rightarrow C' = \{1,2,3,5,10\}$

- Now intersect B' and C'

  $\{1,2,3,9,10\} \cap \{1,2,3,5,10\} = \{1,2,3,10\}$

- Finally, intersect A' with $(B' \cap C')$

  $\{1,6,7,8,9,10\} \cap \{1,2,3,10\} = \{1,10\}$

The answer is $\{1,10\}$.

SUPPLEMENTARY EXERCISE 1.4

Given $U = \{1,2,3,...,10\}$, $A = \{1,2,4,8\}$, $B = \{1,3,7,9,10\}$, $C = \{3,4,5,6,7,8\}$, and $D = \{4,6,8\}$, evaluate the sets in problems 1-20.

1. $A \cap B$
2. $A \cap C$
3. $B \cap C$
4. $B \cap D$
5. $A \cup C$
6. $B \cup C$
7. A'
8. D'
9. $B' \cap C$
10. $C' \cap A$
11. $A' \cup D'$
12. $B \cup B'$
13. $B \cap \emptyset$
14. $A' \cup \emptyset$
15. $A \cap (B \cup C)$
16. $A \cup (B \cap D)$
17. $C \cup (B \cup D)$
18. $B \cap (A \cup C)$
19. $A' \cap (B' \cup D)$
20. $((B \cap C)' \cup D)'$

1.5 PICTURES OF SETS (VENN DIAGRAMS)

1.5.1 Venn Diagrams

A Venn diagram is a pictorial representation of a set. The universal set is shown as a rectangle, and any sets that are subsets of the universal set are shown as circles within the rectangle. These circles partition the rectangle into 2^n regions, where n is the number of sets under discussion. A Venn diagram with one set has 2^1 or 2 regions; a Venn diagram with two sets has 2^2 or 4 regions; and a Venn diagram with three sets has 2^3 or 8 regions. Venn diagrams with one, two, and three sets, and their respective regions are shown below.

SETS   13

Venn Diagram with One Set

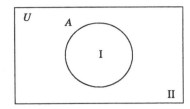

- U consists of regions I and II
- A consists of region I

Venn Diagram with Two Sets

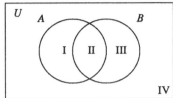

- U consists of regions I, II, III, and IV
- A consists of regions I and II
- B consists of regions II and III
- $A \cap B$ consists of region II
- $A \cup B$ consists of regions I, II, and III

Venn Diagram with Three Sets

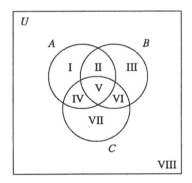

- U consists of all eight regions
- A consists of regions I,II,IV,V
- B consists of regions II,III,V,VI
- C consists of regions IV,V,VI,VII
- $A \cap B$ consists of regions II, V
- $A \cup B$ consists of regions I,II,III,IV,V,VI
- $A \cap C$ consists of regions IV,V
- $A \cup C$ consists of regions I,II,IV,V,VI,VII
- $B \cap C$ consists of regions V, VI
- $B \cup C$ consists of regions II,III,V,VI,VII
- $A \cap B \cap C$ consists of region V

1.5.2 Shading Venn Diagrams

Set expressions that involve the operations of union, intersection, and complement can be pictorially represented via shaded Venn diagrams. In your textbook, Professor Setek uses lines drawn at various angles to distinguish these set operations. This method works well for simple set expressions. If the expressions are more complicated, however, the results of this method are more difficult to interpret. As a result, we will use the procedure outlined below to represent the set expressions of examples 1, 2, and 3, which are given at the end of this section.

14 STUDENT STUDY GUIDE

Procedure for Shading Venn Diagrams

1. Construct a completely labeled one, two, or three set Venn diagram (depending on the problem).
2. Perform all complements, intersections, and unions using the region numbers of the labeled Venn diagram.
3. Shade the region(s) that correspond to the region numbers of the final result.

1.5.3 Cardinality

The **cardinality** of a set A, denoted $n(A)$, is a number that indicates the number of elements contained in set A. For example, if $A = \{2,8,12,19\}$, then $n(A) = 4$.

When using Venn diagrams to represent sets we will indicate the number of elements of a set by writing its Arabic numeral within the set. This number should not be confused with the Roman numerals, which are used to represent the regions of a Venn diagram.

To find the cardinality of a set using a Venn diagram, we follow the following procedure.

Procedure for Determining a Set's Cardinality

1. Determine the regions that make up the set.
2. Identify the number of elements contained within the regions of the set.
3. Find the sum of these elements.

This procedure is demonstrated in Example 4 at the end of this section.

1.5.4 One-To-One Correspondence

If two sets are equivalent (i.e., they contain the same number of elements), then there exists a **one-to-one correspondence** among the elements of the two sets. As a result, each element of one set can be paired with a unique element of the second set. Since the elements of a set are not ordered, if two equivalent sets have more than one element each, then there is more than one way in which a one-to-one correspondence can be established. The number of ways in which this can be done is equal to $n(n-1)(n-2)...1$, where n represents the cardinality of each set. As an illustration, if two sets have four elements each, then the number of one-to-one correspondences is equal to 4(3)(2)(1) or 24.

Example 1 (page 28, #1b,f)

Illustrate the given set expressions using a Venn diagram.

Solution

b. $A' \cap B'$

Using the procedure for shading Venn diagrams, the labeled Venn diagram and the evaluation of the given expression using region numbers are shown below.

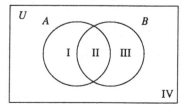

- $A =$ I and II $\to A' =$ III, IV
- $B =$ II and III $\to B' =$ I, IV

The Venn diagram that represents $A' \cap B'$ is a two-set Venn diagram with region IV shaded.

Continued on next page ...

SETS   15

f. $(A' \cup B')'$

The labeled Venn diagram and the evaluation of the given expression using region numbers are shown below.

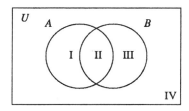

- A = I and II → A' = III, IV
- B = II and III → B' = I, IV
- $A' \cup B'$ = I, III, IV

$(A' \cup B')'$ consists of a two-set Venn diagram with region II shaded.

Example 2 (page 28, #3c,e)

Use a Venn diagram to represent the given set expression.

Solution

c. $(A \cap B) \cup C$

A labeled three-set Venn diagram and the evaluation of the given set expression are shown below.

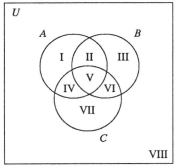

- A = I, II, IV, V
- B = II, III, V, VI
- $A \cap B$ = II, V
- C = IV, V, VI, VII

$(A \cap B) \cup C$ consists of a three-set Venn diagram with regions II, IV, V, VI, and VII shaded.

e. $(A \cup B) \cap (A \cup C)$

A labeled three-set Venn diagram and the evaluation of the given set expression are shown below.

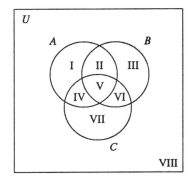

- A = I, II, IV, V
- B = II, III, V, VI
- $A \cup B$ = I, II, III, IV, V, VI
- C = IV, V, VI, VII
- $A \cup C)$ = I, II, IV, V, VI, VII

$(A \cup B) \cap (A \cup C)$ consists of a three-set Venn diagram with regions I, II, IV, V, VI shaded.

Example 3 (page 29, #5b,e)

Use Venn diagrams to show that each of the given statements is true.

Solution

b. $A \cup B = (A' \cap B')'$

If these two sets are equal then the regions that represent them also must be equal. A labeled two-set Venn diagram and the evaluation of these set expressions using the regions of this diagram confirm that the two expressions are equal.

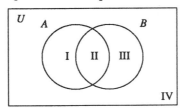

$A \cup B$	$(A' \cap B')'$
$A = \text{I, II}$	$A = \text{I, II} \rightarrow A' = \text{III, IV}$
$B = \text{II, III}$	$B = \text{II, III} \rightarrow B' = \text{I, IV}$
	$(A' \cap B') = \text{IV}$
$A \cup B = \text{I, II, III}$	$(A' \cap B')' = \text{I, II, III}$

Both sets are represented by the same regions (I, II, and III).

e. $A \cap (B \cup C) = (A \cap B) \cup (A \cap C)$

If these two sets are equal then the regions that represent them also must be equal. A labeled three-set Venn diagram and the evaluation of these set expressions using the regions of this diagram confirm that the two expressions are equal.

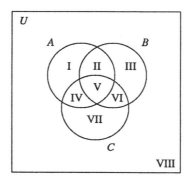

$A \cap (B \cup C)$	$(A \cap B) \cup (A \cap C)$
$B = \text{II, III, V, VI}$	$A = \text{I, II, IV, V}$
$C = \text{IV, V, VI, VII}$	$B = \text{II, III, V, VI}$
$(B \cup C) = \text{II, III, IV, V, VI, VII}$	$(A \cap B) = \text{II, V}$
$A = \text{I, II, IV, V}$	$C = \text{IV, V, VI, VII}$
	$A \cap C = \text{IV, V}$
$A \cap (B \cup C) = \text{II, IV, V}$	$(A \cap B) \cup (A \cap C) = \text{II, IV, V}$

Both sets are represented by regions II, IV, and V.

Example 4 (page 29, #7a,d,f,h)

Determine the cardinality of each set given using the Venn diagram below.

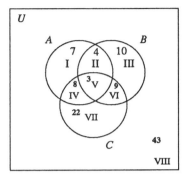

Solution

a. $n(A)$

Set A consists of regions I, II, IV, and V. The cardinality of each of these regions is 7, 4, 8, and 3, respectively. Therefore $n(A)$ is equal to $7 + 4 + 8 + 3$, or 22.

d. $n(U)$

The universal set consists of regions I, II, III, IV, V, VI, VII, and VIII. The cardinality of each of these regions is 7, 4, 10, 8, 3, 9, 22, and 43, respectively. Thus, $n(U) = 106$.

f. $n(B \cap C)$

The set $(B \cap C)$ consists of regions V and VI. Therefore, $n(B \cap C) = 12$.

h. $n(A \cup B \cup C)$

The set $(A \cup B \cup C)$ consists of regions I, II, III, IV, V, VI, and VII. Therefore, $n(A \cup B \cup C) = 63$.

Example 5 (page 29, #9)

In how many ways can a one-to-one correspondence be established between $A = \{i,o,u\}$ and $B = \{x,y,z\}$?

Solution

Sets A and B are equivalent; they both have exactly the same number of elements, namely, three. Therefore there are $(3)(2)(1)$ or 6 possible one-to-one correspondences.

SUPPLEMENTARY EXERCISE 1.5

In problems 1-10 use a Venn diagram to pictorially represent the given set expression. Express your final answer using region numbers.

1. $A \cap B'$
2. $(A \cap B')'$
3. $(A \cap B) \cup C'$
4. $A' \cup (B \cap C)'$
5. $A' \cap (B \cup C)'$
6. $(A' \cap B') \cup C'$
7. $(A \cap B)' \cap C$
8. $(A \cup B)' \cup C$
9. $(A \cup B)' \cap (A' \cup C)'$
10. $(A' \cap B)' \cup (B \cup C')'$

Continued on next page ...

18   STUDENT STUDY GUIDE

In 11 through 18 use the following Venn diagram to determine the number of elements contained in the given set.

11. $n(A \cap B)$

12. $n(B \cup C)$

13. $n((A \cap B) \cup C)$

14. $n((A \cup C) \cap B)$

15. $n((A \cup B)' \cap C)$

16. $n(A' \cap (B \cup C)')$

17. $n((A' \cap C') \cup B')$

18. $n((A' \cup B)' \cap (A' \cup C))$

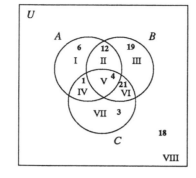

19. If $A = \{\emptyset\}$ then $n(A) = ?$ 20. If $B = \emptyset$ then $n(B) = ?$

In 21 through 26 use $A = \{x, y\}$ and determine whether or not a one-to-one correspondence can be established between each of the given sets and set A.

21. $B = \{\#, \%\}$ 22. $C = \{Y, X\}$ 23. $D = \{12, 3\}$ 24. $E = \emptyset$ 25. $F = \{28\}$ 26. $G = \{\emptyset, \emptyset\}$

1.6 AN APPLICATION OF SETS AND VENN DIAGRAMS

The type of problems solved in this section are known as survey problems. Our goal is to solve these type of problems using Venn diagrams. The following procedure, which is illustrated fully in Example 1, is presented as a guide to assist you in achieving this goal. You should refer to this example for assistance when necessary. (Note: The problems from the textbook that are used as examples here will not be given; only their solutions will be presented. Thus it will be necessary for you to read the problem first from the textbook.)

Procedure for Solving Survey Problems Using Venn Diagrams

1. Construct a Venn diagram to represent the sets under discussion.

2. Extract the data from the problem and identify the appropriate regions of the Venn diagram that correspond to the data.

3. Assign the data to the eight regions of the Venn diagram using the following order:

    a. Region V

    b. Region II, IV, and VI

    c. Regions I, III, VII

    d. Region VIII

4. Answer the questions that are asked in the problem.

Example 1 (page 35, #3)

1. We first construct a Venn diagram to represent the sets. (**Note:** Set A represents Accounting, set P represents Psychology, and set S represents Statistics.)

Continued on next page ...

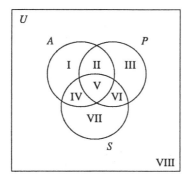

2. Next, the data are extracted from the problem and the regions that correspond to the data are identified.

   a. 75 Total (Regions I, II, III, IV, V, VI, VII, VIII)
   b. 27 Accounting (Regions I, II, IV, V)
   c. 26 Psychology (Regions II, III, V, VI)
   d. 41 Statistics (Regions IV, V, VI, VII)
   e. 12 Accounting and Psychology (Regions II, V)
   f. 13 Accounting and Statistics (Regions IV, V)
   g. 17 Psychology and Statistics (Regions V, VI)
   h. 4 Accounting, Psychology, and Statistics (Region V)

3. We now place the data into the regions of the Venn diagram following the order given in the procedure.

   - Region V gets assigned 4 (See 2h above.)
   - Region II gets assigned 8

     (2e above indicates that regions II and V must have a total of 12. Since region V presently has 4, region II must be assigned 8.)

   - Region IV gets assigned 9

     (2f above indicates that regions IV and V must have a total of 13. Since region V presently has 4, region IV must be assigned 9.)

   - Region VI gets assigned 13

     (2g above indicates that regions V and VI must have a total of 17. Since region V presently has 4, region VI must be assigned 13.)

   - Region I gets assigned 6

     (2b above indicates that regions I, II, IV, and V must have a total of 27. Since regions II, IV, and V have a collective total of 21, region I must be assigned 6.)

   - Region III gets assigned 1

     (2c above indicates that regions II, III, V, and VI must have a total of 26. Since the collective total of regions II, V, and VI is 25, region III must be assigned 1.)

Continued on next page ...

20 STUDENT STUDY GUIDE

- Region VII gets assigned 15

  (2d above indicates that regions IV, V, VI, and VII must have a total of of 41. Since the collective total of regions IV, V, and VI is 26, region VII must be assigned 15.)

- Region VIII gets assigned 19

  (2a above indicates that regions I, II, III, IV, V, VI, VII, and VIII must have a total of 75. Since the collective total of the first seven regions is 56, region VIII must be assigned 19.)

The completed Venn diagram is shown below.

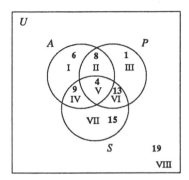

4. Finally, we answer the questions that are asked in the problem.

   a. The number of students *taking only psychology* is located in region III. Therefore, 1 student was taking only psychology.

   b. The number of students *not taking any of the three subjects* can be found in Region VIII. Thus this answer is 19.

   c. The region that represents the number of students taking *accounting and statistics, but not psychology* is region IV. The answer to this question then is 9.

   d. To determine the number of students *taking exactly one of these subjects* we must sum the numbers from regions I, III, and VII. As a result, the answer here is 22.

Example 2 (page 36, #9)

The data extracted from this problem, the regions the data represent, and the completed Venn diagram are shown below. (**Note:** Set *BG* represents backgammon, set *CS* represents chess, and set *CK* represents checkers.)

- 130 surveyed (regions I through VIII)
- 57 played backgammon (Regions I, II, IV, V)
- 89 played checkers (Regions II, III, V, VI)
- 76 played chess (Regions IV, V, VI, VII)
- 35 played checkers and backgammon (Regions II, V)
- 50 played chess and checkers (Regions V, VI)
- 30 played backgammon and chess (Regions IV, V)
- 20 played all three (Region V)

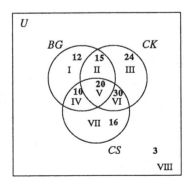

Continued on next page ...

Answers to Questions

a. Region VII represents those that played *only chess*: 16

b. Regions IV, V, VI, and VII represent the total number of students that played chess. The remaining regions (I, II, III, and VIII) represent the number of students that *did not play chess*: 54

c. The concept of *at least one* means one or more. Thus all the regions except region VIII represent the number of students that *played at least one game*: 127

d. The use of the word *or* in this question does not restrict us from playing other games. Therefore, we must consider regions I, II, III, IV, V, and VI, since the number of students represented in these regions all *played backgammon or checkers*: 111

Note: If we wanted to exclude those people that played chess, but still played backgammon or checkers, then the question would be restated as, *How many students played backgammon or checkers, but not chess?* If this were the case, then only regions I, II, and III would be considered.

Example 3 (page 37, #13)

The data extracted from this problem, the regions the data represent, and the completed Venn diagram are shown below.

(**Note:** Set C represents compact cars, set TD represents cars with two door, and set ST represents cars with standard-transmission. These are the primary sets under discussion. The large, four-door, automatic-transmission cars are shown in region VIII, which represents cars that are not compact, have two doors, or standard transmissions.)

- 22 cars are C, TD, and ST (Region V)
- 50 cars have ST (Regions IV, V, VI, VII)
- 28 cars are C with ST (Regions IV, VI)
- 30 cars have ST and TD (Regions V, VI)
- 47 cars have TD (Regions II, III, V, VI)
- 31 cars are C with TD (Regions II, V)
- 44 cars are C (Regions I, II, IV, V)
- 15 cars are large, four-door, automatic-transmission cars (Region VIII)

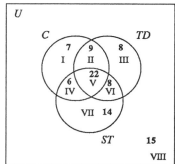

Answers to Questions

a. The total number of cars on the lot can be found by summing all eight regions: 89

b. Region IV represents the number of compact cars with standard transmissions that are not two-door: 6

c. Region VI represents the number of two-door cars with standard transmissions that are not compact: 8

d. The number of cars with standard transmissions that are not compact can be found by adding the data found in regions VI and VII: 22

SUPPLEMENTARY EXERCISE 1.6

1. To provide better facilities to their members a health club determined the following information: of the 100 members, 35 used the racquet ball courts, 36 used the Nautilus machines, and 68 used the swimming pool; 16 of the members used both the racquet ball courts and Nautilus machines, 21 used both the Nautilus machines and the swimming pool, and 15 used both the swimming pool and the racquet ball courts; 10 people used all three of these facilities.

    a. How many members used none of the three facilities?

    b. How many members used the swimming pool and the racquet ball courts, but not the Nautilus machines?

    c. How many members used only the Nautilus machines?

    d. How many members used only one facility?

2. An ice cream parlor wanted to know what flavors were the most popular. A survey of 50 customers revealed: 46 customers liked vanilla, 45 liked chocolate, and 42 liked strawberry; 43 customers liked vanilla and chocolate, 41 liked vanilla and strawberry, and 40 liked chocolate and strawberry; 40 customers liked all three of these flavors.

    a. How many customers liked vanilla or chocolate?

    b. How many customers liked none of these flavors?

    c. How many customers liked only one flavor?

    d. How many customers did not like vanilla?

3. At a college dance, 7 couples danced to soul, country, and rock songs, 12 danced to soul and country songs, 15 danced to soul and rock songs, 11 danced to rock and country songs, 18 couples danced to country songs, 42 danced to soul songs, and 39 danced to rock songs; 80 couples attended the dance.

    a. How many couples did not dance at all?

    b. How many couples danced to country songs?

    c. How many couples danced to rock and country songs, but not soul songs?

4. In analyzing the food preferences of 64 individuals at the salad bar of a school cafeteria it was observed that: 44 people selected carrots, 28 people selected celery, and 41 people selected tomatoes; 19 people selected carrots and celery, 23 selected celery and tomatoes, and 27 people selected tomatoes and carrots; 17 people selected all three vegetables.

    a. How many people selected only tomatoes?

    b. How many people selected none of these three vegetables?

    c. How many people selected exactly two vegetables?

    d. How many people selected carrots or tomatoes?

5. To decide what new shows to include in a new television season, a TV network surveyed 10,000 viewers: 4,370 viewers liked soap operas, 4,200 viewers liked westerns, and 4,432 viewers liked comedies; 397 viewers liked soap operas and westerns, 2,186 viewers liked comedies and westerns, and 678 viewers liked soap operas and comedies; 86 people liked all three types of shows.

    a. How many viewers liked soap operas and westerns, but not comedies?

    b. How many viewers did not like soap operas?

    c. How many viewers liked only one type of show?

    d. How many viewers liked soap operas or westerns, but not comedies?

6. Purina Cat Chow conducted a test using 70 cats: 65 cats ate Special Dinners, 61 cats ate Meow Mix, and 63 cats ate Country Blend; 60 cats ate Special Dinners and Meow Mix, 57 cats ate Meow Mix and Country Blend, and 58 cats ate Country Blend and Special Dinners; 56 cats ate all three varieties.

   a. How many cats would not eat any of the three varieties?

   b. How many cats ate exactly two of the varieties?

   c. How many cats ate only Country Blend?

   d. How many cats did not eat Meow Mix?

7. Swimming, bicycling, and jogging are popular forms of exercise. A magazine survey revealed that of 100 people: 2 people swam, bicycled, and jogged; 12 people swam and bicycled, 9 people bicycled and jogged, and 17 people swam and jogged; 20 people only swam, 30 people only bicycled, and 12 people only jogged.

   a. How many people swam?

   b. How many people did not do any of these exercises?

   c. How many people swam and bicycled, but did not jog?

8. While reviewing his records, a travel agent noticed that of 60 clients: 56 clients traveled to Hawaii, 51 clients traveled to Europe, and 51 clients traveled to Mexico; 50 clients traveled to Hawaii and Europe, 51 clients traveled to Hawaii and Mexico; 48 clients traveled to all three places.

   a. How many clients did not travel to any of these three places?

   b. How many clients traveled only to Hawaii?

   c. How many clients traveled to exactly one of these places?

   d. How many clients did not travel to Mexico?

1.7 CARTESIAN PRODUCTS

Thus far in this chapter we have discussed three set operations: complement, which is a unary operation because it involves a single set, and intersection and union, which are binary operations because they involve two sets. We now discuss a fourth operation, Cartesian product.

A **Cartesian product** is a binary operation, and, as is the case with the previous set operations, yields an answer that is also a set. We denote a Cartesian product as $A \times B$ (read *A cross B*). The members of a Cartesian product consist of **ordered pairs**, denoted (a,b). The first element of the ordered pair, a, is selected from the first set, and the second element of the ordered pair, b, is selected from the second set. Using set-builder notation, an ordered pair can be defined as

$$A \times B = \{(a,b) \,/ (br\ a \in A \text{ and } b \in B\}$$

The ordered pairs of a Cartesian product are formed by pairing each element of the first set with every element of the second set. To illustrate using general terms, if set A were to have the two elements a_1 and a_2, and set B were to have the three elements b_1, b_2, and b_3, the the Cartesian product would be the set

$$A \times B = \{(a_1, b_1), (a_1, b_2), (a_1, b_3), (a_2, b_1), (a_2, b_2), (a_2, b_3)\}$$

24 STUDENT STUDY GUIDE

1.7.1 Notes on Cartesian Product

1. The elements of each ordered pair are separated by commas.
2. The two elements of each ordered pair are enclosed within parentheses.
3. The ordered pairs are separated by commas and are within braces.
4. $A \times B \neq B \times A$ since the elements will have different ordered pairs. (The first and second elements of each ordered pair will be reversed.)
5. $n(A \times B) = n(A) \times n(B)$.
6. A Cartesian product can be pictorially represented as an **array**, which is also known as a **lattice**, or by a **tree diagram**. (See Examples 2 and 3 at the end of this section.)

Example 1 (page 41, #1)

If $A = \{a,b,c\}$ and $B = \{10,20\}$, find $A \times B$, $B \times A$, and $n(A \times B)$.

Solution

- To find $A \times B$ we pair each element of A with every element of B. Thus element a from set A is paired with elements 10 and 20 from set B; element b from set A is paired with 10 and 20; and element c from set A is paired with 10 and 20. The roster form representation of this set is:

$$A \times B = \{(a,10), (a,20), (b,10), (b,20), (c,10), (c,20)\}$$

- To find $B \times A$ we pair each element of set B with every element of set A. Thus element 10 from set B is paired with elements a, b, and c from set A, and element 20 from set B is paired with elements a, b, and c from set A. The roster form representation of this set is:

$$B \times A = \{(10,a), (10,b), (10,c), (20,a), (20,b), (20,c)\}$$

- $n(A \times B) = n(A) \times n(B) \rightarrow (3) \times (2) = 6$

Example 2 (page 41, #7)

Let $A = \{4,5,6\}$ and $B = \{x,y\}$. Make a lattice showing $A \times B$.

Solution

To make a lattice to represent a Cartesian product we place the elements of the first set on a horizontal line segment, and the elements of the second set on an adjoining vertical line segment. The ordered pairs are then represented as dots.

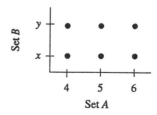

Example 3 (page 41, #9)

If $A = \{x,y\}$ and $B = \{a,e,i,o,u\}$, make a tree diagram showing $A \times B$.

Continued on next page ...

SETS 25

Solution

To construct a tree diagram we physically show each element of the first set (Set A) being paired to every element of the second set (Set B) via a *branch* (a line segment that connects one element to another).

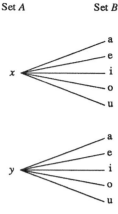

Example 4 (page 41, #11d)

Given $A = \{a,b\}$, $B = \{b,c,d\}$, and $C = \{c,d,e\}$, find $(A \cap B) \times C$.

Solution

- First evaluate $A \cap B$

$$A = \{a,b\} \text{ and } B = \{b,c,d\} \rightarrow A \cap B = \{b\}$$

- Now evaluate the Cartesian product

$$(A \cap B) \times C$$
$$= \{b\} \times \{c,d,e\}$$
$$= \{(b,c), (b,d), (b,e)\}$$

SUPPLEMENTARY EXERCISE 1.7

In 1 through 4 find $A \times B$, $B \times A$ and $n(A \times B)$.

1. $A = \{\text{Pepper, Spike}\}$ and $B = \{\text{Kit, Claudia}\}$
2. $A = \{\text{he, me, fee}\}$ and $B = \{\text{see, be, E}\}$
3. $A = \{AB, C\}$ and $B = \{x,y,z\}$
4. $A = \{R,U,E,Z,?\}$ and $B = \{\text{yes,no}\}$
5. Let $A = \{x \text{ /(br } x \text{ is a vowel in the English alphabet}\}$ and $B = \{x \text{ /(br } x \text{ is a digit from 8 to 10}\}$. Make a lattice and a tree diagram showing $B \times A$.

Let $A = \{2,4\}$, $B = \{3,8\}$, and $C = \{1,3,8\}$. Use these sets to evaluate the expressions given in problems 6 through 10.

6. $A \times (B \cap C)$ 7. $(B \cup C) \times A$ 8. $(A \cap B) \times C$ 9. $(A \cup C) \times (B \cap C)$ 10. $n(A \times B \times C)$

1.8 CHAPTER 1 TEST

In 1 through 12 answer true or false.

1. $U' = \emptyset$
2. $A \cup \emptyset = \emptyset$
3. $A' \cup B' = (A \cup B)'$
4. $A \subset (A \cap B)$
5. $\emptyset$ has exactly one subset

6. $\{x \mid x \text{ is a counting number not less than } 6\} = \{6,7,8,9,...\}$

7. $\{x \mid x \text{ is a positive number}\}$ is not well defined

8. $2 \notin \{1,4,8,10\}$
9. Equal sets are equivalent sets
10. Equivalent sets are equal sets
11. $\{3,4\} = (4,3)$
12. $\{2\} \in \{x \mid x \text{ is an even number}\}$

13. Determine the number of subsets of $M = \{e,f,g\}$ and list all of them.

In 14 and 15 label the set as infinite or finite.

14. $A = \{0,1,2,3,...,1000000000\}$
15. $B = \{x \mid x \text{ is a fraction between 1 and 2}\}$

In 16 and 17 draw a Venn diagram for each set and state the regions that make up the final answer.

16. $P' \cap Q$
17. $(M \cup R)' \cap N$

Given $U = \{1,2,3,...,10\}$, $A = \{2,3,4,5,6\}$, $B = \{5,6,7,8\}$, $C = \{1,9\}$, and $D = \{4,5\}$, evaluate the sets in 18 through 23.

18. $(A \cap B)'$
19. $(A \cap B) \cap D$
20. $(A \cup B) \cap C'$
21. $(B' \cap C) \cup D$
22. $n(A \cup B)$
23. $B \times D$

24. In the *cola wars* a taste test survey revealed the following information:
    - 31 chose Pepsi, 31 chose RC cola, and 24 chose Coke;
    - 13 chose Pepsi & RC cola, 11 chose RC cola & Coke, 8 chose Coke & Pepsi;
    - 11 did not like any of the three
    - 5 chose all three

    a. Construct a Venn diagram to represent the results of this survey.
    b. How many people took part in this survey?
    c. How many people liked Coke or Pepsi, but not RC cola?
    d. How many people did not like Pepsi?

2. LOGIC

2.1 INTRODUCTION

Logic is a science that deals with methods of reasoning. Logic provides us with a means of proving mathematical theorems, drawing conclusions from experiments, and assisting us in solving various everyday problems. In this chapter we will develop some of the basic tools of logic and use these tools to investigate the nature of statements and arguments.

2.2 STATEMENTS AND SYMBOLS

Much of our verbal and written communication consists of sentences that are either true or false. In logic such sentences are referred to as **statements** and can be assigned a **truth value**. Sentences that cannot be assigned a truth value (i.e., are neither true nor false) are not discussed in logic. To illustrate the difference between logical statements and sentences that cannot be assigned a truth value consult the chart below.

Examples of Sentences That Cannot Be Assigned a Truth Value	Examples of Sentences That Can Be Assigned a Truth Value
1. Please get a haircut. (Command) 2. How old are you? (Question) 3. Stop nagging me! (Exclamation)	1. Tim got a haircut. 2. Shannon is 14 years old. 3. My mother is a nag.

Statements in logic are typically represented by symbols. Normally, the symbols P,Q,R, and S (and other capital letters if needed) are used for this purpose. Symbols that are used to represent statements are known as **logical variables**.

There are two general classifications of statements in logic: simple and compound (or complex). **Simple statements** are complete sentences that can be assigned a truth value, but contain no *connecting* words. The three statements in the chart above (right hand side) are examples of simple statements.

When a simple statement is combined with other simple statements, or involves the use of a connecting word, the resulting statement is referred to as a **compound statement**. The connecting words (herein referred to as the **connectives**) include the words *not*, *and*, *or*, *If...then*, and *If and only if*, which is abbreviated *iff*. The symbols used to represent these connectives are given in the chart below.

Connective	Symbol
negation	~
and	$\wedge$
or	$\vee$
if...then	$\rightarrow$
iff	$\leftrightarrow$

2.2.1 Negation

A simple statement, P, that has been negated (i.e., its truth value has been reversed), is known as a **negation**, denoted ~P. The expression ~P is commonly read as *not P*. Additionally, phrases such as *it is not the case that*, *it is false that*, and *it is not true that* represent a negation.

2.2.2 Conjunction

If two statements P and Q are connected by the word *and* then the resulting statement is referred to as a **conjunction**, denoted $P \wedge Q$. This expression is read as *P and Q*.

2.2.3 Disjunction

If two statements P and Q are connected by the word *or* then the resulting statement is referred to as a **disjunction**. There are two types of disjunctions: inclusive and exclusive. An **inclusive disjunction**, denoted $P \vee Q$ (read *P or Q*), means either P or Q, or both P and Q. An **exclusive disjunction**, denoted $P \underline{\vee} Q$ (read *P exclusive or Q*), means either P or Q, but not both.

2.2.4 Conditional

If two statements P and Q are connected in such a manner that the first statement is preceded by the word *if* and the second statement is preceded by the word *then* then the resulting statement is referred to as a **conditional statement**, denoted $P \rightarrow Q$. This expression is read as *if P then Q*, *if P, Q*, or *Q, if P*. P, which is the statement following the word *if* is called the **antecedent**, and Q, which is the statement following the word *then* is called the **consequent**.

2.2.5 Biconditional

A **biconditional statement** is the conjunction of two conditional statements where the antecedent and consequent of the first conditional statement have been reversed in the second statement. A biconditional is denoted $P \leftrightarrow Q$ and read as *if P then Q and if Q then P*. A more convenient (and common) way in which to read a biconditional however is *P if and only if Q*.

Example 1 (page 58, #1a-h)

Identify each of the following sentences as a simple statement, compound statement, or neither. Classify each compound statement as a negation, conjunction, disjunction, conditional, or biconditional.

Solution

a. Today is Friday.

   Simple; this sentence can be assigned a truth value and has no connectives.

b. It is false that Scott is in class.

   Compound/Negation; the phrase *it is false that* negates the simple statement *Scott is in class*.

Continued on next page ...

c. You may play tennis here iff you are a member of the club.

Compound/Biconditional; *iff* indicates *if and only if*, which represents a biconditional statement.

d. Joey is not going to Canada during vacation.

Compound/Negation; the word *not* negates the simple statement *Joey is going to Canada during vacation*.

e. If Addie went swimming, then Julia went sailing.

Compound/Conditional; the use of the words *if...then* indicates that this is a compound statement.

f. Close the door when you leave.

Neither; the statement is a command and cannot be assigned a truth value.

g. Both Ruth and Horace are members of the council.

Compound/Conjunction; the word *and* connects the two simple statements *Ruth is a member of the council* and *Horace is a member of the council*. (Note how the word *both* is used to signify a conjunction.)

h. Either Bill is here or he did not go to school.

Compound/Disjunction; the word *or* connects the simple statement *Bill is here* with the compound statement *he did not go to school*. (Note that the second statement is compound because it is the negation of the simple statement *he did go to school*. Note further how the word *either* is used in a disjunction.)

Example 2 (page 58, #5a-f)

Let P = *Algebra is difficult* and Q = *Logic is easy*. Write each of the following statements in symbolic form.

Solution

a. Algebra is difficult or logic is easy.

Two simple statements are connected by the word *or*. The statement is symbolized P ∨ Q.

b. Logic is not easy and algebra is difficult.

Since Q = logic is easy, ~Q = logic is not easy. The overall statement is a conjunction. The symbolic representation is ~Q ∧ P.

c. It is false that logic is not easy.

The phrase *it is false that* represents a negation, and the statement *logic is not easy* is symbolized as ~Q. The symbolic representation of this statement is ~~Q, or more simply, Q.

d. Logic is easy iff algebra is difficult.

This is a straightforward biconditional. The symbolic representation is Q ↔ P.

e. If algebra is difficult then logic is easy.

This is a straight forward conditional where *algebra is difficult* is the antecedent and *logic is easy* the consequent. The symbolic representation is P → Q.

Continued on next page ...

f. Neither is algebra difficult nor is logic easy.

   A statement of the form *neither P nor Q* means *not P and not Q*. The given statement is rewritten as *algebra is not difficult and logic is not easy*. As a result, the symbolic representation is ~P ∧ ~Q.

Example 3 (page 58, #7a-f)

Write the following statements in symbolic form using the letters given in the parentheses.

Solution

To solve these problems the following procedure is used:

1. Identify the *type* of statement given.

2. Identify the simple statements that make up any given compound statement and assign a logical variable to them.

3. Symbolize the given statement using the variables from step 2 and the correct connective determined from step 1.

a. Either I sink this putt or I lose the match. (P,M)

   1. The statement is a *disjunction* (∨).

   2. Let P = *I sink this putt* and M = *I lose the match*.

   3. The symbolized form is P ∨ M.

   Note: In this problem we could have let M = *I will win the match*. If we had done so, then ~M would represent *I did not win the match*, which effectively has the same meaning as *I will lose the match*. If this were the case then the symbolized form for this problem would be P ∨ ~M. As you can see it is imperative that we clearly indicate what the logical variables represent.

b. Five is greater than zero and five is positive. (G,P)

   1. The statement is a *conjunction* (∧).

   2. Let G = *Five is greater than zero* and P = *five is positive*.

   3. The symbolized form is G ∧ P.

c. If you do not attend class then you will be dropped from the course. (A,D)

   1. The statement is a *conditional* (→).

   2. Let A = *you do attend class*. Thus ~A represents *you do not attend class*. This is the antecedent. Now let D = *you will be dropped from the course*. D is the consequent.

   3. The symbolized form is ~A → D.

d. Either the bus is late, or my watch is not working correctly. (B, W)

   1. The statement is a *disjunction* (∨).

   2. Let B = *the bus is late* and W = *my watch is working correctly*. Since W = *my watch is working correctly*, ~W = *my watch is not working correctly*.

   3. The symbolized form is B ∨ ~W.

Continued on next page ...

e. Two equals one iff three is greater than four.
   1. The statement is a *biconditional* (↔).
   2. Let T = *Two equals one* and let F = *three is greater than four*.
   3. The symbolized form is T ↔ F

f. Smith will raise taxes if he is elected (R,E).
   1. The statement is a *conditional* (→).
   2. Let R = *Smith will raise taxes* and let E = *he is elected.*
   3. The symbolized form is E → R.
      (Recall that the antecedent follows the word *if*.)

Example 4 (page 58, #9a,b,c,f)

Let P = *I like algebra* and Q = *I like geometry*. Write each of the following statements in words.

Solution

To solve these problems the following procedure is used:
1. Identify the *type* of statement given.
2. Read the symbolized version of the statement.
3. Translate the symbolic reading of the statement from step 2 into words.

a. P ∧ Q
   1. The statement is a *conjunction*.
   2. Symbolically, this statement is read *P and Q*.
   3. Translating, we have, *I like algebra and I like geometry*.

b. P → ~Q
   1. The statement is a *conditional*.
   2. Symbolically, this statement is read *If P then not Q*.
   3. Translating, we have *If I like algebra then I do not like geometry*.

c. P V Q
   1. The statement is a *disjunction*.
   2. Symbolically, this statement is read *P or Q*.
   3. Translating, we have *I like algebra or I like geometry*. Alternatively, we could have written *I like either algebra or geometry*.

f. P ↔ Q
   1. The statement is a *biconditional*.
   2. Symbolically, this statement is read *P if and only Q*.
   3. Translating, we have *I like algebra if and only if I like geometry*.

SUPPLEMENTARY EXERCISE 2.2

In 1 through 8 write each statement in symbolic form using P = *We go out to dinner* and Q = *We go to a movie*.

1. If we go out to dinner then we will not go to a movie.
2. Either we will go to a movie or we will go out to dinner.
3. We went neither to dinner nor to a movie.
4. It is false that we will go out to dinner or to a movie.
5. We went to a movie, but we did not go out to dinner.
6. We will go out to dinner if and only if we go to a movie.
7. We will go to a movie if we do not go out to dinner.
8. Either we go out to dinner, or we don't go to a movie.

In 9 through 16 let P = *Shannon loves chicken* and Q = *Pepper needs to lose weight*. Write each statement in words.

9. P ∧ ~Q
10. ~(P → Q)
11. ~P ↔ Q
12. ~P ∨ Q
13. ~Q → P
14. ~(P ∧ Q)
15. Q ∨ P
16. Q ↔ P

2.3 DOMINANCE OF CONNECTIVES

Compound statements that involve multiple connectives can be difficult to interpret without the use of punctuation marks (most notably the comma), or without some order of dominance of the connectives. In logic parentheses are used as punctuation marks. There also is an agreed upon dominant order of the connectives. From strongest to weakest this order is given below.

1. Biconditional ↔
2. Conditional →
3. Conjunction ∧, disjunction ∨
4. Negation ~

In the examples that follow we identify, symbolize, and interpret compound statements with multiple connectives. As an aid in doing this the following observations are noted:

1. Let us assume we have a statement that contains two connectives, one of which is more dominant than the other. If the more dominant connective is placed within parentheses, it automatically becomes less dominant than the connective outside the parentheses. For example, in its present form, the statement ~P → Q is a conditional since → is more dominant than ~. However, the statement ~(P → Q) is a negation since the →, which is now within parentheses, lost its dominance when compared to ~, which is outside the parentheses.

2. The connectives *and, or, if...then, and iff* are **binary** operations. That is they connect two statements. The connective *not*, however, is a **unary** operation; it operates on a single statement.

Example 1 (page 62, #1a-e)

Add parentheses in each statement to form the type of compound statement indicated. If no parentheses are needed then indicate this fact.

Solution

    a. Convert the statement ~P ∧ Q → R into a negation.

        The given statement is a conditional since the arrow is the strongest connective of the three connectives given. To convert the statement into a negation it is necessary to include the entire statement, except the negation symbol, within parentheses. By doing so the negation symbol becomes the most dominant connective. The answer is: ~(P ∧ Q → R).

    b. Convert the statement ~P ∧ Q → R into a conditional.

        The statement in its present form is already a conditional (see above).

        Note: It is not incorrect to group the statements on either side of the arrow to highlight the fact that the arrow is the most dominant connective. Thus (~P ∧ Q) → R and (~P ∧ Q) → (R) are also correct.

    c. Convert the statement ~P ∧ Q → R into a conjunction.

        To make this statement a conjunction it is necessary to enclose the statements on either side of the ∧ symbol within parentheses. Doing so we have (~P) ∧ (Q → R).

        Note: Since ∧ is more dominant than ~, we do not need to mask the ~ from the ∧. An equally correct response is ~P ∧ (Q → R).

    d. Convert the statement P ∧ Q ↔ R into a biconditional.

        Of the two connectives present in the given statement ↔ is the stronger. The statement is a biconditional in its present form.

    e. Convert the statement ~P ∨ Q ∧ R into a disjunction.

        Of the connectives given ∧ and ∨ are of equal dominance and ~ is the weakest. To convert this statement into a disjunction we must *mask* the ∧ symbol from the ∨ symbol. We do this by enclosing that part of the statement within parentheses.
        The correct answer is ~P ∨ (Q ∧ R).

Example 2 (page 62, #3a-c,e)

Let P = *Algebra is difficult*, Q = *Logic is easy*, and R = *Latin is interesting*. Use appropriate connectives and parentheses to symbolize each statement.

Solution

    To solve these problems the following procedure is used:

        1. Identify the type of statement given.

        2. Symbolize the statement using the logical variables given in the problem.

        3. Insert parentheses, if necessary, so that the symbolized statement corresponds to the type of statement as identified in step 1.

Continued on next page ...

a. If logic is easy and algebra is difficult, then Latin is interesting.
   1. The statement is a *conditional*.
   2. The symbolized form is: Q ∧ P → R
   3. No parentheses are needed since the arrow is the stronger connective. The answer is Q ∧ P → R.

b. Latin is interesting and algebra is difficult, or logic is easy.
   1. The statement is a *disjunction* since the placement of the comma is immediately before the word *or*. (**Note:** if the comma is placed immediately before the word *and*, then the statement is a conjunction.)
   2. The symbolized form is: R ∧ P ∨ Q
   3. Parentheses are needed to mask the *and* symbol from the *or* symbol. The correct symbolized form is (R ∧ P) ∨ Q.

c. It is false that logic is easy and algebra is difficult.
   1. The statement is a *negation*.
   2. The symbolized form is: ~Q ∧ P
   3. Parentheses are needed so that the stronger connective is the negation symbol. The correct symbolized form is ~(Q ∧ P).

e. Algebra is difficult iff Latin is interesting and logic is easy.
   1. The statement is a *biconditional*.
   2. The symbolized form is: P ↔ R ∧ Q
   3. No parentheses are needed since the stronger connective is the biconditional symbol. The correct symbolized form is P ↔ R ∧ Q.

Example 3 (page 63, #7a,b,e,f)

Let P = *Algebra is difficult*, Q = *Logic is easy*, and R = *Latin is interesting*. Write each symbolic statement in words.

Solution

To solve these problems the following procedure is used:

1. Identify the type of statement given.
2. Read the statement via its symbolic form.
3. Translate the symbolic reading from step 2 into words using the value of the logical variables given in the problem.

a. P ∧ (Q ∨ R)
   1. The dominant connective is *and*, thus the statement is a *conjunction*.
   2. Symbolically, this statement is read as *P, and Q or R*.
   3. Using the values of the logic variables, the translation of the symbolic reading from step 2 is: *Algebra is difficult, and logic is easy or Latin is interesting*.

Continued on next page ...

b. P ∧ Q → R

   1. The dominant connective is the arrow, thus the statement is a *conditional*.

   2. Symbolically, this statement is read *If P and Q then R*.

   3. Translating, we have *If algebra is difficult and logic is easy, then Latin is interesting.*

e. ~(P ∧ Q)

   1. The dominant connective is the negation symbol, thus the statement is a *negation*.

   2. Symbolically, this statement is read *It is false that P and Q*.

   3. Translating, we have *It is false that algebra is difficult and logic is easy.*

f. ~P ↔ Q ∧ ~R

   1. The dominant connective is the double arrow, thus the statement is a *biconditional*.

   2. Symbolically, this statement is read *Not P if and only if Q and not R*.

   3. Translating, we have *Algebra is not difficult if and only if logic is easy and Latin is not interesting.*

SUPPLEMENTARY EXERCISE 2.3

In 1 through 3 insert parentheses in each statement to form the type of compound statement requested. If no parentheses are needed then indicate so.

1. Conjunction: $P \to \sim Q \land R$

2. Biconditional: $P \leftrightarrow Q \lor \sim R$

3. Disjunction: $\sim P \lor Q \land R$

In 4 through 11 let P = *Alice stayed home*, Q = *Joan went swimming*, and R = *Sarah did her homework*. Symbolize each statement completely.

4. Either Joan went swimming, or if Sarah did her homework then Alice did not stay home.

5. Alice stayed home or Joan went swimming iff Sarah did her homework.

6. It is not the case that Alice stayed home and Sarah did her homework.

7. If Joan did not go swimming and Sarah did her homework, then Alice did not stay home.

8. Alice stayed home or Joan went swimming, and Sarah did her homework.

9. It is false that Joan went swimming iff Alice did not stay home.

10. If Sarah did her homework then Alice stayed home and Joan went swimming.

11. Neither Alice stayed home nor did Joan go swimming.

In 12 through 16 let P = *It rained today*, Q = *I got wet*, and R = *July is a hot month*. Write each symbolic statement in words.

12. P ∨ ~R ↔ Q   13. ~(P ∨ Q)   14. ~P ∨ Q → ~R
15. (P ∨ Q) ∧ R   16. ~P ∨ (Q → ~R)

2.4 TRUTH TABLES

This section is combined with section 2.5.

2.5 MORE TRUTH TABLES — CONDITIONAL AND BICONDITIONAL STATEMENTS

2.5.1 Basic Truth Tables

A **truth table** is a tool we use to determine the truth value of a statement. A completed truth table enables us to investigate under what conditions a given statement is true or false. The basic truth tables for negation, conjunction, disjunction, conditional, and biconditional statements are given below.

Negation		Conjunction		
P	~P	P	Q	P ∧ Q
T	F	T	T	T
F	T	T	F	F
		F	T	F
		F	F	F

Inclusive Disjunction			Exclusive Disjunction		
P	Q	P ∨ Q	P	Q	P v̲ Q
T	T	T	T	T	F
T	F	T	T	F	T
F	T	T	F	T	T
F	F	F	F	F	F

Conditional			Biconditional		
P	Q	P → Q	P	Q	P ↔ Q
T	T	T	T	T	T
T	F	F	T	F	F
F	T	T	F	T	F
F	F	T	F	F	T

2.5.2 Constructing Truth Tables

When constructing a truth table for a given statement that involves multiple connectives, keep in mind that every compound statement is composed of simple statements or other compound statements, or both. The secret to properly constructing a truth table is to recognize the type of statement given and its parts. If the parts of the statement consist of additional compound statements then we must identify them and their parts. In essence, we *decompose* a statement until it is reduced to a collection of simple statements.

To illustrate this concept, we will decompose the statement (P → ~Q) ∨ ~P → Q ∧ P. As you follow along it will be helpful to recall that of the five types of compound statements, all but the negation involve binary operations. Thus every compound statement except the negation contains *two parts*; the negation contains only *one* part.

<u>Decompose the statement (P → ~Q) V ~P → Q ∧ P</u>

Examining the given statement we make the following observations:

1. The statement is a *conditional*.

2. The two parts that make up this conditional are the statements (P → ~Q) V ~P (which is the antecedent and is on the left side of the arrow), and Q ∧ P (which is the consequent and is on the right side of the arrow).

3. Working with the consequent, we observe that the statement Q ∧ P is a *conjunction* and consists of the two simple statements Q and P. Thus the right side of the original statement has now been fully decomposed.

4. Working with the antecedent, we observe that the statement (P → ~Q) V ~P is a *disjunction*, which also consists of two parts: the statements P → ~Q and ~P. We further observe that the first statement P → ~Q is a *conditional*, which can be further decomposed as the simple statement P and the negation ~Q, and ~Q can be decomposed to the simple statement Q. The other statement, ~P can be reduced to the simple statement P.

The simplest way to understand the above explanation is to use a *tree diagram* to represent this decomposition.

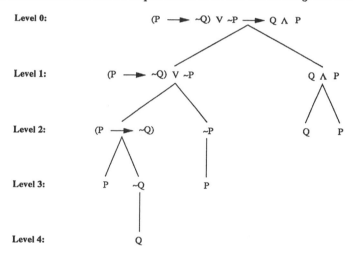

From the tree diagram above note that the given statement is decomposed into five levels. The top level, level 0, is the original statement. Level 1 represents the left and right sides (or parts) of the original statement: the left side is a disjunction, and the right side is a conjunction. At level 2, the disjunction and conjunction (from level 1) are decomposed; the disjunction is decomposed into its two parts (a conditional and a negation), and the conjunction is decomposed into the two simple statements Q and P. At level 3, the conditional and negation (from level 2) are decomposed; the conditional statement is decomposed into its two parts (the simple statement P and a negation), and the negation is decomposed into the simple statement P. Finally at the fifth level, level 4, the negation (from level 3) is decomposed into the simple statement Q.

Note further that it is relatively easy to reconstruct the original statement. All we need to do is follow the tree upwards from its lowest level (level 4) to its highest level (level 0), connecting the statements encountered using the proper connective at each subsequent (higher) level.

Once a given statement is decomposed in this manner the resulting structure lends itself to the completion of a truth table. The tree diagram can aid us in determining the order in which the columns of a truth table are to be completed. By working our way up from the bottom of the tree we complete those columns headed by the weaker connectives (those at lower levels), and move towards those columns headed by the stronger connectives (those at higher levels). Moreover, by moving in a left-to-right direction at each level, we can trace the branches of the tree to determine which statements are to be compared and under which connective.

In this study guide it is our decision to construct truth tables using tree diagrams. Please note that this method is neither discussed nor demonstrated in your textbook. To illustrate this method we complete a truth table for the given statement in the following manner.

Step 1

Assign the truth values (Ts and Fs) to the simple statements P and Q. Label these columns 1 and 2, respectively. (See the table below.) (**Note:** The pattern we use for these truth values is discussed in your textbook.)

P	Q
T	T
T	F
F	T
F	F
1	2

Step 2

Beginning up the tree from level 4, the first action we take is to negate Q. Thus we complete a column for ~Q by negating column 2. Label this column 3. (See the table below.)

P	Q	~Q
T	T	F
T	F	T
F	T	F
F	F	T
1	2	3

Step 3

Moving from level 3 to level 2 we must perform two actions:

1. We must complete a column for P → ~Q by combining columns 1 and 3 using the rules of a conditional
2. We must complete a column for ~P by negating column 1.

As a result, we now have two new columns, columns 4 and 5. (See the table below.)

P	Q	~Q	P → ~Q	~P
T	T	F	F	F
T	F	T	T	F
F	T	F	T	T
F	F	T	T	T
1	2	3	4	5

Continued on next page ...

Step 4

Moving from level 2 to level 1 we once again perform two actions:

1. We need to complete a column for (P → ~Q) V ~P by combining columns 4 and 5 using the rules of a disjunction.

2. We need to complete a column for Q ∧ P by combining columns 2 and 1 using the rules of a conjunction.

As a result we have two more columns, columns 6 and 7. (See the table below.)

P	Q	~Q	P →~Q	~P	(P → ~Q) V ~P	Q ∧ P
T	T	F	F	F	F	T
T	F	T	T	F	T	F
F	T	F	T	T	T	F
F	F	T	T	T	T	F
1	2	3	4	5	6	7

Step 5

Finally, moving from level 1 to level 0 we complete a column for the given statement by combining columns 6 and 7 using the rules of a conditional. Hence, the last column, column 8, represents the truth assignment for the given statement. (See the table below.)

P	Q	~Q	P → ~Q	~P	(P → ~Q) V ~P	Q ∧ P	(P → ~Q) V ~P → Q ∧ P
T	T	F	F	F	F	T	T
T	F	T	T	F	T	F	F
F	T	F	T	T	T	F	F
F	F	T	T	T	T	F	F
1	2	3	4	5	6	7	8

Note how the construction of the truth table follows directly from the tree diagram. Note further that the method we employed to construct the truth table involved a *building process*; we started with something simple and worked our way towards something more complicated.

2.5.3 Truth Table Vocabulary

1. If the final column of a truth table yields all Ts then we say the statement is a **tautology**.

2. If the final column yields all Fs then we say the statement is a **contradiction**.

3. Lastly, if the final columns of two truth tables are identical, then the statements that are represented by the truth tables are said to be **logically equivalent**.

Example 1 (page 68, #3 from section 2.4)

Construct a truth table for ~P ∧ ~Q.

Solution

- Decompose the statement using a tree diagram.

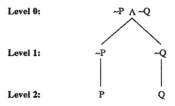

- Construct a truth table using the tree diagram as an aid.

P	Q	~P	~Q	~P ∧ ~Q
T	T	F	F	F
T	F	F	T	F
F	T	T	F	F
F	F	T	T	T
1	2	3	4	5

The truth value of the given conjunction is in column 5.

Example 2 (page 68, #11 from section 2.4)

Construct a truth table for the statement ~P ∨ (P ∧ ~Q).

Solution

- Decompose the statement using a tree diagram.

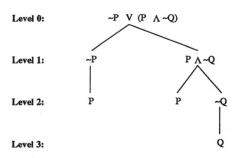

- Construct a truth table using the tree diagram as an aid.

P	Q	~Q	~P	P ∧ ~Q	~P ∨ (P ∧ ~Q)
T	T	F	F	F	F
T	F	T	F	T	T
F	T	F	T	F	T
F	F	T	T	F	T
1	2	3	4	5	6

The truth value of the given disjunction is in column 6.

LOGIC 41

Example 3 (page 74, #3 from section 2.5)

Construct a truth table for the statement ~P → ~Q.

- Decompose the statement using a tree diagram.

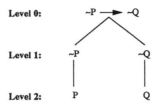

- Construct a truth table using the tree diagram as an aid.

P	Q	~P	~Q	~P → ~Q
T	T	F	F	T
T	F	F	T	T
F	T	T	F	F
F	F	T	T	T
1	2	3	4	5

The truth value of the given conditional is in column 5.

Example 4 (page 74, #9 from section 2.5)

Construct a truth table for the statement P V Q → ~Q.

Solution

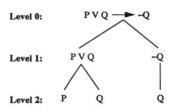

- Decompose the statement using a tree diagram.
- Construct a truth table using the tree diagram as an aid.

P	Q	P V Q	~Q	P V Q → ~Q
T	T	T	F	F
T	F	T	T	T
F	T	T	F	F
F	F	F	T	T
1	2	3	4	5

The truth value of the given conditional is in column 5.

Example 5 (page 75, #11 from section 2.5)

Construct a truth table for the statement $(P \rightarrow Q) \vee P \rightarrow Q$.

- Decompose the statement using a tree diagram.

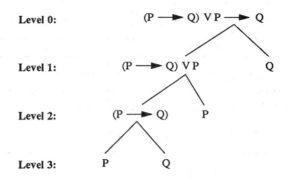

- Construct a truth table using the tree diagram as an aid.

P	Q	$P \rightarrow Q$	$(P \rightarrow Q) \vee P$	$(P \rightarrow Q) \vee P \rightarrow Q$
T	T	T	T	T
T	F	F	T	F
F	T	T	T	T
F	F	T	T	F
1	2	3	4	5

The truth value of the given conditional is in column 5.

Example 6 (page 75, #17 from section 2.5)

Construct a truth table for the statement $(P \wedge Q) \vee (P \wedge R)$

Solution

- Decompose the statement using a tree diagram.

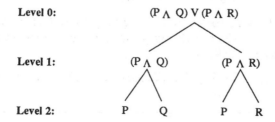

- Construct a truth table using the tree diagram as an aid.

Continued on next page ...

P	Q	R	P ∧ Q	P ∧ R	(P ∧ Q) V (P ∧ R)
T	T	T	T	T	T
T	T	F	T	F	T
T	F	T	F	T	T
T	F	F	F	F	F
F	T	T	F	F	F
F	T	F	F	F	F
F	F	T	F	F	F
F	F	F	F	F	F
1	2	3	4	5	6

The truth value of the given disjunction is in column 6.

Example 7 (page 75, #25a,e from section 2.5)

Use the fact that ~P V Q is logically equivalent to P → Q to rewrite each of the following statements.

Solution

In solving these problems we observe the following procedure:

1. Symbolize the given statement.

2. Express the symbolized form of the statement in its logically equivalent symbolized form.

3. Translate the new symbolized form into English.

a. If the tide is out then we can go clamming.

    1. Let T = *the tide is out* and C = *we can go clamming*. The symbolized form is T → C.

    2. The expression T → C is logically equivalent to the expression ~T V C. (Negate the antecedent and change the arrow to the *or* symbol.)

    3. Translating the expression ~T V C: *The tide is not out or we can go clamming.*

e. Bob did not pass the test, or he is unhappy about something else.

    1. Let P = *Bob did pass the test* and H = *he is happy about something else*. The symbolized form of the given statement is ~P V ~H.

    2. The expression ~P V ~H can be converted to a logically equivalent form by negating the first part of the statement (~P) and changing the *or* symbol to an arrow. Thus ~P V ~H is logically equivalent to P → ~H.

    3. Translating the expression P → ~H: *If Bob did pass the test then he is unhappy about something else.*

SUPPLEMENTARY EXERCISE 2.4-2.5

In 1 through 12 construct a truth table for each statement.

1. ~Q → P   2. ~(P V Q)   3. P ↔ ~Q ∧ P   4. (~P → Q) V P
5. P V (Q ∧ ~P)   6. P ∧ ~Q → ~P V Q   7. (P ↔ Q) V ~P   8. ~(~P → Q)
9. ~(P V Q) ∧ P   10. P ∧ Q ↔ ~Q   11. ~P V Q → R   12. P ∧ ~Q ↔ ~(R V P)

2.6 DEMORGAN'S LAW AND EQUIVALENT STATEMENTS

Any disjunction, conjunction, or the negation of a disjunction or conjunction can be changed to an equivalent statement using DeMorgan's Law. To do this we perform the following three steps:

1. Negate the entire statement. This involves enclosing the statement within parentheses and placing the symbol for negation (~) in front of the expression.//
2. Negate each statement that makes up the original disjunction or conjunction.
3. If the given statement is a conjunction, then change it to a disjunction; if the given statement is a disjunction, then change it to a conjunction.

The examples that follow demonstrate this procedure.

Example 1 (page 79, #1)

Use DeMorgan's Law to create an equivalent statement to $\sim(\sim P \wedge Q)$.

Solution

- Negate the entire statement.

$$\sim\sim(\sim P \wedge Q) = \sim P \wedge Q$$

- Negate each individual statement.

$$\sim\sim P \wedge \sim Q = P \wedge \sim Q$$

- Change the conjunction to a disjunction.

$$P \vee \sim Q$$

Thus an equivalent statement to $\sim(\sim P \wedge Q)$ is $P \vee \sim Q$.

Example 2 (page 79, #9)

Use DeMorgan's Law to write an equivalent statement to $\sim(P \rightarrow \sim Q)$

Solution

The given statement, $\sim(P \rightarrow \sim Q)$ is the negation of a conditional statement. DeMorgan's Law can only be applied to disjunctions or conjunctions. As a result we must first express the conditional statement $P \rightarrow \sim Q$ as an equivalent disjunction or conjunction. From our previous work (see example 7 in the previous section of this study guide) we learned that a conditional statement can be expressed as an equivalent disjunction by negating the antecedent and changing $\rightarrow$ to $\vee$. Applying this procedure to the conditional part of the given statement we have $P \rightarrow \sim Q \equiv \sim P \vee \sim Q$. Therefore, the given statement can be reexpressed as $\sim(\sim P \vee \sim Q)$. We can now use DeMorgan's Law.

- Negate the entire statement.

$$\sim\sim(\sim P \vee \sim Q) = (\sim P \vee \sim Q) = \sim P \vee \sim Q$$

- Negate each individual statement.

$$\sim\sim P \vee \sim\sim Q = P \vee Q$$

- Change the disjunction to a conjunction.

$$P \wedge Q$$

Thus an equivalent statement to $\sim(P \rightarrow \sim Q)$ is $P \wedge Q$.

LOGIC 45

Example 3 (page 79, #13)

Rewrite the following statement using DeMorgan's Law.

I did not pass the test, or I studied too much.

Solution

- First, symbolize the given statement.

  Let P = *I did pass the test* and S = *I studied too much*

  The symbolized form of the statement is: ~P V S

- Next, use DeMorgan's Law to write an equivalent statement.

  ~P V S ≡ ~(P ∧ ~S)

- Translate ~(P ∧ ~S) back to an English statement.

  It is false that I did pass the test and I did not study too much.

Example 4 (page 79, #21)

Rewrite the statement *If the wind doesn't come up, then we can't sail* using DeMorgan's Law.

Solution

- Symbolize the statement.

  Let W = *the wind does come up* and S = *we can sail*.

  The symbolized form is ~W → ~S.

- Rewrite the conditional statement as an equivalent disjunction.

  ~W → ~S ≡ W V ~S

- Use DeMorgan's Law to express the disjunction as an equivalent conjunction.

  W V ~S ≡ ~(~W ∧ S)

- Translate back to English.

  It is false that the wind did not come up and we did sail.

Example 5 (page 79, #25)

Use DeMorgan's Law to create an equivalent expression to the set expression A′ ∩ B′.

Solution

To solve this problem we must first express the given set expression as an equivalent statement in logic. To do this we use the following information.

- Set complement (′) corresponds to *not* (~)
- Set intersection (∩) corresponds to *and* (∧)
- Set union (∪) corresponds to *or* (V)

As a result, A′ ∩ B′ can be rewritten as ~A ∧ ~B. Using DeMorgan's Law this statement can be expressed as ~(A V B). Finally, converting back to set expression form, we have (A ∪ B)′.

SUPPLEMENTARY EXERCISE 2.6

In 1 through 5 use DeMorgan's Law to write an equivalent statement to each given statement.

1. ~P ∧ Q   2. ~(P ∧ ~Q)   3. ~P → Q   4. ~P V Q   5. ~[(P ∧ Q) V ~R]

In 6 through 10 use DeMorgan's Law to rewrite each statement.

6. It is false that Ann and Sue went to the beach.

7. I didn't go to the store or I played tennis.

8. It is not the case that if Fred cooks dinner, John will wash the dishes.

9. Neither Margaret nor Rita took the test.

10. If I don't wash my car then I'll bake peach muffins.

2.7 THE CONDITIONAL (OPTIONAL)

A conditional is a statement of the form *If P then Q* and is denoted $P \to Q$. The statement P that follows the word *If* is called the **antecedent** and the statement Q that follows the word *then* is called the **consequent**. Closely related to a conditional are three additional statements: the converse, inverse, and contrapositive.

- The **converse** of a conditional is formed by interchanging the antecedent and consequent of the original statement. The converse of P → Q is Q → P.

- The **inverse** of the conditional statement is formed by replacing P and Q of the original statement with their respective negations. The inverse of P → Q is ~P → ~Q.

- The **contrapositive** of a conditional statement is formed by interchanging the antecedent and consequent of the original statement and then replacing them with their respective negations. The contrapositive of P → Q is ~Q → ~P.

Of these three statements, only the contrapositive is logically equivalent to the original conditional statement from which it is derived.

2.7.1 Terminology Related to the Conditional Statement

The conditional statement, P → Q, can be expressed equivalently in many different ways. These expressions include, but are not limited to, those listed below.

1. *If* P, *then* Q
2. *If* P, Q
3. Q *if* P
4. P *implies* Q
5. Q *is implied by* P
6. Q *whenever* P
7. P *only if* Q
8. P *is sufficient* for Q
9. Q *is necessary for* P

2.7.2 Terminology Related to the Biconditional Statement

The biconditional statement, *P if and only if Q* (P ↔ Q), has two different but equivalent expressions.

1. P *if and only if* Q   2. P *is necessary and sufficient for* Q

Example 1 (page 87, #1a,e)

Use the suggested notation to write the converse, inverse, and contrapositive for each of the following statements in symbolic form.

Solution

a. If Jim studies, then he will pass. (S,P)

Let S = *Jim studies* and P = *He will pass*. The original statement is symbolized as S → P. We can now determine the converse, inverse, and contrapositive of this statement.

- The converse of S → P is: P → S (*If Jim passes then he will study.*)

- The inverse of S → P is: ~S → ~P (*If Jim does not study then h will not pass.*)

- The contrapositive of S → P is: ~P → ~S (*If Jim does not pass then he does not study.*)

e. If today is not Thursday then yesterday was not Wednesday. (T,W)

Let T = *Today is Thursday* and W = *Yesterday was Wednesday*. The original statement is symbolized as ~T → ~W.

- The converse is ~W → ~T (*If yesterday was not Wednesday then today is not Thursday.*)

- The inverse is T → W (*If today is Thursday then yesterday was Wednesday.*)

- The contrapositive is W → T (*If yesterday was Wednesday then today is Thursday.*)

Example 2 (page 87, #3a,c,d)

Let P = *It is a logic course* and Q = *It is interesting*. Write each of the following statements in symbolic form.

a. It is a logic course only if it is interesting.

The phrase *only if* indicates that the statement that precedes *only if* is the antecedent and the statement that follows *only if* is the consequent.

$$\begin{array}{ccc} \text{It is a logic course} & \textbf{only if} & \text{it is interesting} \\ P & \rightarrow & Q \end{array}$$

The symbolized form of the given statement is P → Q.

c. The fact that it is a logic course is sufficient for it to be interesting.

The phrase *is sufficient for* indicates that the statement that precedes it is the antecedent and the statement that follows it is the consequent.

$$\begin{array}{ccc} \text{it is a logic course} & \textbf{is sufficient for} & \text{it to be interesting.} \\ P & \rightarrow & Q \end{array}$$

The symbolized form of the given statement is P → Q.

Continued on next page ...

d. Only if it is interesting is it a logic course.

Once again, the phrase *only if* indicates that the statement that follows it is the consequent. The remaining statement must be the antecedent.

$$\text{Only if} \quad \underset{\text{(Consequent)}}{\underset{Q}{\text{it is interesting}}} \quad \underset{\text{(Antecedent)}}{\underset{P}{\text{is it a logic course.}}}$$

→

The symbolized form of the given statement is P → Q.

SUPPLEMENTARY EXERCISE 2.7

In 1 through 3 use the suggested notation to write the converse, inverse, and contrapositive of the given statement.

1. If I go shopping then I will buy a dress. (S,B)
2. Sally will not bake cookies if Margie bakes a cake. (M,S)
3. If the sun isn't shining then Rachel won't got to the beach. (S,R)

In 4 through 8 let P = *John likes peanut butter* and Q = *George likes apples*. Write each given statement in symbolic form.

4. George likes apples only if John likes peanut butter.
5. George likes apples, if John likes peanut butter.
6. For John to like peanut butter it is necessary and sufficient for George to like apples.
7. If John doesn't like peanut butter then George doesn't like apples.
8. For George to like apples it is sufficient for John to like peanut butter.

In 9 through 13 let P = *She will eat breakfast* and Q = *She has time*. Write each given statement in symbolic form.

9. She will eat breakfast iff she has time.
10. If she doesn't eat breakfast, then she doesn't have time.
11. Her having time is necessary for her eating breakfast.
12. She will eat breakfast only if she has time.
13. Her having time is sufficient for her eating breakfast.

2.8 VALID ARGUMENTS

An **argument** consists of a set of statements called **premises**, which when connected (or joined) by a conjunction, either imply or does not imply a statement called a **conclusion**. If the premises imply the conclusion (i.e., the conclusion follows logically from the premises), then the argument is said to be **valid**; otherwise the argument is **invalid**.

By joining the premises under a conjunction and implying the conclusion we in effect have a conditional statement. The conjuncted premises represent the antecedent and the implied conclusion represents the consequent. If we construct a truth table for this conditional statement we will be able to determine whether the argument is valid or invalid. If the truth table yields a tautology (the final column consists of all *trues*), then the argument is valid; otherwise it is invalid.

In the following examples we symbolize each argument using the suggested notation. A truth table also is constructed to determine whether the argument is valid or invalid.

Example 1 (page 95, #3)

Symbolize the following argument and determine whether the argument is valid or invalid (using a truth table).

> If I pass math, then I will graduate. (P,G)
> I graduated.
> Therefore, I passed math.

Solution

- Symbolize the argument.

$$P \to G$$
$$\underline{G}$$
$$P$$

- Join the premises by conjunction and imply the conclusion.

$$\underbrace{(P \to G)}_{\text{Premise One}} \land \underbrace{G}_{\text{Premise Two}} \to \underbrace{P}_{\text{Conclusion}}$$

- The truth table for this statement shows that the argument is invalid.

P	G	P → G	(P → G) ∧ G	(P → G) ∧ G → P
T	T	T	T	T
T	F	F	F	T
F	T	T	T	F
F	F	T	F	T
1	2	3	4	5

Example 2 (page 95, #5)

Symbolize the following argument and determine whether the argument is valid or invalid (using a truth table).

> I will graduate iff I pass math. (G,P)
> I graduated.
> Therefore, I passed math.

Solution

- Symbolize the argument.

$$G \leftrightarrow P$$
$$\underline{G}$$

Continued on next page ...

- Join the premises by conjunction and imply the conclusion.

$$(G \leftrightarrow P) \land G \rightarrow P$$

Premise One — Premise Two — Conclusion

- The truth table for this statement shows that the argument is valid.

G	P	G ↔ P	(G ↔ P) ∧ G	(G ↔ P) ∧ G → P
T	T	T	T	T
T	F	F	F	T
F	T	F	F	T
F	F	T	F	T
1	2	3	4	5

Example 3 (page 95, #7)

Symbolize the following argument and determine whether the argument is valid or invalid (using a truth table).

Addie and Bill will be at the party. (A,B)
Bill was at the party.
Therefore, Addie was at the party.

Solution

- Symbolize the argument.

$$\frac{\begin{array}{c} A \land B \\ B \end{array}}{A}$$

- Join the premises by conjunction and imply the conclusion.

$$(A \land B) \land B \rightarrow A$$

Premise One — Premise Two — Conclusion

- The truth table for this statement shows that the argument is valid.

A	B	A ∧ B	(A ∧ B) ∧ B	(A ∧ B) ∧ B → A
T	T	T	T	T
T	F	F	F	T
F	T	F	F	T
F	F	F	F	T
1	2	3	4	5

Example 4 (page 95, #15)

Symbolize the following argument and determine whether the argument is valid or invalid (using a truth table).

> Paul did not study, or he is bluffing. (P,B)
> If he is bluffing, then he will cut class. (B,C)
> Therefore, Paul did not study, or he will cut class.

Solution

- Symbolize the argument.

$$\frac{\begin{array}{c}\sim P \vee B \\ B \rightarrow C\end{array}}{\sim P \vee C}$$

- Join the premises by conjunction and imply the conclusion.

$$\underbrace{(\sim P \vee B)}_{\text{Premise One}} \wedge \underbrace{(B \rightarrow C)}_{\text{Premise Two}} \rightarrow \underbrace{\sim P \vee C}_{\text{Conclusion}}$$

- The truth table for this statement shows that the argument is valid.

P	B	C	~P	~P V B	B → C	(~P V B) ∧ (B → C)	~P V C	(~P V B) ∧ (B → C) → ~P V C
T	T	T	F	T	T	T	T	T
T	T	F	F	T	F	F	F	T
T	F	T	F	F	T	F	T	T
T	F	F	F	F	T	F	F	T
F	T	T	T	T	T	T	T	T
F	T	F	T	T	F	F	T	T
F	F	T	T	T	T	T	T	T
F	F	F	T	T	T	T	T	T
1	2	3	4	5	6	7	8	9

SUPPLEMENTARY EXERCISE 2.8

In 1 through 10 symbolize each of the following arguments using the suggested notation and use a truth table to determine whether the argument is valid or invalid.

1. Tom will walk or ride his bicycle. (W,B)
   Tom didn't ride his bicycle.
   Therefore, Tom walked.

2. If I wreck my car, my father will be mad at me. (C,M)
   My father is mad me.
   Therefore, I wrecked my car.

3. Shannon will run into the kitchen iff she hears the electric can opener. (R,H)
   Shannon doesn't run into the kitchen.
   Therefore, Shannon heard the electric can opener.

4. If Laura doesn't hurry, she'll be late for work. (H,L)
   Laura hurries.
   Therefore, she's not late for work.

52   STUDENT STUDY GUIDE

5. If 3 + 2 ≠ 5, then 4 + 4 = 8. (N,E)
    4 + 4 = 8.
    Therefore, 3 + 2 = 5.

6. Patti and John liked the movie. (P,J)
    John liked the movie.
    Therefore, Patti liked the movie.

7. If Alan doesn't call soon, Carol will be worried. (C,W)
    Carol isn't worried.
    Therefore, Alan called.

8. Paul will buy a shirt or a pair of shoes. (S,P)
    Paul didn't buy a shirt.
    Therefore, Paul didn't buy a pair of shoes.

9. If Robert has pizza for lunch, then he won't eat dinner. (P,D)
    Robert has pizza for lunch.
    Therefore, he won't eat dinner.

10. I'll go in the ocean iff the water is warm. (O,W)
    I don't go in the ocean.
    Therefore, the water isn't warm.

2.9 PICTURING STATEMENTS WITH VENN DIAGRAMS (OPTIONAL)

In addition to representing sets pictorially, Venn diagrams can be used in logic. Two such uses include:

- determining whether two statements are **inconsistent** (they cannot be true together; i.e., they contradict each other), or **consistent** (they can be true together); and
- determining whether a **syllogism** (which is the topic of section 2.9), is valid or invalid.

2.9.1 Quantified Statements

To apply Venn diagrams to these two areas of logic, certain types of statements, known as **quantified** statements, must be considered. We also must be able to symbolize (i.e., picture) these statements via Venn diagrams. Four such statements and their corresponding Venn diagram representations are summarized below.

The Universal Affirmative: All A's are B's

The quantifier *All* in the statement *All A's are B's* indicates that all the elements in Set A are contained in Set B. This implies that region I is empty. This is shown by *blanking out* region I in the Venn diagram.

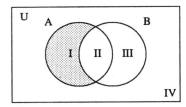

The Universal Negative: No A's are B's

The quantifier *No* in the statement *No A's are B's* indicates that Set A does not contain any elements that are also in Set B. This implies that region II is empty. This is shown by *blanking out* region II in the Venn diagram.

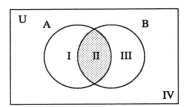

The Particular Affirmative: Some A's are B's

The statement *Some A's are B's* indicates that there is at least one element from Set A that is also contained in Set B. This implies that region II is not empty and contains at least one element. This is shown by placing an **x** in region II.

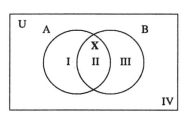

The Particular Negative: Some A's are not B's

The statement *Some A's are not B's* indicates that there is at least one element from Set A that is not contained in Set B. This implies that region I is not empty and contains at least one element. This is shown by placing an **x** in region I.

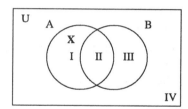

Example 1 (page 102, #1)

Identify and represent via a Venn diagram the statement *No heroes are imps*.

Solution

Let H = *Heroes* and I = *Imps*. The quantifier *No* indicates that the statement is a universal negative. Thus we must blank out the intersection of the two sets, namely region II.

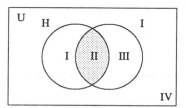

Example 2 (page 102, #3)

Identify and represent via a Venn diagram the statement *Some girls are not imps*.

Solution

Let G = *Girls* and I = *Imps*. The words *Some..are not...* indicate that the statement is a particular negative. Thus we symbolize this statement by placing an **x** in the region that represents *only girls*, namely region I.

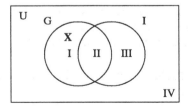

Example 3 (page 102, #5)

Identify and represent via a Venn diagram the statement *Some taxpayers are happy*.

Solution

Let T = *Taxpayers* and H = *Happy*. The words *Some...are...* indicate that the statement is a particular affirmative. Thus we must place an **x** in the intersection of the two sets, namely region II.

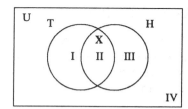

Example 4 (page 102, #21)

Use a Venn diagram to determine if the given pair of statements is consistent or inconsistent.

Some kind people are clever.
No clever people are kind.

Solution

- We first construct a Venn diagram for each statement.

    Some kind people are clever

    Let K = *Kind People* and C = *Clever People*. The statement is of the form *Some A's are B's*, which is a particular affirmative. Thus we place an **x** in region II.

    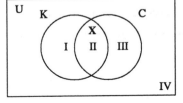

    No clever people are kind

    Let K = *Kind People* and C = *Clever People*. The statement is of the form *No A's are B's*, which is a universal negative. Thus we blank out region II.

    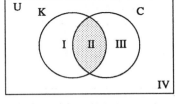

- Now we combine the two Venn diagrams. Note that the two statements are inconsistent. The first statement says that there is at least one kind person that is clever, but the second statement says that no people are clever and kind.

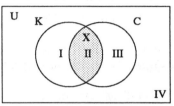

Example 5 (page 102, #25)

Use a Venn diagram to determine if the given pair of statements is consistent or inconsistent.

Some gamblers are losers.
Some losers are not gamblers.

Solution

- Construct a Venn diagram for each statement.

    Some gamblers are losers

    Let G = *Gamblers* and L = *Losers*. The statement is of the form *Some A's are B's*, which is a particular affirmative. Thus we place an **x** in region II.

    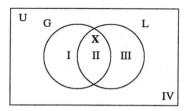

    Some losers are not gamblers

    Let G = *Gamblers* and L = *Losers*. The statement is of the form *Some A's are not B's*, which is a particular negative. Thus we place an **x** in region III. (**Note:** The x is placed in region III because of the way the sets of the Venn diagram are labeled.)

    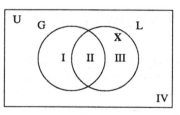

Continued on next page ...

LOGIC 55

- Now we combine the two Venn diagrams. The statements are consistent since it is possible for them to coexist.

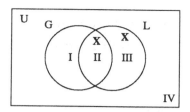

Example 6 (page 102, #29)

Use a Venn diagram to determine if the given pair of statements is consistent or inconsistent.

All politicians are sly.
No sly people are politicians.

Solution

Letting P = *Politicians* and S = *Sly People* we find that the two statements are **consistent**.

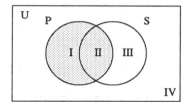

SUPPLEMENTARY EXERCISE 2.9

In 1 through 8 identify each statement as a *universal affirmative, universal negative, particular affirmative, particular negative*, and then represent the statement using a Venn diagram.

1. All Floridians like orange juice.
2. Some cats sleep all day.
3. Some flowers aren't red.
4. No cars are allowed on the beach.
5. Some cats don't like milk.
6. All dogs eat bones.
7. Some bicycles are twelve-speeds.
8. No dogs climb trees.

In 9 through 16 use a Venn diagram to determine whether each pair of statements is consistent or inconsistent.

9. Some cats are white.
   No cats are white.
10. All dogs bit people.
    Some dogs don't bite people.
11. Some books are too long.
    Some things that are too long are not books.
12. Some movies are funny.
    Some movies aren't funny.
13. Some rainy days are cold.
    Some cold days are rainy.
14. All women like jewelry.
    No one who likes jewelry is a woman.
15. Some cats are not black.
    No black animals are cats.
16. No candy is good for you.
    Some things that are good for you are not candy.

2.10 VALID ARGUMENTS AND VENN DIAGRAMS (OPTIONAL)

A **syllogism** is a special kind of argument that contains a major premise, a minor premise, and a conclusion. Venn diagrams are very useful in determining whether or not a syllogism is valid. (Recall that an argument is valid if the conclusion follows logically from the premises.) To determine the validity of a syllogism using Venn diagrams, the following procedure is used.

56 STUDENT STUDY GUIDE

Procedure for Determining the Validity of a Syllogism Using Venn Diagrams

1. Construct a Venn diagram with three overlapping circles. Each circle represents a unique set that is discussed in the premises and conclusion of the syllogism.

2. Diagram each premise using the technique discussed in the previous section.

3. *Do not diagram the conclusion.* Rather, check to see if the conclusion is pictured in the Venn diagram as a result of having diagramed the premises. If so, the syllogism is valid; otherwise the syllogism is invalid.

Example 1 (page 107, #1)

Determine if the following syllogism is valid using Venn diagrams.

No logic students are gullible.
Some gullible people are superstitious.
Hence, no logic students are superstitious.

Solution

- Construct a Venn Diagram to represent the syllogism. Let L = *Logic Students*, G = *Gullible People*, and S = *Superstitious People*.

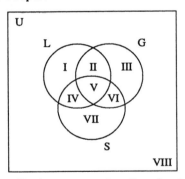

- Diagram each premise of the syllogism separately. Premise 1 indicates there are no elements contained in the intersection of *Logic Students* and *Gullible People*. So we blank out regions II and V. Premise 2 indicates there is at least one element contained in the set of *Gullible People* and *Superstitious People*. The intersection of these two sets is represented by regions V and VI. Since region V has been blanked out as a result of Premise 1, we have to place an x in region VI.

Premise 1
No logic students are gullible

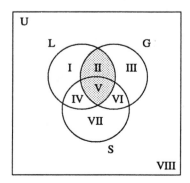

Premise 2
Some gullible people are superstitious

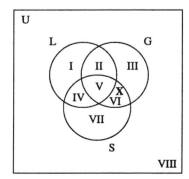

Continued on next page ...

- Determine if the conclusion is supported by the premises.

The conclusion, *No logic students are superstitious* implies that there are no elements contained in the intersection of the sets L and S (regions IV and V). If we were to diagram this statement regions IV and V would be blanked out. Note that the completed Venn diagram, which is obtained by combining the two premises, does not support this; region IV is not blanked out. Since the Venn diagram of the premises does not yield the conclusion, the syllogism is invalid.

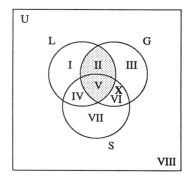

Example 2 (page 107, #3)

Determine if the following syllogism is valid using Venn diagrams.

All coins are valuable.
Some coins are old.
Therefore, some old things are valuable.

Solution

- Construct a Venn Diagram to represent the syllogism. Let C = *Coins*, O = *Old Things*, and V = *Valuable Things*.

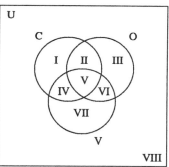

- Diagram each premise of the syllogism separately. Premise 1 indicates that elements contained in set C are also contained in set V. This implies that regions I and II must be blanked out. Premise 2 indicates that there is at least one element from set C contained in set O. We are forced to place an **x** in region V.

Premise 1
All coins are valuable

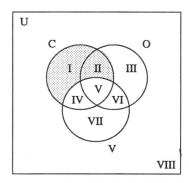

Premise 2
Some coins are old

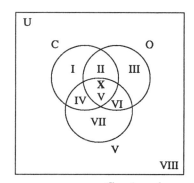

Continued on next page ...

58 STUDENT STUDY GUIDE

- Determine if the conclusion is supported by the premises.

The conclusion, *Some old things are valuable* implies there is at least one element contained in the intersection of sets O and V (regions V and VI). If we were to diagram this statement we would place an **x** in either region V or VI. Note that the completed Venn diagram, which is obtained by combining the two premises, supports this; region V has an **x**. Since the Venn diagram of the premises yields the conclusion, the syllogism is valid.

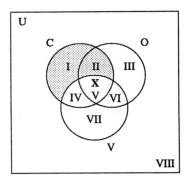

Example 3 (page 107, #5)

Determine if the following syllogism is valid using Venn diagrams.

All cars have four wheels.
No bikes have four wheels.
Therefore, no cars are bikes..

Solution

- Construct a Venn Diagram to represent the syllogism. Let C = *Car*, F = *Four Wheels*, and B = *Bikes*.

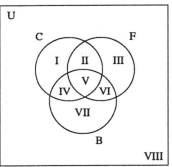

- Diagram each premise of the syllogism separately. Premise 1 indicates that every car must have wheels. So we blank out regions I and IV. Premise 2 is represented by blanking out the intersection of sets B and F.

Premise 1
All cars have four wheels

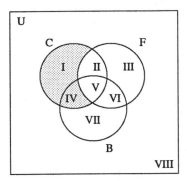

Premise 2
No bikes have four wheels

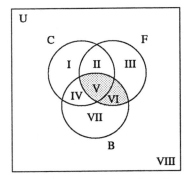

Continued on next page ...

- Determine if the conclusion is supported by the premises.

    The conclusion, *No cars are bikes* would be diagramed by blanking out regions IV and V. Note that the completed Venn diagram, which is obtained by combining the two premises, supports this; regions IV and V are blanked out. Since the Venn diagram of the premises yields the conclusion, the syllogism is valid.

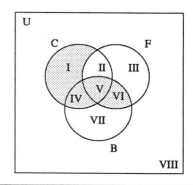

Example 4 (page 108, #7)

Determine if the following syllogism is valid using Venn diagrams.

All books are readable.
Some readable things are not interesting.
So, some books are not interesting.

Solution

- Construct a Venn Diagram to represent the syllogism.

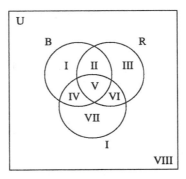

- Diagram each premise of the syllogism separately.

    Premise 1 indicates that regions I and IV must be blanked out since all the elements of set B are also contained in set R.

Premise 1

All books are readable

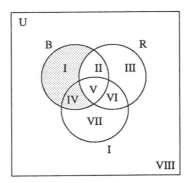

Continued on next page ...

Premise 2 indicates that there is at least one element from set R that is not contained in set I. Note that we have a choice of where to place the **x** when diagraming this statement. We can place the **x** in region II (as shown in the Venn diagram for Premise 2A) or in region III (as shown in the Venn diagram for Premise 2B).

Premise 2A
Some readable things are not interesting

Premise 2B
Some readable things are not interesting

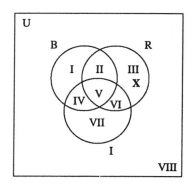

To decide which region gets the **x** we need to consider the conclusion, *Some books are not interesting*. If we disregard set R for a moment, the conclusion can be shown by placing an **x** in either region I or II. Since region I is empty, though (as a result of Premise I), we are forced to place the **x** in region II. This supports the conclusion and makes the syllogism valid. (See Possible Conclusion 1.) On the other hand, if we were to place the **x** in region III, then the completed Venn diagram would not support the conclusion, and the syllogism is invalid. (See Possible Conclusion 2.)

Possible Conclusion 1

Possible Conclusion 2

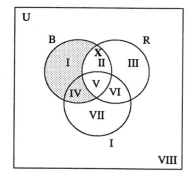 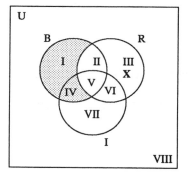

As a result, we have ambiguity. In one case, we can diagram the premises so the conclusion is supported. Yet in a second case the premises can be diagramed so the conclusion is not supported. In such instances, we opt for the latter. *If it is at all possible to diagram the premises without showing the conclusion, then the syllogism is invalid.*

LOGIC 61

SUPPLEMENTARY EXERCISE 2.10

In 1 through 10 use Venn diagrams to determine whether the following syllogisms are valid or invalid.

1. Some cats are black.
   Some black things have green eyes.
   Therefore, some cats have green eyes.

2. All tomatoes are red.
   Some apples are red.
   So, some apples are tomatoes.

3. All dogs are playful.
   No cats are playful.
   Hence, no dogs are cats.

4. All cars have power steering.
   Some trucks have power steering.
   Therefore, some trucks are not cars.

5. No mink coats are cheap.
   Some cheap things last a long time.
   Hence, no mink coats last a long time.

6. No oranges are vegetables.
   No vegetables are fruits.
   Therefore, no oranges are fruits.

7. No shoes are comfortable.
   Some beds are comfortable.
   Therefore, some beds are not shoes.

8. All surfers are good swimmers.
   All good swimmers are agile.
   Some agile people are surfers.

9. All birds can fly.
   Some things that cannot fly are not alive.
   Therefore, some birds are not alive.

10. No women are bodybuilders.
    Some men are bodybuilders.
    Hence, some women are not men.

2.11 SWITCHING NETWORKS

A statement that involves a conjunction, disjunction, or negation can be expressed via a **network diagram**. Typically, the value *true* (T) is interpreted as a closed switch (electrical current can pass through it) and *false* (F) is interpreted as an open switch (electric current cannot pass through it).

2.11.1 A Switching Network for a Conjunction

If two switches P and Q are connected in **series** then the network corresponds to the conjunction $P \wedge Q$. Electrical current will pass through the network if both switches are closed (T), but not otherwise (F). An illustration of this is shown below where a light bulb is lit when both switches are closed, but is unlit when at least one of the two switches is open.

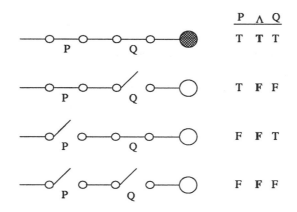

2.11.2 A Switching Network for a Disjunction

If two switches P and Q are connected in **parallel** then the network corresponds to the disjunction $P \vee Q$. Electrical current will pass through the network when either switch P or Q, or both P and Q are closed, but not otherwise. The light bulb is lit when at least one switch is closed.

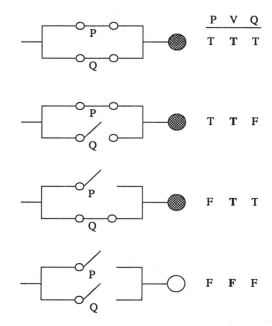

	P	V	Q
	T	T	T
	T	T	F
	F	T	T
	F	F	F

2.11.3 A Switching Network for a Negation

If switch P is open then switch P′ is closed. Inversely, if switch P is closed then switch P′ is open. Shown below are two different switching networks (series and parallel), which depict complementary switches. Note that in the series network electrical current will never pass through the network since one of the two switches will always be open. However, in the parallel network electrical current will always pass through it.

Series Circuit for a Negation

	P	∧	~P
	T	F	F
	F	F	T

Parallel Circuit for Negation

	P	V	~P
	T	T	F
	F	T	T

LOGIC 63

In the examples that follow the notation —P— is used to represent a switch whose position is not known.

Example 1 (page 117, #1)

Write a symbolic statement for the network shown below.

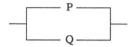

Solution

The switches P and Q are in parallel and represent the disjunction P ∨ Q.

Example 2 (page 117, #3)

Write a symbolic statement for the network shown below.

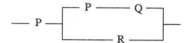

Solution

- The top wire contains switches P and Q connected in series. Since wires connected in series imply a conjunction, the top wire is represented by P ∧ Q.

- The bottom wire contains switch R. This wire is connected to the top wire in parallel. Since wires connected in parallel imply disjunction, the top and bottom wires are represented by the statement (P ∧ Q) ∨ R.

- Finally, the lone wire on the left side of the given network contains switch P. Since this wire is connected in series with the rest of the circuit, the symbolic statement that represents the given network is the conjunction P ∧ [(P ∧ Q) ∨ R].

Example 3 (page 117, #7)

Write a symbolic statement for the network shown below.

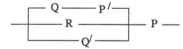

Solution

- The top wire contains switches Q and P′ connected in series. This wire is represented by Q ∧ ~P.

- The middle wire contains switch R and switch P. At this junction we only consider switch R Switch P is incroprated at the end. Since the middle wire is connected in series with the top wire, switch R is combined with Q ∧ ~P via a disjunction. These two wires are represented by the disjunction (Q ∧ ~P) ∨ R.

- The bottom wire consists of switch Q′. Since this wire is connected in parallel with the first and second wires, switch Q′ is combined with the statement (Q ∧ ~P) ∨ R via a disjunction. This leads to the statement (Q ∧ ~P) ∨ R ∨ ~Q.

- Now incorporate switch P. Switches Q, P′, R, and ~Q are in series with switch P. This implies that the statement (Q ∧ ~P) ∨ R ∨ ~Q is combined with P via a conjunction. The complete circuit is represented by the statement [(Q ∧ ~P) ∨ R ∨ ~Q] ∧ P.

Example 4 (page 118, #15)

Construct the switching network that corresponds to the statement (P V Q) ∧ (~P V R).

Solution

- Construct the switching network for P V Q.

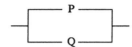

- Construct the switching network for ~P V R.

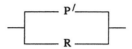

- Since the given statement is a conjunction, place the two networks in series.

Example 5 (page 118, #17)

Construct the switching network that corresponds to the statement P V [Q ∧ (P V ~Q)].

Solution

- Construct the switching network for P V ~Q.

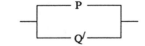

- Place switch Q in series with P V ~Q.

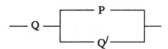

- Since the given statement is a disjunction, the network is placed in parallel with switch P.

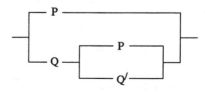

SUPPLEMENTARY EXERCISE 2.11

In 1 through 5 write a symbolic statement for the networks shown.

1.

2.

3.

4.

5.

In 6 through 10 construct the switching networks that correspond to the given statements.

6. (P ∧ Q) V R
7. ~P → Q
8. (P V ~Q) ∧ (R V ~P)
9. [P V (R ∧ Q)] ∧ P
10. (~P ∧ Q) V [(~R V P) ∧ ~Q]

2.12 CHAPTER 2 TEST

In 1 through 8 determine whether the given statement is true or false.

1. P → Q is the inverse of Q → P.
2. P ↔ ~P is a contradiction.
3. P → Q ≡ P V ~Q.
4. ~(P ∧ Q) ≡ ~P ∧ ~Q.
5. P → Q is the contrapositive of ~Q → ~P.
6. *Don't make noise* is a simple statement.
7. *Q only if P* means Q → P.
8. ~P V ~Q → R ∧ ~S is a conditional.

In 9 through 13 let P = *Pam does the laundry* and Q = *Sue does the ironing*. Write each of the given statements in symbolic form.

9. Pam does the laundry whenever Sue does the ironing.
10. If Sue doesn't do the ironing then Pam doesn't do the laundry.
11. Pam does the laundry, but Sue doesn't do the ironing.
12. Either Pam does the laundry or Sue doesn't do the ironing.
13. It is false that if Pam does the laundry then Sue does the ironing.

Continued on next page ...

In 14 through 16 translate the given statement where P = *You use sunscreen* and Q = *You get a sunburn*.

14. $P \rightarrow \sim Q$ 15. $\sim P \wedge Q$ 16. $\sim(Q \rightarrow \sim P)$

In 17 and 18 construct a truth table for the given statement.

17. $P \leftrightarrow \sim Q$ 18. $\sim(\sim P \vee Q)$

19. Determine whether the statement $P \vee \sim Q$ is logically equivalent to the statement $Q \rightarrow \sim P$ using a truth table.

20. Symbolize the following argument and then use a truth table to determine whether the argument is valid or invalid.

Louise or Marcia will sew the dress. (L,M)
Marcia did not sew the dress.
Therefore, Louise sewed the dress.

In 21 through 23 use a Venn diagram to determine whether the given arguments are valid or invalid.

21. *All liars are devious.*
    Some liars are not clever.
    Therefore, some clever people are not liars.

22. *Some girls are good cooks.*
    All good cooks are well organized.
    Therefore some girls are well organized.

23. *All movie stars are rich.*
    No janitors are rich.
    Therefore no movie stars are janitors.

24. Let P = *Tom plays tennis* and Q = *Tom rides his bicycle*. Write the converse, inverse, and contrapositive of the statement, *If Tom plays tennis, then he will not ride his bicycle.*

In 25 and 26 use a Venn diagram to determine whether each pair of statements is consistent or inconsistent.

25. No horses are green.   26. All rainy days are depressing.
    Some horses are green   Some rainy days are not depressing.

3. PROBABILITY

3.1 INTRODUCTION

There are many situations in life in which it is impossible to determine *exactly* a specific result. For example, it is impossible to know exactly what the weather will be like on a particular day. In such instances an experiment is usually conducted using some sort of model. Since the results of the experiment cannot be determined at the time the experiment is conducted, it is necessary to predict the results. This prediction is based on the likelihood of occurrence of the possible results of the experiment. The likelihood of a specific result is assigned a number between zero and one and represents the *probability* of that result occurring. In this chapter we will define probability and discuss how it is calculated.

3.2 DEFINITION OF PROBABILITY

3.2.1 The Vocabulary of Probability

Formally, the result of an experiment is called an **outcome** of the experiment. The set of all possible outcomes is called a **sample space**. If every member (i.e., outcome) of the sample space has an equal chance of occurring, then the outcomes are referred to as **equally likely outcomes**. Each individual outcome of an experiment, or any grouping of the outcomes, is called an **event**. Thus an event is any subset of a sample space.

As an example consider the experiment of rolling a die and observing the number of *dots* that appear on the top face of the die when it comes to rest. The sample space for this experiment consists of {1,2,3,4,5,6}. Some possible events and their corresponding subsets are listed in the table below.

Possible Events	Subsets
Observe a one	{1}
Observe a five	{5}
Observe an even number	{2,4,6}
Observe a number less than five	{1,2,3,4}
Observe a number greater than six	∅

3.2.2 Definition of Probability

If the outcomes of an experiment are equally likely, then the probability of event A, denoted $P(A)$, is defined to be

$$P(A) = \frac{\text{The number of successful outcomes}}{\text{The total number of possible outcomes}}$$

The sum of the probabilities of the individual outcomes in a sample space is one (1). Moreover, the probability of an event occurring added to the probability of the same event not occurring is equal to one. That is, $P(A) + P(\text{not } A) = 1$.

68   STUDENT STUDY GUIDE

3.2.3 Procedure for Solving Simple Probability Problems

To solve the probability problems discussed in this section, the following procedure will be followed.

1. Determine the sample space. That is, determine the total number of possible outcomes a particular experiment has. This will be the denominator of the answer.

2. From the sample space, determine how many of the outcomes satisfy the event being discussed. That is, determine the number of successful outcomes. This will be the numerator of the answer.

Example 1 (page 131, #1a,b,d,f)

On a single toss of one die find the probability of obtaining (a) a 4, (b) an odd number, (d) a number less than 4, and (f) an odd or even number.

Solution

For each part of this problem the sample space consists of the set $\{1,2,3,4,5,6\}$, which represents the total number of possible outcomes. Thus the total number of possible outcomes is 6.

a. $P(4) = ?$

   Since there is only a single 4 in the sample space, the number of successful outcomes is one. Thus, $P(4) = \frac{1}{6}$.

b. $P(\text{odd number}) = ?$

   In the sample space, there are only three odd numbers: 1, 3, and 5. Since the total number of successful outcomes is three, $P(\text{odd number}) = \frac{3}{6}$ or $\frac{1}{2}$.

d. $P(\# < 4) = ?$

   The sample space contains three numbers less than four: 1, 2, and 3. Thus, $P(\# < 4) = \frac{3}{6}$ or $\frac{1}{2}$.

f. $P(\text{odd or even}) = ?$

   There are three odd numbers (1,3,5) and three even numbers (2,4,6) in the sample space. Thus, $P(\text{odd or even}) = \frac{6}{6}$ or 1.

Example 2 (page 132, #5b,d)

On a single draw from a shuffled standard deck of 52 cards, find the probability of obtaining (b) a picture card or a heart, and (d) a jack and a heart.

Solution

The sample space for this problem is a standard deck of cards. Therefore the total number of possible outcomes is 52.

b. $P(\text{picture card or heart}) = ?$

   - There are three picture cards for each suit of cards: The jack, queen, and king of diamonds, hearts, clubs, and spades. This yields a total of 12 successful outcomes for *picture card*.

Continued on next page ...

- There are 13 hearts (ace through king). However, note that the jack, queen, and king of hearts were included in the outcomes for *picture card*. So the total number of successful outcomes for *heart* is 10.

Therefore, $P(\text{picture card or heart}) = \dfrac{12 + 10}{52} = \dfrac{22}{52} = \dfrac{11}{26}$

d. $P(\text{jack and heart}) = ?$

There are four jacks: The jack of hearts, diamonds, clubs, and spades. Of these four jacks only one of them is also a heart. Therefore, $P(\text{jack and heart}) = \dfrac{1}{52}$.

Example 3 (page 132, #7a,c,e)

Gloria Glove keeps all of her mittens on the top shelf of her hall closet. On the shelf are four blue mittens, six brown mittens, and four green mittens. Gloria reaches up and pulls a mitten out at random. Find the probability that the mitten chosen is (a) blue or green, (c) not red, and (e) neither blue nor green.

Solution

The sample space for the problem consists of 14 mittens (4 blue, 6 brown, and 4 green).

a. $P(\text{blue or green}) = \dfrac{4 \text{ blue} + 4 \text{ green}}{14} = \dfrac{8}{14} = \dfrac{4}{7}$

c. $P(\text{not red}) = \dfrac{4 \text{ blue} + 6 \text{ brown} + 4 \text{ green}}{14} = \dfrac{14}{14} = 1$

e. $P(\text{Neither blue nor green}) = P(\text{brown}) = \dfrac{6}{14} = \dfrac{3}{7}$

SUPPLEMENTARY EXERCISE 3.2

In problems 1 through 5 assume a die is tossed once. Find the probability of obtaining:

1. a number greater than 2
2. an odd number less than 5
3. a number divisible by 4
4. a number greater than 6
5. a 1 or a 3

In problems 6 through 15 assume that a single card is drawn from a shuffled standard deck of 52 cards. Find the probability of obtaining:

6. the five of clubs
7. a seven
8. a black picture card
9. a heart
10. a heart or a diamond
11. a nine or a diamond
12. a nine and a diamond
13. a picture card or a spade
14. a heart and a diamond
15. a red king

Use the following story to answer questions 16 through 20:

Aunt Sarah baked four different kinds of cookies: butter, chocolate chip, oatmeal, and sugar. All that are left in her cookie jar are four butter, two chocolate chip, seven oatmeal, and five sugar. If she reaches into the jar and picks one cookie out at random, what is the probability that she picks:

16. a butter cookie?
17. a sugar cookie?
18. a chocolate chip or oatmeal cookie?
19. anything but a butter cookie?
20. neither a sugar cookie nor an oatmeal cookie?

3.3 SAMPLE SPACES

(This section is combined with section 3.4.)

3.4 TREE DIAGRAMS

Frequently the total number of possible outcomes of a particular experiment is not readily apparent. This is especially true when an experiment is repeated more than once, or if an experiment consists of two or more parts (each with different outcomes). In such cases we use the **Fundamental Counting Principle (FCP)** to determine the total number of possible outcomes.

3.4.1 The Fundamental Counting Principle

Assume we are conducting an experiment with more than one part to it. Further assume that the first part of the experiment consists of M different outcomes, the second part consists of N different outcomes, ... , and the last part consists of T different outcomes. The total number of outcomes of the experiment is determined by multiplying the number of outcomes for each part of the experiment (i.e., $M \times N \times ... \times T$).

A tree diagram (see section 1.7) can also be used to construct the sample space of a multiple part experiment.

Example 1 (page 138, #3a-f from section 3.3)

A box contains a one-dollar bill, a five-dollar bill, a ten-dollar bill, a twenty-dollar bill and a fifty-dollar bill.. Two bills are chosen at random in succession. The first bill is not replaced before the second bill is drawn. Answer the questions that follow.

Solution

a. How many outcomes are possible?

   This experiment consists of two parts: Choose one bill, do not replace it, and then choose another bill. The total number of outcomes for the first part of the experiment is five (there are five bills from which to choose). Once a bill is chosen, only four remain. Thus the second part of the experiment has only four possible outcomes. Using the FCP, the total number of outcomes for this experiment is $5 \times 4 = 20$.

b. Construct a sample space for this experiment. (A tree diagram is used.)

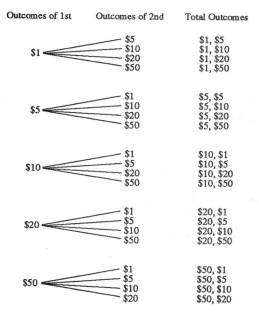

Note that the outcomes of the second part of the experiment is dependent upon the result of the first part. Also, the total number of outcomes shown is 20 and represents the sample space for this experiment.

c. P (first bill is even) = ?

Looking at the sample space in (b) above, there are 12 instances where the first bill is even (i.e., the first bill is $10, $20, or $50). Thus, P (first bill is even) = $\frac{12}{20} = \frac{3}{5}$.

d. P (second bill is even) = ?

The sample space contains 12 instances where the second bill is even.
P (second bill is even) = $\frac{12}{20} = \frac{3}{5}$.

e. P (both bills are even) = ?

The sample space contains six instances where both bills are even.
P (both bills are even) = $\frac{6}{20} = \frac{3}{10}$.

f. P (neither bills are even) = ?

There are two instances in which both bills are not even (i.e., odd).
P (neither bills are even) = $\frac{2}{20} = \frac{1}{10}$.

Example 2 (page 142, #3a-f from section 3.4)

Use a tree diagram to list the sample space showing the possible arrangements of boys and girls with exactly three children. (Questions a-f are answered following the tree diagram.)

Solution

The sample space for this problem is shown below.

1st Child	2nd Child	3rd Child	Total Outcomes
B	B	B	BBB
B	B	G	BBG
B	G	B	BGB
B	G	G	BGG
G	B	B	GBB
G	B	G	GBG
G	G	B	GGB
G	G	G	GGG

There are eight possible outcomes. This is confirmed by the FCP: $2 \times 2 \times 2$.

a. P (all three children are boys) = P (BBB) = $\frac{1}{8}$

b. P (two boys and one girl) = P (BBG, BGB, GBB) = $\frac{3}{8}$

c. P (at least one girl) = $1 - P$ (all boys) = $1 - \frac{1}{8} = \frac{7}{8}$

Continued on next page ...

(**Note:** *At least one girl* means one or more girls. There are seven instances where there is one or more girls. This is equivalent to *one minus the probability of all boys*.)

d. $P(\text{no boys}) = P(\text{all girls}) = P(GGG) = \dfrac{1}{8}$

e. $P(\text{all three are the same sex}) = P(BBB, GGG) = \dfrac{2}{8} = \dfrac{1}{4}$

f. $P(\text{fourth child is a girl}) = P(\text{girl}) = \dfrac{1}{2}$

(**Note:** Since the sex of the first three children was determined, the probability that the fourth child is a girl is based on a sample space of {boy, girl}.)

SUPPLEMENTARY EXERCISE 3.3-3.4

1. A nickel is flipped and a dime is flipped.

   a. Use a tree diagram to list the sample space for this experiment.

   b. Find F (at least one head)

In problems 2 through 4 use a tree diagram to list the sample space showing the possible arrangements of heads and tails when three coins are tossed.

2. Find the probability that all three coins are tails.

3. Find the probability that exactly two coins are tails.

4. Find the probability that at most two coins are tails.

In problems 5 through 8 a car dealer has the following selections available.

Category I	Category II	Category III
Convertible	Automatic	Black
Sedan	Standard Shift	Blue
Station Wagon		Red
		White

A customer selects one option from each category.

5. What is the probability that a customer chooses a station wagon?

6. What is the probability that a customer selects a blue convertible with an automatic transmission?

7. What is the probability that a customer selects a white, standard shift sedan or white, standard shift station wagon?

8. What is the probability that a customer chooses either a red or black sedan with a standard shift?

9. There are 20 mathematics instructors at Brevard Community College. Eight of these instructors teach calculus, 16 teach algebra, and 14 teach trigonometry. Seven teach calculus and algebra, 11 teach trigonometry and algebra, and six teach calculus and trigonometry. Five teach all three courses. Using a Venn diagram find the probability that a mathematics instructor teaches:

   a. only calculus b. only algebra
   c. only trigonometry d. calculus and algebra, but not trigonometry
   e. algebra or trigonometry f. none of these courses

Continued on next page ...

10. A florist took a survey of 50 customers to determine which flowers they preferred. Forty-five customers liked roses, 38 liked lilacs, and 35 liked carnations. Thirty-five customers liked roses and lilacs, 31 liked lilacs and carnations, and 33 liked carnations and roses. Thirty customers liked all three kinds of flowers. Use a Venn diagram to find the probability that a customer likes:

   a. only roses b. only lilacs c. exactly one of these flowers
   d. none of these flowers   e. carnations and roses, but not lilacs

3.5 ODDS AND EXPECTATIONS

3.5.1 Odds

Odds can occur in two ways: odds in favor and odds against an event occurring. The formulas needed to calculate either types of odds are given below.

3.5.1.1 Formulas Used To Calculate Odds

$$\text{Odds in favor of an event } A \text{ occurring} = \frac{\text{Probability that } A \text{ will occur}}{\text{Probability that } A \text{ will not occur}}$$

$$\text{Odds against an event } A \text{ occurring} = \frac{\text{Probability that } A \text{ will not occur}}{\text{Probability that } A \text{ will occur}}$$

Notice that *odds in favor* and *odds against* are reciprocals of each other. In order to calculate odds we must first know the probability of an event occurring. Once this probability is known we can then find the probability of the same event not occurring by using the formula $P(notA) = 1 - P(A)$, (since $P(A) + P(notA) = 1$). The method used for calculating probability is the same as before, namely, $P(A) = \frac{success}{total}$.

3.5.2 Expectations

A related topic of probability and odds is **mathematical expectation**. Mathematical expectation represents the expected value (i.e., what you would expect to win or lose) of an experiment conducted over a long period of time. Expected value can be calculated by the formula $E = P \times M$ where $P = $ *the probability that an event will occur* and $M = $ *the amount that will be won or lost if the event were to occur*. Expected value can also be regarded as a fair price to pay to play a game.

Example 1 (page 151, #1a,d)

In a single toss of a pair of dice find the odds for parts a and d.

Solution

a. odds in favor of obtaining a 7 = ?

   - The probability of obtaining 7 is $\frac{6}{36} = \frac{1}{6}$
   - The probability of not obtaining a 7 is $1 - \frac{6}{36} = \frac{30}{36} = \frac{5}{6}$

Continued on next page ...

- The odds in favor of obtaining a 7 are equal to $\dfrac{P(7)}{P(\text{not } 7)}$, which is equal to

$$\dfrac{\frac{1}{6}}{\frac{5}{6}} = \dfrac{1}{6} \div \dfrac{5}{6} = \dfrac{1}{6} \times \dfrac{6}{5} = \dfrac{6}{30} = \dfrac{1}{5}$$

The odds are 1 to 5.

d. Odds against obtaining a 12 = ?

- The probability of obtaining a 12 is $\dfrac{1}{36}$
- The probability of not obtaining a 12 is $\dfrac{35}{36}$
- The odds against obtaining a 12 are equal to $\dfrac{P(\text{not } 12)}{P(12)}$, which is equal to

$$= \dfrac{\frac{35}{36}}{\frac{1}{36}} = \dfrac{35}{36} \div \dfrac{1}{36} = \dfrac{35}{36} \times \dfrac{36}{1} = \dfrac{35}{1} \text{ or } 35:1$$

The odds are 35 to 1.

Example 2 (page 151, #5a,b,e)

In a single toss of two coins make a sample space and find the odds for parts a, b, and e.

Solution

The sample space for this problem is shown below.

```
      1st Toss    2nd Toss    Total Outcomes
                     H           HH
          H <
                     T           HT

                     H           TH
          T <
                     T           TT
```

a. Odds in favor of two tails = ?

- $P(TT) = \dfrac{1}{4}$
- $P(\text{not } TT) = \dfrac{3}{4}$
- *Odds in favor of* $(TT) = \dfrac{1}{4} \div \dfrac{3}{4} = \dfrac{1}{4} \times \dfrac{4}{3} = 1:3$

b. Odds against two heads = ?

- $P(HH) = \dfrac{1}{4}$
- $P(\text{not } HH) = \dfrac{3}{4}$
- *Odds against* $(HH) = \dfrac{3}{4} \div \dfrac{1}{4} = \dfrac{3}{4} \times \dfrac{4}{1} = 3:1$

Continued on next page ...

e. Odds in favor of at least one tail = ?

- $P(\text{at least one tail}) = P(HT, TH, TT) = \dfrac{3}{4}$

- $P(\text{not at least one tail}) = \dfrac{1}{4}$

- $\text{Odds in favor of at least one tail} = \dfrac{3}{4} \div \dfrac{1}{4} = \dfrac{3}{4} \times \dfrac{4}{1} = 3:1$

Example 3 (page 151, #7a,c)

Find the probability that event B will happen given the odds in parts a and c.

Solution

a. 8:3 in favor of B.

   To find the probability given the odds, note the following:

   1. The sum of the individual digits that make up the odds represents the total number of possible outcomes of the experiment, which is the denominator of the corresponding probability. In this problem the odds are 8:3 so there are a total of 11 possible outcomes (8 + 3 = 11). The **denominator** is 11.

   2. The individual digits that make up the odds represent the numerator of the corresponding probability. However, we must be sensitive to which probability we need: in favor or against. In this problem the odds (8:3) are given as odds in favor. Thus there are 8 favorable outcomes and 3 unfavorable outcomes (i.e., against). Therefore the **numerator** of the probability is 8.

   Since the formula for probability is
   $$P(A) = \dfrac{\text{\# of favorable outcomes}}{\text{\# of possible outcomes}}$$
   the probability that event B will happen if the odds are 8:3 in favor of B is $\dfrac{8}{11}$.

c. 7:5 against B.

   The solution to part c is similar to that given for part a. The odds against formula gives the ratio of unfavorable outcomes to favorable outcomes. The sum of the number of unfavorable outcomes and the number of favorable outcomes equals the number of possible outcomes. So, 7 + 5 = 12. Furthermore, the number of favorable outcomes is 5. Using the probability formula, the probability that event B will happen given that the odds are 7:5 against is $\dfrac{5}{12}$.

Example 4 (page 151, #13)

The Association for Conservation is awarding a $1,000 cash prize to the winner of a raffle. A total of 5,000 tickets are sold for $2 each. What is the mathematical expectation for a person who buys five tickets?

Solution

Mathematical expectation (E) is equal to $P \times M$ where P = the probability the event will occur and M = the amount that will be won if an event occurs. In this problem, the probability that the event will occur translates to the probability of buying the winning ticket, and the amount to be won (M) is equal to $1,000. We first calculate the probability.

Continued on next page ...

$$P\text{(buying winning ticket)} = \frac{\#\ of\ favorable\ outcomes}{\#\ of\ possible\ outcomes}$$

$$= \frac{5}{5000}$$

$$= \frac{1}{1000}$$

With $P = \frac{1}{1000}$ and $M = \$1,000$, we now calculate expectation.

$$E = \frac{1}{1000} \times \$1,000$$

$$= \$1$$

The mathematical expectation is $1 for five tickets or $0.20 for one ticket. This implies that a fair price to play this game is 20¢. So the actual cost to play the game ($2) is too much to pay.

SUPPLEMENTARY EXERCISE 3.5

In problems 1 through 5 a pair of dice is tossed once. Find the odds

1. in favor of getting 9
2. in favor of getting an even number less than 8
3. against getting a 4
4. against getting a 2 or 3
5. against getting a sum other than 8

In problems 6 through 10 a single card is drawn from a shuffled deck of 52 cards. Find the odds

6. in favor of drawing a red jack
7. in favor of drawing a 7
8. in favor of drawing a 5 or 6
9. against drawing a red picture card
10. against drawing a black card or ace

In problems 11 through 14 find the probability that event A will happen if the odds are

11. 3 to 4 in favor of A
12. 6 to 5 in favor of A
13. 5 to 1 against A
14. 4 to 7 against A

15. In order to win a game Amanda must select a picture card from a 52 card deck of shuffled cards. What are the odds for and against this event?

16. The odds in favor of Lisa winning a cooking contest are 2 to 5. What is the probability that Lisa will win?

17. Eau Gallie High School is raffling a motorcycle worth $2000. A total of 1,000 tickets are sold for $5 each. If Margaret buys two tickets what is her mathematical expectation?

18. George will win $10 if he throws a six on the first toss of a pair of dice. What are the odds in favor of his winning? What is a fair price for him to play this game?

19. On a game show there are six different prizes, each hidden behind a door. Two of the prizes are worth $500, two are worth $1,000, one is worth $2,000, and the last is valued at $5,000. If a contestant picks a door at random, what is the contestant's mathematical expectation?

20. In a certain lottery a player selects any four-digit number from 0000 to 9999. The cost of a ticket to play this game is $2. If the player wins the payoff is $2,000. What is the probability a player will win? What is a fair price to pay for a ticket?

3.6 COMPOUND PROBABILITY

Compound probability involves finding the probability of compound events. Recall that events are outcomes (or results) of an experiment. Simple events represent individual outcomes, whereas compound events represent any grouping of the individual outcomes. To illustrate, in an experiment we toss a die and let event A = *a one occurs* and event B = *an odd number occurs*. Event A is simple since it has only one possible outcome, namely 1. However, event B is compound since it involves a grouping of the individual outcomes, namely, 1,3,5.

When dealing with compound events we must identify the simpler events (of which the compound events are composed) as being independent, dependent, or mutually exclusive events. Definitions of these various type of events are now provided.

3.6.1 Definitions

Independent Events

Two or more events are said to be independent if the occurrence of one event has no effect upon the occurrence or non-occurrence of the other event(s).

Dependent Events

Two or more events are said to be dependent if the occurrence of one event has an effect upon the occurrence or non-occurrence of the other event(s).

Mutually Exclusive Events

Two events are mutually exclusive if the occurrence of one event precludes the occurrence of the other. That is, one event or the other can occur, but both cannot occur simultaneously.

Compound probability problems can be usually reworded into *and* or *or* type problems. For example,

- Given two events A and B, the probability that both events occur is

$$P(A \text{ and } B) = P(A) \times P(B)$$

  (**Note:** This formula can be extended for more than two events. Also, we always assume that event A occurs before event B.)

- Given two mutually exclusive events A and B, the probability that either one of these two events occur is

$$P(A \text{ or } B) = P(A) + P(B)$$

- Given two nonmutually exclusive events A and B, the probability that either one of these events occur is

$$P(A \text{ or } B) = P(A) + P(B) - P(A \text{ and } B)$$

The key to deciding which formula to use is a function of the use of the words *and* or *or* in the problem.

[1] If the problem is stated as finding the probability of event A and B occurring, use the first formula. You must, however, identify the individual events as independent or dependent when you calculate the individual probabilities.

[2] If the problem is stated as finding the probability of event A *or* B occurring, then you must first determine whether or not the events are mutually exclusive. Once this is done you can then use the second or third formula accordingly.

Example 1 (page 158, #1a,d)

A bag contains three red marbles, two green marbles, and one blue marble. One marble is drawn and then replaced, after which a second marble is drawn. Find the probability for parts a and d.

Solution

(**Note:**) The individual events are independent since we have replacement, and there are a total of six possible outcomes.)

a. P (first is red and the second is blue)

$$= P(\text{red}) \times P(\text{blue})$$
$$= \frac{3}{6} \times \frac{1}{6}$$
$$= \frac{3}{36} = \frac{1}{12}$$

d. P (both are red)

$$= P(\text{red}) \times P(\text{red})$$
$$= \frac{3}{6} \times \frac{3}{6}$$
$$= \frac{9}{36} = \frac{1}{4}$$

Example 2 (page 158, #3a,e)

In a certain political science class there are 20 students: 11 males, of which four are freshman and seven are sophomores; nine females, of which three are freshman and six are sophomores. A panel consisting of three students is to be selected from the class. Find the probability for parts a and e.

Solution

(Note: The events are dependent since there is no replacement.)

a. P (all freshman) = ?

There are a total of seven freshman: four males and three females. If we let F = *freshman* then

$$\begin{aligned} P(\text{all freshman}) &= P(F \text{ and } F \text{ and } F) \\ &= P(F) \times P(F) \times P(F) \\ &= \frac{7}{20} \times \frac{6}{19} \times \frac{5}{18} \\ &= \frac{210}{6840} = \frac{7}{228} \end{aligned}$$

e. P (a freshman and then two sophomores) = ?

There are 7 freshman and 13 sophomores. Note that order is important here. We first choose a freshman, and then we choose two sophomores. Let F = *freshman* and S = *sophomore*.

Continued on next page ...

PROBABILITY 79

$$P(F \text{ and } S \text{ and } S \text{ and } S) = P(F) \times P(S) \times P(S)$$
$$= \frac{7}{20} \times \frac{13}{19} \times \frac{12}{18}$$
$$= \frac{1092}{6840} = \frac{91}{570}$$

Example 3 (page 158, #5a,b)

One card is randomly selected from a shuffled deck of 52 cards and then a quarter is flipped. Find the probability for parts a and b.

Solution

a. P (red card and heads)
$$= P(\text{red card}) \times P(\text{heads})$$
$$= \frac{26}{52} \times \frac{1}{2}$$
$$= \frac{26}{104} = \frac{1}{4}$$

b. P (a red card or heads) = ?

Since this is an *or* problem we must first determine whether or not the two events are mutually exclusive. In this case the answer is *no*, the two events are not mutually exclusive because they can occur at the same time. (Note that in part (a) above we calculated the probability of these two events occurring at the same time.) Therefore, P (red card or heads)

$$= P(\text{red card}) + P(\text{heads}) - P(\text{red card and heads})$$
$$= \frac{26}{52} + \frac{1}{2} - \frac{1}{4}$$
$$= \frac{52}{52} - \frac{1}{4}$$
$$= \frac{3}{4}$$

Example 4 (page 159, #9a,b,c,d,f)

Two cards are selected at random from a shuffled standard deck of 52 cards, without replacement. Find the probability for parts a through d and f.

Solution

(**Note:** The events are dependent. Once a card is picked it cannot be picked again.)

a. P (both cards are picture cards)
$$= P(\text{picture card}) \times P(\text{picture card})$$
$$= \frac{12}{52} \times \frac{11}{51}$$
$$= \frac{132}{2652} = \frac{11}{221}$$

Continued on next page ...

80 STUDENT STUDY GUIDE

b. P(first card is an ace and second card is a picture card)

$$= P(\text{ace}) \times P(\text{picture card})$$
$$= \frac{4}{52} \times \frac{12}{51}$$
$$= \frac{48}{2652} = \frac{4}{221}$$

c. P(first card is the ace of spades and second card is a picture card)

$$= P(\text{ace of spades}) \times P(\text{picture card})$$
$$= \frac{1}{52} \times \frac{12}{51}$$
$$= \frac{12}{2652} = \frac{1}{221}$$

d. P(both cards are kings)

$$= P(\text{king}) \times P(\text{king})$$
$$= \frac{4}{52} \times \frac{3}{51}$$
$$= \frac{12}{2652} = \frac{1}{221}$$

f. P(both cards are of same denomination) = ?

"Both cards are of same denomination" means 2 aces or 2 threes or 2 kings, etc. From part d above, the P(both cards are kings) is $\frac{1}{221}$. Since there are 13 similar cases,

$$P(\text{both cards are of same denomination}) = 13 \times \frac{1}{221} = \frac{1}{17}$$

Example 5 (page 159, #11a,d,f)

An ice chest contains 5 cans of ginger ale, 3 cans of orange soda, and 4 cans of root beer. If 3 cans are drawn at random from the chest without replacement, find the probability of getting

Solution

a. three cans of root beer

Note that the order makes no difference, and there are a total of 12 cans of soda.

$$P(3 \text{ cans of root beer})$$
$$= P(\text{1st is root beer}) \text{ and } P(\text{2nd is root beer}) \text{ and } P(\text{3rd is root beer})$$
$$= \frac{4}{12} \times \frac{3}{11} \times \frac{2}{10}$$
$$= \frac{4 \times 3 \times 2}{12 \times 11 \times 10}$$
$$= \frac{24}{1320} = \frac{1}{55}$$

Continued on next page ...

d. A can of orange, then a can of root beer, then a can of ginger ale

Note that order makes a difference, and there are a total of 12 cans of soda.

$$P(\text{orange, root beer, ginger ale})$$
$$= P(\text{1st is orange}) \text{ and } P(\text{2nd is root beer}) \text{ and } P(\text{3rd is ginger ale})$$
$$= \frac{3}{12} \times \frac{4}{11} \times \frac{5}{10}$$
$$= \frac{60}{1320} = \frac{1}{22}$$

f. no root beer

"No root beer" implies either an orange or ginger ale.

$$P(\text{no root beer})$$
$$= P(\text{1st is not root beer}) \text{ and } P(\text{2nd is not root beer}) \text{ and } P(\text{3rd is not root beer})$$
$$= \frac{8}{12} \times \frac{7}{11} \times \frac{6}{10}$$
$$= \frac{336}{1320} = \frac{14}{55}$$

Example 6 (page 159, #13)

A track coach must decide who is going to run on the school's relay team. He must choose four runners from a list of six equally fast runners: Nenno, Nelson, Gilligan, Neanderthal, Clar, and Connelly. He decides that the easiest way to make the decision is to draw names from a hat in succession. (Note that there is no replacement since the coach needs four different runners.) Find the probability that

Solution

a. he chooses Neanderthal first

$$P(\text{Neanderthal first}) = \frac{favorable\ pick}{possible\ picks} = \frac{1}{6}$$

b. he chooses Neanderthal first and Nenno second (These events are dependent.)

$$P(\text{Neanderthal first and Nenno second})$$
$$= P(\text{Neanderthal first}) \times P(\text{Nenno second, given Neanderthal first})$$
$$= \frac{1}{6} \times \frac{1}{5} = \frac{1}{30}$$

c. he chooses Nelson or Gilligan first (These events are mutually exclusive.)

$$P(\text{Nelson or Gilligan first})$$
$$= P(\text{Nelson first}) + P(\text{Gilligan first})$$
$$= \frac{1}{6} + \frac{1}{6} = \frac{1}{3}$$

Continued on next page ...

d. his third choice is a person whose name begins with N, given that Connelly and Clar were already chosen

Since Clar and Connelly have already been chosen, 4 names remain and 3 of them begin with N.

P (name begins with N, given that Clar and Connelly were already chosen) $= \dfrac{3}{4}$

e. Nenno is chosen first, Neanderthal is chosen second, Nelson is chosen third, and Gilligan is chosen to run the anchor leg (fourth)

The events are dependent.

P (Nenno first, Neanderthal second, Nelson third, Gilligan fourth)

$= P$ (Nenno first) $\times P$ (Neanderthal second) $\times P$ (Nelson third) $\times P$ (Gilligan fourth)

$= \dfrac{1}{6} \times \dfrac{1}{5} \times \dfrac{1}{4} \times \dfrac{1}{3} = \dfrac{1}{360}$

SUPPLEMENTARY EXERCISE 3.6

In problems 1 through 4 two cards are selected at random, without replacement, from a shuffled deck of 52 cards. Find the probability that

1. both cards are red
2. both cards are nines
3. the first is a diamond and the second is the queen of hearts
4. the first is a spade and the second card is red

In problems 5 through 8 a bowl contains 10 peanuts, 16 pecans, 8 walnuts and 12 cashews. A nut is drawn at random and then replaced. A second nut is then drawn at random. Find the probability that

5. the first is a pecan and the second is a walnut
6. both are cashews
7. the first is a peanut and the second is a pecan
8. both are walnuts

In problems 9 through 12 one card is randomly selected from a shuffled deck of 52 cards and then a quarter is flipped. Find the probability of obtaining

9. a club and tails
10. a club or tails
11. a red king and heads
12. a six or heads

In problems 13 through 16 a jar contains 5 white, 6 red, and 8 blue marbles. Three marbles are picked randomly with no replacement after each pick. Find the probability that

13. all three are red.
14. the first two are blue and the third is white.
15. the first is red, the second white, the third blue.
16. the first is white, the second blue, the third white.

3.7 COUNTING, ORDERED ARRANGEMENTS, AND PERMUTATIONS (OPTIONAL)

Counting problems address the question *How many?*. One type of counting problem deals with **permutations**. Permutations represent an ordered arrangement of items. That is, the order in which the items under discussion are arranged is important. A permutation can consist of a set of N distinct items arranged in some specific order, or a set of N distinct items from which R of these items are to be arranged in a specific order. Two methods are used to calculate permutations: the *slot* method and the formula method.

Calculating Permutations By The Slot Method

1. Determine the total number of items (from the given set of items) that are to be arranged and make a *slot* for each.

2. Fill in each *slot* by asking the questions *From the given set of items, how many can be placed in slot 1, how many in slot 2, ..., how many in slot n?*

3. Multiply the numbers from step 2. The product is the total number of possible arrangements.

Calculating Permutations By The Formula Method

Alternatively (and equivalently), we can use the $_nP_r$ formula for calculating permutations. The number of permutations of n distinct items taken r at a time can be found by

$$_nP_r = \frac{n!}{(n-r)!}$$

The following examples use the *slot* method.

Example 1 (page 170, #1)

Given the set of digits {4,5,6,7,8,9}, how many three digit numbers can be formed if no digit is to be repeated? How many three digit numbers can be formed if repetition is allowed?

Solution

Solving For No repetition

- Since we want to form three-digit numbers, the total number of items to be arranged is three. Thus we must fill in three slots.

- The given set of digits has six items. Since repetition is not allowed there are six candidates for the first slot, five candidates for the second slot, and four candidates for the third slot.

$$\underline{6} \quad \underline{5} \quad \underline{4}$$

There are $6 \times 5 \times 4 = 120$ possible numbers.

Solving With Repetition

This problem is exactly as the previous problem except repetition is now allowed. This means that all three slots have six possible candidates.

$$\underline{6} \quad \underline{6} \quad \underline{6}$$

There are $6 \times 6 \times 6 = 216$ possible numbers.

Example 2 (page 170, #7)

a. How many license plates can be made if each plate must consist of two letters followed by four digits, if we assume no repetition?

b. How many plates are possible if the first digit cannot be zero?

c. How many plates are possible if the first letter cannot be O and the first digit cannot be zero?

Continued on next page ...

84 STUDENT STUDY GUIDE

Solution

a. There are 26 letters from which to choose, of which two are to be arranged. There are ten digits from which to choose, of which four are to be arranged.

$$\underset{\text{Letter}}{26} \times \underset{\text{Letter}}{25} \times \underset{\text{Digit}}{10} \times \underset{\text{Digit}}{9} \times \underset{\text{Digit}}{8} \times \underset{\text{Digit}}{7} = 3{,}276{,}000$$

b. If the first digit cannot be zero then there are only 9 digits that can be used for the first digit slot. The second digit slot also has 9 possible digit candidates since zero can now be included. (Remember, we used one digit for the first digit slot and there is no repetition.)

$$\underset{\text{Letter}}{26} \times \underset{\text{Letter}}{25} \times \underset{\text{Digit}}{9} \times \underset{\text{Digit}}{9} \times \underset{\text{Digit}}{8} \times \underset{\text{Digit}}{7} = 2{,}948{,}400$$

c. If the first letter cannot be O then there are 25 letters from which to chose for the first letter slot. The second letter slot also has 25 letter candidates because the letter O can now be included. (Remember, we used one letter for the first letter slot and there is no repetition.)

$$\underset{\text{Letter}}{25} \times \underset{\text{Letter}}{25} \times \underset{\text{Digit}}{9} \times \underset{\text{Digit}}{9} \times \underset{\text{Digit}}{8} \times \underset{\text{Digit}}{7} = 2{,}835{,}000$$

Example 3 (page 170, #9)

The Rochester Tennis Club is having a mixed doubles tournament. If eight women and their husbands sign up for the tournament, (a) how many mixed doubles teams are possible, and (b) how many teams can be formed if no woman is paired with her husband?.

Solution

a. A mixed doubles team consists of two people, namely, one man and one woman. There are 8 male candidates for the man's slot, and 8 women candidates for the woman's slot.

$$\underset{\text{Man}}{8} \times \underset{\text{Woman}}{8} = 64$$

b. If no woman is to be paired with her husband then the set of women has one less person than the set of men. (When a man is selected, his wife cannot be considered for that pairing.)

$$\underset{\text{Man}}{8} \times \underset{\text{Woman}}{7} = 56$$

Example 4 (page 171, #19)

In how many distinct ways can the letters of the two words given be arranged?

Solution

a. *ALGEBRA*

There are 7 letters of which one letter occurs twice (A).

$$= \frac{7!}{2!} = \frac{7 \times 6 \times 5 \times 4 \times 3 \times 2 \times 1}{2 \times 1} = 7 \times 6 \times 5 \times 4 \times 3 = 2{,}520$$

Continued on next page ...

b. *STATISTICS*

There are 10 letters, of which the S occurs three times, the T occurs three times, and the I occurs twice.

$$= \frac{10!}{3!\ 3!\ 2!} = \frac{10 \times 9 \times 8 \times 7 \times 6 \times 5 \times 4 \times 3 \times 2 \times 1}{(3 \times 2 \times 1)(3 \times 2 \times 1)(2 \times 1)} = 50{,}400$$

SUPPLEMENTARY EXERCISE 3.7

1. Lisa has five skirts and seven sweaters. How many outfits does she have to choose from? (Assume every sweater *goes with* every skirt.)

2. How many five digit numbers can be formed from the set of digits {1,3,5,6,7,9} if (a) no digit can be repeated? and (b) repetition is allowed?

3. A dinner dance is attended by nine couples. In how many ways can these couples dance if husband and wife are not to dance together?

4. In a restaurant, a meal consists of an appetizer, entree, and dessert. A menu has five appetizers, eight entrees, and four desserts. In how many ways can a meal be selected from this menu?

5. In a club of 15 girls and 10 boys a girl is to be chosen as president, a boy as vice president, and either a boy or a girl as secretary-treasure. In how many ways can this be done?

6. A large office building has eight exits. In how many ways can a man enter the building and leave by a different exit?

7. At a college there are six instructors who teach freshman English, four instructors who teach freshman history, and three instructors who teach freshman science. If a freshman takes English, history, and science, how many different sets of instructors are possible?

8. In how many different ways can the letters of the word *CALIPER* be arranged?

In problems 9 through 13 evaluate the given expression.

9. $7!$ 10. $\dfrac{9!}{3!}$ 11. $_7P_3$ 12. $\dfrac{10!}{8!}$ 13. $_8P_1$

14. In how many distinct ways can the letters of the word *CALIFORNIA* be arranged?

15. In Florida, a license plate consists of three letters followed by three digits. How many license plates can be made if the first digit cannot be zero?

16. In how many ways can a basketball coach select a first string of five players from a squad of 14?

17. In how many distinct ways can the letters of the word *TELEPHONE* be arranged?

18. In how many different ways can the five finalists in a beauty pageant be arranged in line?

19. A club of 30 people is going to elect a president and secretary. In how many different ways can this be done?

20. Given the set of digits {0,1,2,3,4,5,6}, how many different numbers consisting of five digits can be formed when repetition is allowed? (Assume a number cannot begin with zero.)

3.8 COMBINATIONS (OPTIONAL)

Another type of counting problem deals with **combinations**. Combinations represent an unordered arrangement of items. That is, the order in which the items occur is not important. The general formula for the number of combinations of n items taken r at a time is

$$_nC_r = \frac{n!}{(n-r)!r!}$$

The expression $_nC_r$ can be interpreted as n *choose* r. For example, if we have nine items from which two items are to be selected without regard to order, then the number of ways this can be done is *9 choose 2*, which is

$$_9C_2 = \frac{9!}{(9-2)!2!} = \frac{9!}{7!2!} = \frac{9 \times 8 \times 7!}{2! \times 7!} = \frac{9 \times 8}{2} = 36$$

When solving counting problems always determine first whether the problem is one of permutation (order is important) or combination (order is not important). To distinguish between these two types of counting problems it is sometimes helpful to think of combinations as involving a number of selections, and permutations as involving a number of arrangements.

Example 1 (page 176, #3)

If the Xerox Corporation has to transfer four of its ten junior executives to a new location, in how many ways can the four executives be chosen?

Solution

This is a problem in combinations since we do not care about the order in which the executives are selected. Accordingly, the problem is *10 choose 4*.

$$_{10}C_4 = \frac{10!}{(10-4)!4!} = \frac{10!}{6!4!} = \frac{10 \times 9 \times 8 \times 7 \times 6!}{4! \times 6!} = 210$$

Example 2 (page 176, #5)

Alice has a penny, a nickel, a dime, a quarter, and a half dollar. She may spend any three coins. (a) In how many ways can Alice do this? and (b) What is the most money she can spend using just three coins?

Solution

a. This is a combination problem since the order is not important. Alice has five coins from which she may choose three.

$$_5C_3 = \frac{5!}{(5-3)!3!} = \frac{5!}{2!3!} = \frac{5 \times 4 \times 3!}{2! \times 3!} = 10$$

b. If Alice were to select the three highest valued coins (namely, the half dollar, the quarter, and the dime), then her selection would result in the most money she can spend. These coins total 85¢. (Note that the order in which these coins are chosen has no affect on this total.)

Example 3 (page 177, #11)

A baseball squad consists of eight outfielders and seven infielders. If the baseball coach must choose three outfielder and four infielders, in how many ways can this be done?

Solution

- Outfielders

$$_8C_3 = \frac{8!}{(8-3)!3!} = \frac{8!}{5!3!} = \frac{8 \times 7 \times 6 \times 5!}{3! \times 5!} = 56$$

- Infielders

$$_7C_4 = \frac{7!}{(7-4)!4!} = \frac{7!}{3!4!} = \frac{7 \times 6 \times 5 \times 4!}{3! \times 4!} = 35$$

- Total Number of Ways

$$56 \times 35 = 1960$$

SUPPLEMENTARY EXERCISE 3.8

In problems 1 through 3 evaluate the given expression.

1. $_9C_6$ 2. $_9C_4$ 3. $_6C_1$

4. A club of 50 people is to select three delegates to attend to a convention. In how many ways can this be done?

5. Roxanne has eight necklaces and 13 bracelets. In how many ways can she select one necklace and two bracelets?

6. A bakery has seven different kinds of cookies. In how many ways can David select three kinds of cookies?

7. In how many ways can 12 magazines be divided equally among four airline passengers?

8. A student senate consists of five freshmen, five sophomores, ten juniors, and 10 seniors. How many different committees can be formed that consist of two freshmen, two sophomores, three juniors, and three seniors?

9. Four sections of trigonometry are being offered. In how many different ways can eight students make their selections so that three students are in one section, two are in a second section, two are in a third section, and one is in the fourth section?

10. At a Chinese restaurant a dinner for four people consists of three selections from column A and two selections from column B. Column A has 12 different dishes listed and column B has ten. How many different meals can be chosen?

3.9 MORE PROBABILITY (OPTIONAL)

The Fundamental Counting Principle, permutations, and combinations are now brought together as a means for providing an alternative method of solution for previously considered probability problems, and for solving more involved probability problems.

Example 1 (page 180, #1)

Find the probability of being dealt two queens when you are dealt two cards from a shuffled deck of 52 cards.

Solution

This problem can be solved using methods learned in section 3.6.

$$P(2Q) = P(Q \text{ and } Q) = P(Q) \times P(Q) = \frac{4}{52} \times \frac{3}{51} = \frac{12}{2652} = \frac{1}{221}$$

Alternatively, we can use combinations.

- First find the total number of outcomes. Since we want to select two cards from 52 cards, we have *52 choose 2*, which is equal to 1326.
- Next find the total number of ways in which two queens can be selected. Since there are only four queens in a standard deck of cards, we have *4 choose 2*, which is equal to 6.
- The probability of being dealt two queens is $\frac{6}{1326}$ or $\frac{1}{221}$.

Example 2 (page 181, #5a,c)

On the track team of the York Athletic Club there are eight sprinters and ten distance runners. A relay team consisting of four people must be chosen at random, without regard to whom runs first, second, third, or fourth. Find the probability of parts a and c.

Solution

Note: In solving the probability problems in this problem we first must determine the total number of ways in which a relay team can be selected. (We need this for our denominator in the formula $P(A) = \frac{success}{total}$.) Since the order is not important, we have *18 choose 4* ($_{18}C_4$), which is equal to 3060.

a. P (team consists of only sprinters) = ?

- Determine the number of ways a team of all sprinters can be selected. Since there are eight sprinters and a team must consist of four people, $_8C_4 = 70$.
- Since the total number of ways in which a team can be selected is 3060 (see the note above), P (all sprinters) = $\frac{70}{3060} = \frac{7}{306}$

c. P (team consists of two sprinters and two distance runners) = ?

- The number of ways in which two sprinters can be selected is $_8C_2 = 28$.
- The number of ways in which two distance runners can be selected is $_{10}C_2 = 45$.
- The number of ways in which two sprinters and two distance runners can be selected is $28 \times 45 = 1260$.
- The probability of selecting a team with two sprinters and two distance runners is $\frac{1260}{3060} = \frac{7}{17}$

Example 3 (page 181, #13)

An urn contains six orange balls, four blue balls, and three red balls. If three balls are drawn at random indicate (but do not evaluate) the following probabilities.

Solution

Once again, as with any probability problem, we need to know the total number of possible outcomes. The total number of ways in which three balls can be selected from a total of 13 balls (6 orange, 4 blue, and 3 red), is $_{13}C_3$. This number will now be the denominator for the probabilities that follow.

a. $P(\text{3 orange balls}) = \dfrac{_6C_3}{_{13}C_3}$

d. $P(\text{2 orange balls and 1 blue ball}) = \dfrac{_6C_2 \times {_4C_1}}{_{13}C_3}$

Example 4 (page 182, #15)

Find the probability of being dealt a flush (five cards of the same suit) in hearts in five-card poker.

Solution

- The total number of ways of being dealt five cards is $_{52}C_5 = 2,598,960$. (See example 2 in your textbook.)
- The number of ways in which five cards of the same suit can be selected is $_{13}C_5 = 1287$. (There are 13 cards for each suit.)
- The probability of being dealt a flush is $\dfrac{1287}{2598960} = \dfrac{33}{66,640}$

SUPPLEMENTARY EXERCISE 3.9

1. Beth has a bag that contains five red marbles, six blue marbles, eight yellow marbles, and six white marbles. She selects four marbles at random. Find the probability that three are blue and one is yellow.

2. You are dealt four cards from a shuffled deck of 52 cards. Find the probability two cards are queens and two cards are sixes.

3. Alice has seven books: four are mysteries and three are romances. If she picks three books at random to take on a vacation, what is the probability that all three books are mysteries?

4. Marcia is on a committee that consists of ten people. Three members are to be chosen at random to attend a convention. What is the probability that Marcia will be chosen?

5. A cat has a litter of five white and three black kittens. If two kittens are chosen at random what is the probability that both are white?

6. You are dealt four cards from a shuffled deck of 52 cards. What is the probability that you have three jacks and a five?

7. A class consists of ten girls and seven boys. A committee of three people is to be chosen at random. Find the probability that the committee will contain more girls than boys.

8. Fifteen girls have reached the finals in trying out for the cheerleading squad. Eight of the girls have brown hair, five have blonde hair, and two have red hair. Seven girls will be eliminated at random. Find the probability that the cheerleading squad will consist of four girls with brown hair and four girls with blonde hair.

Continued on next page ...

9. A picnic basket contains three ham sandwiches, four peanut butter and jelly sandwiches, and four roast beef sandwiches. If two sandwiches are selected at random, what is the probability that one of the sandwiches is peanut butter and jelly and the other is ham?

10. Find the probability of being dealt two cards of the same suit when you are dealt two cards from a shuffled deck of 52 cards.

3.10 CHAPTER 3 TEST

In problems 1 through 6 use the following information to find the indicated probabilities.

A container has eight red, five green, and seven blue balls. One ball is to be selected at random.

1. $P(\text{red})$ 2. $P(\text{blue})$ 3. $P(\text{green})$ 4. $P(\text{red and green})$ 5. $P(\text{green or blue})$ 6. $P(\text{not blue})$

In problems 7 and 8 two cards are drawn (without replacement) from a standard deck of 52 cards. Find:

7. $P(\text{both cards are aces})$ 8. $P(\text{both cards are clubs})$

9. What is the probability of rolling a six or a ten on a pair of dice?

In problems 10 and 11 two coins are tossed. Find each probability.

10. $P(\text{two tails})$ 11. $P(\text{at least one head})$

12. Suppose the probability you catch a cold this winter is $\frac{9}{37}$. What is the probability that you do not catch a cold this winter?

In problems 13 through 16 one card is randomly selected from a shuffled deck of 52 cards, and then a die is rolled. Find the indicated probability.

13. $P(\text{a queen and a one})$ 14. $P(\text{a queen or a one})$ 15. $P(\text{a red card and a one})$ 16. $P(\text{a red card or a one})$

In problems 17 and 18 assume the probability that an event A will happens is $\frac{7}{10}$. Find the indicated odds.

17. odds favoring event A 18. odds against event A

19. If the odds against a horse winning are 3 to 5, what is the probability that the horse will win?

20. What would be your expectation if you were to win a prize of $104 for drawing a king from a shuffled deck of 52 cards?

21. If the probability of gaining $100 is $\frac{1}{4}$ and the probability of gaining $4000 is $\frac{3}{4}$, what is the expectation?

22. A car wash business loses $20 on rainy days and gains $50 on rainless days. If the probability of rain is 0.2, what is the expectation?

23. How many different license plates can be made if each plate is to consist of one letter of the alphabet followed by four digits, where the digits can be repeated?

24. In how many distinct ways can the letters of the word *PARALLEL* be arranged?

In problems 25 and 26, the student council has 10 sophomores, 9 juniors, and 10 seniors. Three students are to be selected to form a committee. Find the probability of selecting

25. two sophomores and then a junior.   26. a senior and then two juniors.

4. STATISTICS

4.1 INTRODUCTION

Statistics is a branch of mathematics that deals with the collection, analysis, interpretation, and presentation of numerical information, called data. Statistics can be an invaluable tool in decision making in areas that have a degree of uncertainty (e.g., product comparisons, political elections, quality control, and so forth). In this chapter we study descriptive statistics, which includes the concepts and techniques used to summarize and describe specific characteristics of a set of data.

4.2 MEASURES OF CENTRAL TENDENCY

When a large amount of data has been collected it is often desirable to have a single number that best describes or represents the data. One such number is referred to as a **measure of central tendency**, or more simply, **average**. An average is a central value (i.e., its location is somewhere in the middle of a set of data) that is characteristic of the data. Four different values are commonly used to represent an average: mean, median, mode, and midrange.

4.2.1 Mean

The mean of a set of data is found by adding up the data and then dividing the sum by the total number of pieces of data. Mean is symbolized by $\bar{x}$.

4.2.2 Median

The median of a set of data is found by first ranking the data, from lowest to highest, and then finding the middle number. Median is symbolized by $\tilde{x}$.

4.2.3 Mode

The mode of a set of data is that number, item, or value that occurs most frequently.

4.2.4 Midrange

The midrange of a set of data is found by adding the largest and smallest data items and dividing the sum by two.

Example 1 (page 200, #7)

Find the mean, median, mode, and midrange for the set of data {11, 99, 77, 88, 66, 44, 55, 22, 33}.

Solution

- Mean

    To find the mean we add the numbers and divide the sum by nine (since there are nine pieces of data).

    $$\bar{x} = \frac{11 + 99 + 77 + 88 + 66 + 44 + 55 + 22 + 33}{9} = \frac{495}{9} = 55$$

Continued on next page ...

- Median

  To find the median we need to first rank the data.

  Unranked: 11 99 77 88 66 44 55 22 33
  Ranked: 11 22 33 44 55 66 77 88 99

  The middle number of the ranked set of data is the median. Thus $\bar{x} = 55$.

- Mode

  There is no mode because no one data item from the set occurs most frequently.

- Midrange

  To find the midrange add 11 (the smallest) and 99 (the largest) and divide the sum by two.

  $$\text{Midrange} = \frac{11 + 99}{2} = \frac{110}{2} = 55$$

Example 2 (page 202, #19)

The mean score on a set of 13 scores is 77. What is the sum of the 13 test scores?

Solution

Since the mean is calculated by dividing the sum of the scores by the total number of scores, we can calculate the sum of the scores by multiplying the total number of scores by the mean.

$$\text{If } \bar{x} = \frac{\text{the sum of the scores}}{\text{the total number of scores}}$$

then $\bar{x} \times (\text{total number of scores}) = \text{the sum of the scores}$

The sum of the 13 test scores is $77 \times 13 = 1001$.

SUPPLEMENTARY EXERCISE 4.2

In problems 1 through 10 find the mean, median, mode, and midrange for each set of data. Round off any decimal answer to the nearest tenth.

1. {3, 3, 4, 5, 5, 5, 7, 8}
2. {1, 3, 5, 7, 9}
3. {9, 12, 8, 4, 8, 3, 15}
4. {6, 1, 8, 1, 3, 9, 2, 8}
5. {5, 6, 9, 3, 7, 9, 8, 4}
6. {67, 43, 90, 56, 35, 74}
7. {83, 17, 25, 16, 62, 32, 41}
8. {803, 749, 190, 20, 107, 600}
9. {618, 101, 3000, 423, 202, 11, 101}
10. {2468, 2310, 1111, 3100, 5322, 4231}

11. Howard Prince, a mathematics teacher, gave a test to his Algebra II class. After grading the papers he recorded the following results. (**Note:** All numbers represent total points earned based on a possible 100 points. A score of 73 means $\frac{73}{100}$, or 73%.)

    73, 80, 49, 62, 81, 85, 91, 98, 19, 43, 66, 71, 83
    84, 88, 96, 91, 86, 68, 57, 29, 91, 40, 53, 82

    Upon receiving their scores the students lodged a formal complaint against Dr. Prince, charging that the test was too difficult since the class average was a failing grade (below 60%). Dr. Prince countered that the class average was not below 60%; rather, the average was 80%.

    a. Which measure of central tendency did the students use?

    b. Which measure of central tendency did Dr. Prince use?

    Continued on next page ...

12. The table below gives the annual wage distribution for Danko Business Machines, Inc. Using the information provided in the table find the mean, median, mode, and midrange annual income.

Annual Income (in dollars) (in dollars)	Number of People Receiving This Income
250000	1
100000	2
55000	13
33000	11
20000	8
16000	5

4.3 MEASURES OF DISPERSION

A measure of dispersion is a numerical value that describes to what extent the individual values of a set of data are spread out. Your textbook discusses two such measures: range and standard deviation.

4.3.1 Range

The range (R) for a set of data is the difference between the largest data value (L) and the smallest data value (S): ($R = L - S$.)

4.3.2 Standard Deviation

The standard deviation for a set of data indicates how widely spread out (i.e., scattered) the individual pieces of data are from the mean. To calculate standard deviation we must perform a number of operations.

Procedure for Calculating Standard Deviation

1. Find the mean ($\bar{x}$).
2. Find the difference between each piece of data and the mean: $(x - \bar{x})$.
3. Square each answer found in the second step: $((x - \bar{x})^2)$.
4. Add the answers found in step 3 and divide this sum by the total number of pieces of data that make up the set.
5. Take the square root of the answer found in step 4.

For n pieces of data, where $x_1, x_2, x_3, \cdots x_n$ represent the individual pieces of data, and $\bar{x}$ is the mean of the set, the standard deviation, denoted σ, is represented by the following general formula.

$$\sigma = \sqrt{\frac{(x_1 - \bar{x})^2 + (x_2 - \bar{x})^2 + (x_3 - \bar{x})^2 + \cdots + (x_n - \bar{x})^2}{n}}$$

Example 1 (page 210, #3)

Find the standard deviation for the set of data {16, 13, 13, 12, 9, 9}.

Solution

Step 1: Find $\bar{x}$

$$\bar{x} = \frac{16 + 13 + 13 + 12 + 9 + 9}{6} = \frac{72}{6} = 12$$

Continued on next page ...

Steps 2 and 3: Set up a table and complete each column

x	$(x - \bar{x})$	$(x - \bar{x})^2$
16	$(16 - 12) = 4$	$4^2 = 16$
13	$(13 - 12) = 1$	$1^2 = 1$
13	$(13 - 12) = 1$	$1^2 = 1$
12	$(12 - 12) = 0$	$0^2 = 0$
9	$(9 - 12) = -3$	$(-3)^2 = 9$
9	$(9 - 12) = -3$	$(-3)^2 = 9$

Step 4

Add the values of the last column from the table above and divide this sum by six, which is the total number of pieces of data.

$$\frac{16 + 1 + 1 + 0 + 9 + 9}{6} = \frac{36}{6} = 6$$

Step 5

Take the square root of the answer found in step 4. (Use Table 2 in Appendix A and round to the nearest tenth if necessary.)

$$\sqrt{6} = 2.449 = 2.4 \text{ (to the nearest tenth)}$$

Thus, $\sigma = 2.4$ (to the nearest tenth).

Alternatively, we can use the formula to find standard deviation directly.

$$\sigma = \sqrt{\frac{(x_1 - \bar{x})^2 + (x_2 - \bar{x})^2 + (x_3 - \bar{x})^2 + \cdots + (x_n - \bar{x})^2}{n}}$$

$$= \sqrt{\frac{(16 - 12)^2 + (13_2 - 12)^2 + (13 - 12)^2 + (12 - 12)^2 + (9 - 12)^2 + (9 - 12)^2}{6}}$$

$$= \sqrt{\frac{(4)^2 + (1)^2 + (1)^2 + (0)^2 + (-3)^2 + (-3)^2}{6}}$$

$$= \sqrt{\frac{16 + 1 + 1 + 0 + 9 + 9}{6}}$$

$$= \sqrt{\frac{36}{6}}$$

$$= \sqrt{6}$$

$$= 2.4 \text{ (to the nearest tenth)}$$

Example 2 (page 211, #11a-f)

A sample of ten bowlers was taken in a tournament. The number of strikes recorded for each bowler in his or her first game was as follows: {2,3,4,5,5,6,3,2,7,3}. For this set of data find the mean, median, mode, range, midrange, and standard deviation.

Solution

a. Mean

$$\bar{x} = \frac{2+3+4+5+5+6+3+2+7+3}{10} = \frac{40}{10} = 4$$

b. Median

Unranked data: 2, 3, 4, 5, 5, 6, 3, 2, 7, 3
Ranked data: 2, 2, 3, 3, 3, 4, 5, 5, 6, 7

The median lies between the fifth and sixth pieces of data.

$$\tilde{x} = \frac{(fifth\ piece\ of\ data) + (sixth\ piece\ of\ data)}{2} = \frac{3+4}{2} = \frac{7}{2} = 3.5$$

c. Mode

The most frequently occurring data value is 3.

d. Range

$$R = L - S = 7 - 2 = 5$$

e. Midrange

$$M = \frac{L+S}{2} = \frac{7+2}{2} + \frac{9}{2} = 4.5$$

f. Standard Deviation

x	$(x - \bar{x})$	$(x - \bar{x})^2$
2	$(2 - 4) = -2$	$(-2)^2 = 4$
2	$(2 - 4) = -2$	$(-2)^2 = 4$
3	$(3 - 4) = -1$	$(-1)^2 = 1$
3	$(3 - 4) = -1$	$(-1)^2 = 1$
3	$(3 - 4) = -1$	$(-1)^2 = 1$
4	$(4 - 4) = 0$	$0^2 = 0$
5	$(5 - 4) = 1$	$1^2 = 1$
5	$(5 - 4) = 1$	$1^2 = 1$
6	$(6 - 4) = 2$	$2^2 = 4$
7	$(7 - 4) = 3$	$3^2 = 9$

$$\sigma = \sqrt{\frac{4+4+1+1+1+0+1+1+4+9}{10}} = \sqrt{\frac{26}{10}} = \sqrt{2.6} = 1.6 \text{ (to nearest tenth)}$$

SUPPLEMENTARY EXERCISE 4.3

In problems 1 through 10 find the standard deviation for each set of data. Round off any decimal answers to the nearest tenth.

1. {4,2,3,5,1}
2. {7,3,2,8,5}
3. {1,4,5,4,3,1}
4. {3,5,2,4,5,2,2,1}
5. {19,45,78,18}
6. {38,43,34,37,28}
7. {20,22,24,26,28}
8. {49,52,50,48,48,53}
9. {59,79,92,77,76,84,80,69}
10. {61,45,59,77,82,73,50,49}

11. While working at an amusement park during the summer Larry made deposits of $65, $43, $55, $97, $62, $88, and $52 into his savings account. For this set of data find the mean, median, mode, range, midrange, and standard deviation.

12. The average monthly temperatures for each month of the year for Cincinnati, Ohio, are (in degrees Fahrenheit) 31, 34, 43, 54, 63, 72, 76, 74, 67, 56, 44, and 34. For this set of data find the mean, median, mode, range, midrange, and standard deviation.

4.4 MEASURES OF POSITION (PERCENTILES)

A measure of position is a numerical value that is used to locate the position of a piece of data relative to the rest of the data. Two measures of position discussed in the textbook are percentile and quartile.

4.4.1 Percentile

A **percentile rank** is defined as the percentage of scores (i.e., data) in a set of scores with data at or below a given score. If a score is identified by its percentile rank, then we refer to the score as a **percentile**. Thus a percentile refers to the location of a specific score relative to the remaining set and is interpreted as the percentage of data that are below the score. For example,

- The 10th percentile, denoted P_{10}, of a set of data refers to that score in which exactly 10% of the data are below it.
- The 80th percentile, denoted P_{80}, of a set of data refers to that score in which exactly 80% of the data are below it.

By multiplying the percentile (expressed as a decimal) that corresponds to a particular score by the total number of items contained in a set of data, we can determine how many pieces of data lie below the given score.

To calculate a percentile for a score we must know the score's position in the set of data and the total number of items that make up the set of data. (Note that it will be necessary to first rank the set of data from lowest to highest.) Once these two values are known we then

1. subtract the score's rank from the total number of items in the set of data, and
2. divide this value by the total number of items in the given set.

4.4.2 Quartile

A quartile partitions a set of data in four distinct parts.

- The first quartile, Q_1, refers to a score that separates the lower 25% of the set of data from the rest of the data. Note that Q_1 is equivalent to P_{25}.
- The second quartile, Q_2, refers to a score that separates the lower 50% of the set of data from the rest of the data. Note that Q_2 is equivalent to P_{50}, which is also equivalent to the median.
- The third quartile, Q_3, refers to a score that separates the lower 75% of the set of data from the rest of the data. Note that Q_3 is equivalent to P_{75}.

Example 1 (page 217, #1)

In a class of 200 students Julia has a rank of twelfth. What is Julia's percentile rank in the class?

Solution

- The total number of students is 200.
- Julia's rank is 12th.
- Julia's percentile rank is

$$= \frac{200 - 12}{200} = \frac{188}{200} = 0.94$$

Thus Julia's percentile rank is 94th, which means that she is at the 94th percentile (i.e, 94% of the students are ranked lower than Julia.)

Example 2 (page 217, #5)

In a statistics class of 30 students an exam was given and Eddie scored at the 80th percentile. How many students scored lower than Eddie?

Solution

Since Eddie scored at the 80th percentile, 80% of the 30 students scored lower than he did. Thus, $0.80 \times 30 = 24$. Twenty-four students scored lower than Eddie.

Example 3 (page 217, #9)

Jessie is ranked at the third quartile in her economics class of 60 students. What is her rank in the class?

Solution

The third quartile is equivalent to the 75th percentile. This means that 75% of the 60 students are ranked lower than Jesse. Alternatively, 25% of the 60 students are at her rank or are ranked higher. Since a class ranking is usually given from highest to lowest, Jesse's class rank is $0.25 \times 60 = 15$.

Example 4 (page 217, #11)

Given the twelve ranked scores below, we will answer parts a and b.

$$62, 72, 74, 75, 75, 82, 85, 86, 86, 87, 89, 90$$

Solution

a. What is the rank (from the top) of a score of 87?

Score:	62	72	74	75	75	82	85	86	86	**87**	89	90
Rank:	12	11	10	9	8	7	6	5	4	**3**	2	1

Note that a score of 87 is third from the top and hence has a rank of three.

Continued on next page ...

b. What is the percentile rank of a score of 87?

- The total number of scores is 12.
- The score 87 has a ranking of 3.
- The percentile rank for 87 is $= \dfrac{12-3}{12} = \dfrac{9}{12} = 0.75$.

Thus a score of 87 has a percentile rank of 75, which means it is located at the 75th percentile (or third quartile).

SUPPLEMENTARY EXERCISE 4.4

1. Twelve hundred people took the General Equivalency Diploma (GED) test and Harriet was ranked 84th. What was her percentile rank?
2. A high school graduating class consists of 420 seniors. If Tim is ranked 63rd what is his percentile rank?
3. Five hundred people competed in a mini-triathalon contest. If Robert placed 125th what was his percentile rank?
4. Julia was ranked at the first quartile in her college senior class of 500 students. What was her rank in the class (from the top)?
5. Jane's score on her Graduate Record Exam (GRE) was ranked at the 97th percentile. If 800 students took the exam what was Jane's rank (from the top)?
6. A college course has an enrollment of 150 students. On his first exam Tom's score was ranked at the 84th percentile. What was Tom's rank (from the top)?
7. In a mathematics class of 40 students an exam was given and Shannon scored at the 60th percentile. How many students scored lower than Shannon?

For problems 8 through 14 use the following scores to answer each question.

80, 77, 62, 54, 91, 88, 73, 96, 90, 37, 44, 110, 75, 85, 66, 50

8. What is the rank (from the top) of a score of 77?
9. What is the percentile rank of a score of 77?
10. What is the percentile rank of a score of 96?
11. What score is at the first quartile?
12. What score is at the third quartile?
13. What score is at the 63rd percentile?
14. What score is at the 100th percentile?

15. Claudio is ranked 35th in his senior class of 180 students. Juanita is in the same senior class and has a percentile rank of 80. Of the two seniors who has the higher standing in the class?

4.5 PICTURES OF DATA

The most common types of graphs used to represent data are bar graphs, pictograms, circle graphs, histograms, and frequency polygons (i.e., line graphs). A review of these graphs is presented here.

4.5.1 Bar Graphs

Bar graphs use either vertical or horizontal bars of equal width to represent the relationship within a set of data. Vertical bars are used for vertical bar graphs, and the frequencies are placed along the vertical axis. Horizontal bars are used for horizontal bar graphs, and the frequencies are placed along the horizontal axis. An example of a vertical and horizontal bar graph is shown for the data from the following table.

Holiday	Percentage
Christmas	25
Valentine's Day	40
Easter	25
Pet's Birthday	70

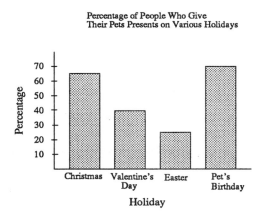

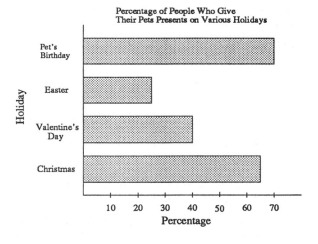

4.5.2 Pictograms

Pictograms use symbols to stand for a certain quantity of a particular item. A *key* also is given that details what the symbol represents. The total quantity is then calculated by counting the number of symbols for an item. An example of a pictogram is shown for the data from the following table.

Holiday	Number of Roses
Christmas	100
Valentine's Day	400
Easter	350
Halloween	25
Thanksgiving	75

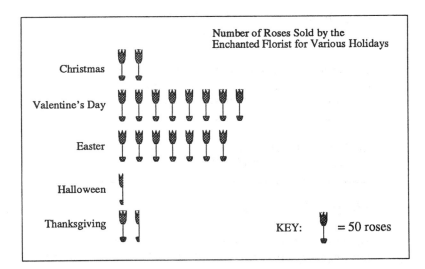

4.5.3 Circle Graph

A circle graph, which is also referred to as a pie chart, consists of a circle partitioned into sections. Each section, formally called a **sector**, represents a percentage of the entire circle. Circle graphs are useful when it is necessary to illustrate how a whole quantity is divided into parts.

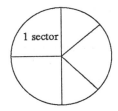

A circle graph with five sectors

4.5.4 Histogram

A histogram is a type of bar graph that pictorially represents a **grouped frequency distribution**. To construct a histogram we first construct a grouped frequency distribution. The following procedure is used to construct a grouped frequency distribution.

Procedure For Constructing A Grouped Frequency Distribution

1. Rank the data.
2. Group the data into *classes*. Appropriate classes should be chosen so that
    a. each piece of data belongs to only one class;
    b. each class is of the same width; and
    c. the number of classes for the set of data ranges from 5 to 12.
3. Record the number of pieces of data that belong to each class.

Once a grouped frequency distribution has been constructed, we can construct the histogram that represents it. The histogram consists of a horizontal line segment (called the *horizontal axis*), and an adjoining vertical line segment (called the *vertical axis*). The frequency is usually measured on the vertical axis, and the scores (i.e., the classes) are indicated on the horizontal axis. The height of each bar represents frequency, and the width of each bar is equal. There also is no space between bars, and every histogram should be appropriately titled.

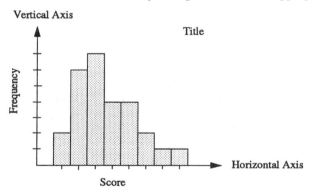

4.5.5 Frequency Polygon

A frequency polygon is a line graph that pictorially represents a grouped frequency distribution. This graph can be constructed easily from a histogram by connecting the midpoints of the top of each bar of a histogram. A frequency polygon of the histogram above is shown below. In effect, every histogram has a corresponding frequency polygon.

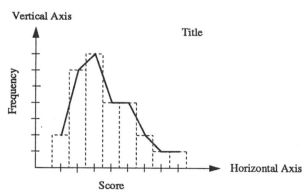

Example 1 (page 228, #3)

The table gives the location and length (in miles) of six notable canals. Construct a vertical bar graph representing this information.

Name	Location	Length in Miles
Amsterdam-Rhine	The Netherlands	45.0
Beaumont-Port Authur	United States	40.0
Houston	United States	43.0
Panama	Canal Zone	50.7
Kiel	Germany	61.3
Welland	Canada	27.5

Solution

Since we are asked to construct a vertical bar graph, the frequencies (in this example, length in miles) are placed along the vertical axis. Note that the lengths range from 27.5 to 61.3. A convenient scale is from 20 to 65 in intervals of five.

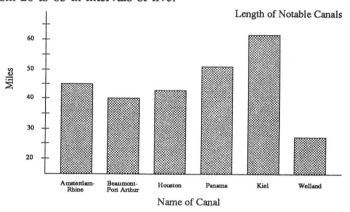

Example 2 (page 228, #7)

The graph shown represents the annual sales of the Ronolog Corporation for the years 1988-1992.

a. Which year had the lowest sales amount?

b. Approximately how much were sales in 1992?

c. In what years did sales decrease?

d. What was the approximate total sales for the five years?

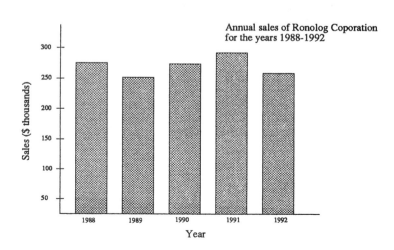

Continued on next page ...

Solution

In this example we are asked to interpret the graph.

a. The year with the lowest sales corresponds to the year that has the shortest bar associated with it. This is 1989.

b. To interpret the sales for the year 1992, we read from the vertical axis the number that approximates the height of the bar that represents the year 1992. This is just a little more than $250,000. A good approximation is $260,000. (Note: The answer in the textbook is $262,000. Given the scale of the vertical axis, this number is very difficult to obtain.)

c. Notice that from 1988 to 1989, and from 1991 to 1992, the bars "drop" in height. This corresponds to a decrease in sales.

d. The approximate total sales for the five years is found by adding the amount of approximate sales corresponding to each bar. (Note: Your answer might be different than that given in the textbook since the scale of the vertical axis is subject to personal interpretation.)

$$\$280{,}000 + \$245{,}000 + \$280{,}000 + \$295{,}000 + \$260{,}000 = \$1{,}360{,}00$$

Example 3 (page 229, #11)

The number of pizzas sold by Phil's Pizza Palace during the last seven days were as follows:

Sunday	90	Thursday	70
Monday	closed	Friday	105
Tuesday	70	Saturday	85
Wednesday	50		

Represent this information by means of a pictogram. Let each symbol O represent 10 pizzas sold.

Solution

Since the symbol given represents 10 pizzas we will need 9 symbols for Sunday, none for Monday, 7 for Tuesday, 5 for Wednesday, 7 for Thursday, $10\frac{1}{2}$ for Friday, and $8\frac{1}{2}$ for Saturday. Do not forget to include a key with the graph.

Pizzas Sold by Phil's Pizza Palace
(Last Seven Days)

Sunday   O O O O O O O O O
Monday   Closed
Tuesday O O O O O O O
Wednesday O O O O O
Thursday O O O O O O O
Friday   O O O O O O O O O O ◖
Saturday O O O O O O O O ◖

Key: O = 10 pizzas sold

Example 4 (page 230, #15a,b,c)

The graph below shows the net income of the Josco Company for the years 1986-1990.

a. What was the net income for 1987?

b. What was the net income for 1988?

c. What was the total net income for 1988 and 1989?

```
Josco Company Net Income

1986   $ $ $
1987   $ $ $ $
1988   $ $ $ $ $
1989   $ $ $ $ $
1990   $ $ $ $ $

    Key: $ = $50,000
```

Solution

We must interpret this graph. Since it is a pictogram, there must be a key associated with it. Note that the symbol given represents $50,000.

a. The year 1987 is represented by $3\frac{1}{2}$ symbols. Thus the total net income for this year is $50,000 \times 3\frac{1}{2}$, which is equal to $175,000.

b. The year 1988 is represented by $4\frac{1}{2}$ symbols. Thus the total net income is $225,000.

c. The total net income for 1988 and 1989 is $225,000 + $250,000 = $475,000.

Example 5 (page 230, #17)

In a survey of 180 students it was determined that 90 students came to school by bus, 60 students came by car, and 30 students walked. Construct a circle graph representing this information.

Solution

The sectors of the circle are determined by multiplying the percentage of students taking a bus, car, or walking by 360° (the total number of degrees in a circle).

- Bus: $\frac{90}{180} = \frac{1}{2}$ and $\frac{1}{2} \times 360° = 180°$

- Car: $\frac{60}{180} = \frac{1}{3}$ and $\frac{1}{3} \times 360° = 120°$

- Walk: $\frac{30}{180} = \frac{1}{6}$ and $\frac{1}{6} \times 360° = 60°$

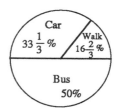

Example 6 (page 231, #25)

A police radar unit measured the speed of 25 cars on a certain street. The resulting data are

23	38	24	26	18
23	52	30	45	27
28	25	28	37	29
33	27	34	36	32
23	18	23	38	21

Construct a grouped frequency distribution of the data by using 15-19 as the first class, and a frequency polygon that represents this data.

Solution

- Rank the data.

18	23	26	30	37
18	23	27	32	38
21	24	28	33	38
23	25	28	34	45
23	26	29	36	52

- Classify the data and record the number of times each score lies within a class. Since we are instructed to use 15-19 as the first class, note that the class width will be 4 (19-15=4). This is the basis by which subsequent classes are obtained.

Classes	Frequency
15 - 19	2
20 - 24	6
25 - 29	7
30 - 34	4
35 - 39	4
40 - 44	0
45 - 49	1
50 - 54	1

- Construct the frequency polygon by first constructing the histogram.

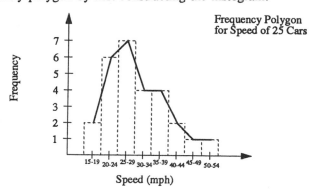

SUPPLEMENTARY EXERCISE 4.5

1. In a survey, people were asked if they trusted members of several occupations. The results were as follows:

Percent that Trusted	Occupation
65	Lawyer
65	Doctor
15	Car Salesman
70	Clergyman
75	Teacher

   Construct a vertical bar graph that represents this information.

2. The number of people attending the movies at Atlantic Plaza Movie Theater during its final six months of operation is given in the table below.

September	1,000
October	1,500
November	1,750
December	2,500
January	2,250
February	2,000

   Represent this information by means of a pictogram. Select a symbol and let this symbol represent 500 people.

3. The graph below shows the net income of the Sinclair Company for the years 1987-1991.

   a. What was the net income for 1988?
   b. What was the net income for 1990?
   c. What was the total net income for 1989 and 1990?
   d. What was the total net income for 1987 through 1991?

Net Income of the Sinclair Company (1987-1991)	
1987	$$$$
1988	$$$$
1989	$$$$$
1990	$$$
1991	$$
Key: $ = $100,000	

4. In a survey of 40 students it was found that ten were majoring in English, six in math, four in human services, nine in psychology, nine in political science, and two in physical education. Construct a circle graph representing this information.

5. The grades of 26 students in an introductory computer science course are given below. Construct a frequency distribution and a histogram to represent this data.

73	63	68	73	83	88	83	78	83
73	68	73	73	63	73	63	88	63
78	83	78	83	83	93	73	83	

Continued on next page ...

6. The ages of 24 people at a nursing home (including staff and residents) are given below.

$$
\begin{array}{cccccccc}
67 & 68 & 65 & 63 & 71 & 77 & 72 & 74 \\
37 & 51 & 58 & 99 & 88 & 80 & 48 & 93 \\
95 & 42 & 55 & 85 & 53 & 65 & 61 & 96
\end{array}
$$

a. Construct a grouped frequency distribution for the data using 37-45 as the first class.

b. Using the frequency distribution from part (a), construct a histogram that represents this data.

7. A survey of 20 students was conducted to determine the number of hours of television they watch per week. The results are given below.

$$
\begin{array}{ccccc}
14 & 22 & 32 & 15 & 21 \\
28 & 11 & 6 & 5 & 13 \\
19 & 30 & 27 & 17 & 25 \\
8 & 26 & 29 & 2 & 24
\end{array}
$$

a. Construct a grouped frequency distribution for this data using 2-6 as the first class.

b. Using the frequency distribution from part (a) construct a frequency polygon that represents this data.

8. A telephone poll was conducted to determine the monthly rent (in dollars) for a one-bedroom apartment. The set of data collected from 24 respondents is given below.

$$
\begin{array}{cccccccc}
255 & 300 & 375 & 350 & 180 & 200 & 520 & 550 \\
325 & 300 & 475 & 220 & 380 & 355 & 330 & 200 \\
580 & 500 & 340 & 350 & 400 & 420 & 375 & 280
\end{array}
$$

a. Construct a grouped frequency distribution for this data using classes of 180-250, 251-321, 322-392, and so forth.

b. Using the frequency distribution from part (a) construct a histogram that represents this data.

4.6 THE NORMAL CURVE

The normal curve is a bell shaped curve in which the mean, median, and mode all have the same value, and all occur exactly at the center of the distribution. For normally distributed data, approximately 68% of the data is within one standard deviation of the mean, approximately 95% of the data is within two standard deviations of the mean, and approximately 99.7% of the data is within three standard deviations of the mean. This information is summarized in the figure below.

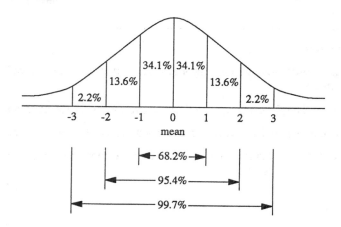

The position of a score in terms of the number of standard deviations it is located from the mean is called a **z-score**, or standard score. A z-score is found by the formula

$$z = \frac{(individual\ score) - (mean)}{standard\ deviation}$$

Symbolically, if we let x = an individual score, $\bar{x}$ = the mean, and σ = the standard deviation, the z-score formula can be expressed as

$$z = \frac{x - \bar{x}}{\sigma}$$

If in a problem the number of standard deviations about the mean (i.e., the z-score) is fractional, it will be necessary to use Table 6, which is on page 225 of your textbook, to determine the percentage of data that lies between the mean and this z-score. (**Note:** A copy of Table 6, which is called *Areas of the Standard Normal Distribution*, is given in appendix B of this study guide.) Also, when doing problems such as those discussed in this section, it is always helpful if a sketch of a normal curve is constructed to depict the information given in a problem.

Example 1 (page 243, #1)

The IQ scores for a certain group of elementary school students are approximately normally distributed with a mean of 100 and a standard deviation of 10. What percentage of the students has IQ scores between (a) 90 and 110? (b) 80 and 120? (c) 70 and 130?

Solution

First construct a normal curve based on the information given in the problem. Note that the mean is 100 and the standard deviation is 10.

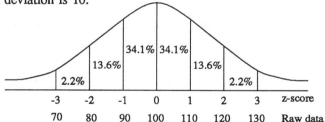

a. Between 90 and 110, there is 34.1% + 34.1%, which is equal to 68.2%.

b. Between 80 and 120, there is 13.66% + 34.1% + 34.1% + 13.6%, which is equal to 95.4%.

c. Between 70 and 130, there is 2.2% + 13.6% + 34.1% + 34.1% + 13.6% + 2.2%, which is equal to 99.7%.

Example 2 (page 245, #9)

The scores on a mathematics exam were approximately normally distributed and the instructor assigned grades as follows.

- All scores more than 1.7 standard deviations above the mean received an A.
- All scores between 1.1 and 1.7 standard deviations above the mean received a B.
- All scores between 1.1 standard deviations above the mean and 1.2 standard deviations below the mean received a C.

Continued on next page ...

- All scores between 1.2 and 1.8 standard deviations below the mean received a D.
- All scores more than 1.8 standard deviations below the mean received a F.

What percentage of the class received each letter grade?

Solution

First draw a normal curve based on the information given in the problem. Note that the standard deviations above and below the mean that are to be considered are fractional. This indicates that we need to use Table 6 in the textbook.

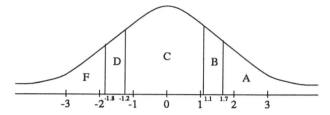

Grade A

Scores that are 1.7 standard deviations above the mean were assigned a grade of A. According to table 6, 45.5% of the data is between the mean and 1.7 standard deviations above it. Since 50% of the data is to the right of the mean we must subtract 45.5% from 50% to find the percentage of data that is to the right of 1.7. 50.0% − 45.5% = 4.5%. Thus 4.5% of the class received a grade of A. This is shown in the figure below.

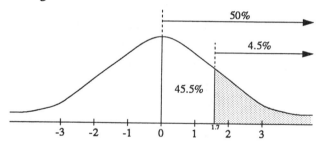

Grade B

Scores that are between 1.1 and 1.7 standard deviations above the mean were assigned a grade of B. According to Table 6 45.5% of the data is between the mean and 1.7 standard deviations above it. Additionally, 36.4% of the data is between the mean and 1.1 standard deviations above it. The area of interest, which is shaded in the figure below, is found by subtracting 36.4% from 45.5%. Thus, 45.5% − 36.4% = 9.1% of the class received a grade of B.

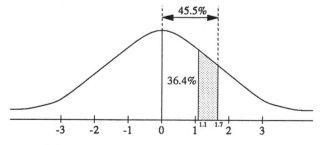

The percentages for those receiving grades of C, D, and F are found in a similar manner and will not be presented here.

Example 3 (page 245, #13a,d)

A survey was taken among students in a college cafeteria. to find out how many hours a week they spent studying for courses for which they were currently enrolled. The data were approximately normally distributed with a mean of 24 hours and a standard deviation of five hours. We will answer questions a and d of this problem.

Solution

a. What percentage of students studied more than 28 hours per week?

Draw a normal curve based on the information given in the problem. Note that the score of 28 does not lie on a standard deviation *boundary*. As a result we must calculate the z-score and use Table 6 in the textbook.

$$z = \frac{x - \bar{x}}{\sigma} = \frac{28 - 24}{5} = \frac{4}{5} = 0.8$$

From Table 6, 28.8% of the data lies between the mean and 0.8 standard deviations above it. Thus 50% − 28.8% = 21.2% of the students studied more than 28 hours per week.

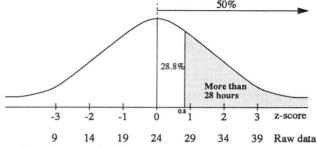

d. What percentage of students studied between 22 and 30 hours per week?

Draw a normal curve based on the information given in the problem. Note that the scores of 22 and 30 do not lie on a *boundary*. Calculate the z-scores and use Table 6.

The z-score for 22 is

$$z = \frac{x - \bar{x}}{\sigma} = \frac{22 - 24}{5} = \frac{-2}{5} = -0.4$$

The z-score for 30 is

$$z = \frac{x - \bar{x}}{\sigma} = \frac{30 - 24}{5} = \frac{6}{5} = 1.2$$

Using Table 6, 15.5% of the data lies between the mean and 0.4 standard deviations below it, and 38.5% of the data lies between the mean and 1.2 standard deviations above it. Therefore, 15.5% + 38.5% = 54% of the students studied between 22 and 30 hours per week.

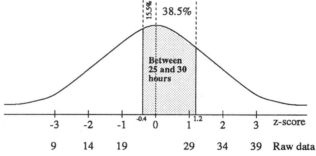

110   STUDENT STUDY GUIDE

SUPPLEMENTARY EXERCISE 4.6

1. A survey of faculty members from colleges in the state of Florida was taken and the rank of each person was recorded. It was found that the number of faculty at each rank was approximately normally distributed in the following manner.

   Full Professor: More than 0.9 standard deviations above the mean
   Associate Professor: 0.3 standard deviations below the mean to 0.9 standard deviations above the mean
   Assistant Professor: 0.3 to 1.3 standard deviations below the mean
   Instructor: More than 1.3 standard deviations below the mean

   What is the percentage of faculty members at each rank?

2. At a local high school 1200 students took a minimum competency exam. The exam had a mean of 300 and a standard deviation of 20, with the results being approximately normally distributed. (Round all answers to the nearest whole number.)

   a. How many students scored above 340?

   b. How many students scored between 340 and 360?

   c. How many students scored between 300 and 340?

   d. How many students scored between 280 and 320?

   e. How many students scored below 240?

3. The average length in time of a long distance telephone call is 28 minutes with a standard deviation of five minutes. Assuming the call times are normally distributed, answer the following questions.

   a. Approximately what percentage of calls are longer than 35 minutes?

   b. Approximately what percentage of calls are less than 35 minutes?

   c. Approximately what percentage of calls are between 15 and 20 minutes long?

   d. Approximately what percentage of calls are at least one-half hour in length?

   e. Approximately what percentage of calls are between 22 and 37 minutes in length?

4. The test scores of the Scholastic Aptitude Test (SAT are normally distributed with a mean of 500 and a standard deviation of 100.

   a. What is the probability that a student selected at random will score higher than 620?

   b. What is the probability that a student selected at random will score between 380 and 510?

   c. What is the probability that a student selected at random will score less than 320?

   d. If 840 students of a school district took the exam and local scholarships were awarded to those students who scored at least 730, approximately how many students would receive a scholarship?

4.7 CHAPTER 4 TEST

1. The graph below represents the monthly rainfall in Orlando, Florida for five consecutive months. Answer the questions in parts a-d.

   a. Which month had the least amount of rain?

   b. How many inches of rain fell in March?

   c. During which months did rainfall increase compared with the previous month?

   d. What was the total rainfall for the five months?

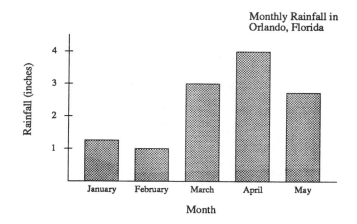

2. The number of baseballs sold by Tom's Sporting Goods Store last week is shown in the graph below. Answer the questions in parts a-c.

   a. How many baseballs were sold during the one week period?

   b. How many baseballs were sold on Tuesday?

   c. Between which two days was there the largest increase in the number of baseballs sold?

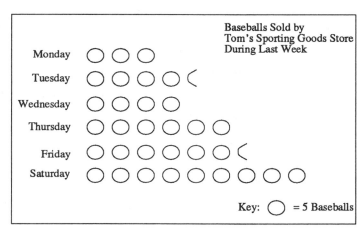

For problems 3 through 10 use the set of data given here to find the following statistics:

$$\{68, 21, 35, 62, 50, 48, 77, 42, 45, 39, 71, 63\}$$

3. the mean
4. the median
5. the midrange
6. the mode (if any)
7. the range
8. the standard deviation
9. the score at the first quartile
10. the score at the third quartile

Continued on next page ...

11. In a class of 70 students Milt ranks 12 and Nancy's rank is at the 80th percentile. Who has the higher class rank?

12. The daily high temperatures for a 25-day period during the month of June at Daytona Beach, Florida were recorded and are listed below (in degrees Fahrenheit).

82	88	86	90	88
88	82	80	80	86
90	88	84	90	86
84	82	84	88	86
88	88	90	86	86

   a. Construct a frequency distribution for the data.

   b. Construct a histogram for the data.

13. It has been found that the breath-holding time for Navy divers is normally distributed with a mean of 130 seconds and a standard deviation of 15 seconds.

   a. Approximately what percent of Navy divers can hold their breath for more than 150 seconds?

   b. Approximately what percent of Navy divers can hold their breath between 110 and 140 seconds?

   c. Approximately what percent of Navy divers cannot hold their breath for more than 90 seconds?

   d. What is the probability that a Navy diver selected at random can hold his or her breath between 145 and 160 seconds?

5. AN INTRODUCTION TO THE METRIC SYSTEM

5.1 INTRODUCTION

The metric system of weights and measures is an international system. The entire world (except for the United States and a few small countries) is either using the metric system predominately, or is actively and totally committed to using it. In this chapter we discuss the metric system of measurement.

5.2 HISTORY OF SYSTEMS OF MEASUREMENT

The metric system of measurement is a decimal based system. Consequently, conversions from one unit to another within the metric system can be accomplished by multiplying or dividing by ten or powers of ten. Each general category of measurement is represented by a *most commonly used unit*.

- The most commonly used unit of measure for length is the **meter**.
- The most commonly used unit of measure for volume is the **liter**.
- The most commonly used unit of measure for mass is the **gram**.
- The most commonly used unit of measure for temperature is one **Celsius** degree.

With the exception of temperature, other units of measure that are smaller or larger than the most commonly used unit are expressed by attaching a prefix to the most commonly used unit. The most commonly used prefixes and their meanings are listed in the table below. (Abbreviations are in parentheses.)

Prefix	Meaning
kilo (k)	1000
hecto (h)	100
deka (da)	10
deci (d)	$\frac{1}{10}$ or 0.1
centi (c)	$\frac{1}{100}$ or 0.01
milli (m)	$\frac{1}{1000}$ or 0.001

By using prefixes, other denominations are formed from the standard units. For example, in measuring length the standard unit is the meter. Using the prefixes shown we can represent lengths that are larger or smaller than the meter.

Larger Than a Meter
$$\begin{cases} 1 \text{ \textbf{kilometer} (km)} = 1000 \text{ meters} \\ 1 \text{ \textbf{hectometer} (hm)} = 100 \text{ meters} \\ 1 \text{ \textbf{dekameter} (dam)} = 10 \text{ meters} \end{cases}$$

Smaller Than a Meter $\begin{cases} 1 \text{ \textbf{decimeter}} \text{ (dm)} = \frac{1}{10} \text{ meter} \\ 1 \text{ \textbf{centimeter}} \text{ (hm)} = \frac{1}{100} \text{ meter} \\ 1 \text{ \textbf{milliliter}} \text{ (mm)} = \frac{1}{1000} \text{ meter} \end{cases}$

Example 1 (page 264, #7a,e)

Given the fact that one meter is approximately equal to 39.37 inches, how many inches are contained in (a) 1 kilometer and (e) 1 centimeter?

Solution

a. Since one kilometer is equal to 1000 meters we must multiply 39.37 by 1000. Note that this can be accomplished easily by moving the decimal point (in 39.37) three places to the right. Thus 1 *km* = 39,370 *in.*.

e. Since one centimeter is equal to $\frac{1}{100}$ of a meter we must multiply 39.37 by $\frac{1}{100}$. This calculation is equivalent to dividing 39.37 by 100, which can be accomplished by moving the decimal point (in 39.37) two places to the left. Thus 1 *cm* = 0.3937 *in.*.

Example 2 (page 264, #11b,d)

Complete each of the following (parts b and d only).

Solution

b. 1 liter = _____ milliliters

The prefix *milli* means $\frac{1}{1000}$. Thus one milliliter is equal to $\frac{1}{1000}$ of a liter. That is,

$$1 \text{ } milliliter = \frac{1}{1000} \text{ } liter$$

If we were to multiply both sides of this equation by 1000, we get

$$1000 \text{ milliliters} = \frac{1000}{1000} \text{ liter}$$

or   1000 milliliters   =   1 liter

d. 1 gram = _____ centigrams

The prefix *centi* means $\frac{1}{100}$. Thus one centigram is equal to $\frac{1}{100}$ of a gram. That is,

$$1 \text{ } centigram = \frac{1}{100} \text{ } gram$$

If we were to multiply both sides of this equation by 100, we get

$$100 \text{ centigrams} = \frac{100}{100} \text{ gram}$$

or   1000 centigrams   =   1 gram

Example 3 (page 264, #13a,c,e)

Complete each of the following (parts a, c, and e only).

Solution

a. 2000 meters = _____ kilometers

The prefix *kilo* means 1000. Thus,

$$1 \text{ km} = 1000 \text{ m}$$
$$\text{or} \quad 2 \text{ km} = 2000 \text{ m} \quad (\textit{Multiply by 20})$$

c. 3 grams = _____ milligrams

The prefix *milli* means $\frac{1}{1000}$. Thus,

$$1 \text{ mg} = \frac{1}{1000} \text{ gm}$$
$$\text{or} \quad 1000 \text{ mg} = 1 \text{ gm} \quad (\textit{Multiply by 1000})$$
$$\text{or} \quad 3000 \text{ mg} = 3 \text{ gm} \quad (\textit{Multiply by 3})$$

e. 3 hectoliters = _____ liters

The prefix *hecto* means 100. Thus,

$$1 \text{ h}l = 100 \text{ } l$$
$$\text{or} \quad 3 \text{ h}l = 300 \text{ } l \quad (\textit{Multiply by 3})$$

SUPPLEMENTARY EXERCISE 5.2

In problems 1 through 6 match each item in column A with the correct item in column B.

A	B
1. deci	a. $\frac{1}{10}$
2. deka	b. 100
3. milli	c. $\frac{1}{100}$
4. centi	d. 10
5. hecto	e. $\frac{1}{1000}$
6. kilo	f. 1000

In 7 through 15 write the word for each abbreviation.

7. l 8. ml 9. g 10. kg 11. dam 12. cl 13. cm 14. hl 15. mg

5.3 LENGTH AND AREA

The most commonly used unit of length measurement in the metric system is the **meter**. Using the prefixes from the previous section we can name multiples and submultiples of the meter. A summary is provided in Table 5.3.1.

Table 5.3.1 Units of Linear Measure

1 kilometer (km)*	=	1000 meters
1 hectometer (hm)	=	100 meters
1 dekameter (dam)	=	10 meters
1 meter (m)*	=	1 meter
1 decimeter (dm)	=	0.1 meter
1 centimeter (cm)*	=	0.001 meter
1 millimeter (mm)*	=	0.001 meter

* Units most commonly used

By rewriting these units horizontally from largest to smallest we can see the decimal relationships among the length measurements in the metric system.

km	hm	dam	m	dm	cm	mm
1	10	100	1000	10000	100000	1000000
0.1	1	10	100	1000	10000	100000
0.01	0.1	1	10	100	1000	10000
0.001	0.01	0.1	1	10	100	1000
0.0001	0.001	0.01	0.1	1	10	100
0.00001	0.0001	0.001	0.01	0.1	1	10
0.000001	0.00001	0.0001	0.001	0.01	0.1	1

Reading each row from left to right, notice that when we move from a larger unit to a smaller unit, we multiply by a power of 10 (e.g., 1 km = 10 hm). Similarly, when we move from a smaller unit to a larger unit (i.e., moving from right to left), we divide by a power of 10 (e.g., 1 m = 0.1 dam). Recall that multiplying or dividing a number by a power of 10 has the effect of moving the decimal point of the number to the right or left, respectively. This concept leads to the following procedure for converting units of linear measure.

Procedure for Converting a Linear Measure from Unit to Another Unit

1. Determine the *direction* of the conversion. (If the required unit is larger than the given unit then the direction will be *left*; otherwise it will be *right*.)

2. Determine the number (n) of *positional moves* needed to make the conversion.

3. Move the decimal point of the given number n places left or right.

We can also convert from metric units to English units, and vice versa. Table 5.3.2 lists the appropriate conversion factors for measures of length and area.

Table 5.3.2 Length and Area English-Metric Conversions

ENGLISH TO METRIC		
When you know	**Multiply by**	**To find**
inches (in.)	2.56	centimeters (cm)
feet (ft.)	30.0	centimeters (cm)
yards (yd.)	0.9	meters (m)
miles (mi.)	1.6	kilometers (km)
square inches ($in.^2$)	6.5	square centimeters (cm^2)
square feet ($ft.^2$)	0.09	square meters (m^2)
square yards ($yd.^2$)	0.8	square meters (m^2)
acres	0.4	hectares (ha)
METRIC TO ENGLISH		
When you know	**Multiply by**	**To find**
millimeters (mm)	0.04	inches (in.)
centimeters (cm)	0.4	inches (in.)
meters (m)	3.3	feet (ft.)
meters (m)	1.1	yards (yd.)
kilometers (km)	0.6	miles (mi.)
square centimeters (cm^2)	0.16	square inches ($in.^2$)
square meters (m^2)	1.2	square yards ($yd.^2$)
hectares (ha)	2.5	acres

Example 1 (page 271, #3a,c)

Complete each of the following (parts a and c only).

Solution

a. 18 cm = _____ dm

- Decimeter is one place to the left of centimeter. Thus the direction is *left* and the number of *positional moves* is 1.

$$\text{km} \quad \text{hm} \quad \text{dam} \quad \textbf{m} \quad \textbf{dm} \quad \text{cm} \quad \text{mm}$$

- Move the decimal point in 18 one place to the left.

$$18 \Rightarrow 18. \Rightarrow 1.8$$

Therefore, 18 cm = 1.8 dm.

c. 3.2 km = _____ cm

- Centimeters is five places to the right of kilometers. Thus the direction is *right* and the number of *positional moves* is five.

$$\textbf{km} \quad \text{hm} \quad \text{dam} \quad \text{m} \quad \text{dm} \quad \textbf{cm} \quad \text{mm}$$

- Move the decimal point in 3.2 five places to the right.

$$3.2 \Rightarrow 320000.$$

Therefore, 3.2 km = 320000 cm.

Example 2 (page 271, #7)

If the distance from Seattle to New Orleans is 2,625 miles, how many kilometers is it?

Solution

Using Table 5.3.2 we multiply the number of miles by 1.6 to find an approximate distance in kilometers. Therefore, 2625 *mi.* = 2625 × 1.6 = 4200 *km*.

Example 3 (page 271, #11)

The speed limit in a certain town is 75 kilometers per hour (kph). If radar records Larry's speed as 50 miles per hour (mph) should he get a ticket?

Solution

Convert 50 miles to an approximate distance in kilometers by multiplying by 1.6.

$$50 \times 1.6 = 80$$

Thus, 50 mph ≅ 80 kph. Since Larry was traveling 50 mph he was exceeding the speed limit of the town by 5 mph. Larry should get a ticket.

Example 4 (page 272, #19)

Which measurement of area is greater in each case (parts and and c only). (**Note:** To solve these problems we must work with equivalent units.)

Solution

a. 1 square inch or 1 square centimeter

   Since 1 inch is approximately equal to 2.56 centimeters (Table 5.3.2), 1 square inch (i.e., 1 in. × 1 in.) must equal 6.5536 square centimeters (2.56 cm × 2.56 cm). Note that 1 sq. in., which is approximately 6.5536 sq. cm, is larger than 1 sq. cm.

c. 1 square yard or 1 square meter

   Using Table 5.3.2, 1 square meter is approximately equal to 1.2 square yards, which is bigger that one square yard. Thus one square meter is larger.

SUPPLEMENTARY EXERCISE 5.3

In problems 1 through 20 convert each length measurement.

1. 23 cm = _____ mm
2. 0.07 cm = _____ mm
3. 8.6 m = _____ mm
4. 6.2 km = _____ mm
5. 2.7 dm = _____ cm
6. 3.8 m = _____ cm
7. 200 mm = _____ cm
8. 6 km = _____ cm
9. 297 mm = _____ dm
10. 687 cm = _____ dm
11. 21.3 m = _____ dm
12. 4.68 km = _____ dm
13. 5.63 cm = _____ m
14. 387 cm = _____ m
15. 29.6 mm = _____ m
16. 0.52 km _____ m
17. 32 hm = _____ km
18. 2586 m = _____ km
19. 370.6 dam _____ km
20. 1,270,309 cm = _____ km

21. Which is less expensive: an item selling for $8 per meter or $7.90 per yard?

22. Cynthia was clocked doing 35 miles per hour in a 50 kilometer per hour speed zone. Did she exceed the speed limit? If so, how much faster was she traveling than the speed limit?

23. The distance between home plate and first base is 90 feet. What is this distance in centimeters?

METRIC SYSTEM 119

5.4 VOLUME

Measures of volume are also referred to as measures of capacity. In the the English system of measurement there are two kinds of capacity measure: dry measure and liquid measure. Distinguishing between these two English capacity measures can occasionally lead to confusion. Consider for example the units of *quarts*, which can measure either dry quarts (e.g., a quart of strawberries) or liquid quarts (e.g., a quart of milk). The metric system, however, uses the same units of measures of capacity, whether they are dry or liquid.

5.4.1 The Liter

The basic unit of capacity measure in the metric system is the **liter**. One liter is defined as the volume of a cube, with dimensions of 10 centimeters. Thus 1 liter (l) is equal to $10 \times 10 \times 10$, or 1000 cubic centimeters (cc).

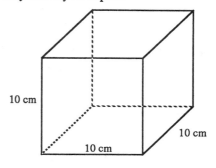

1000 cubic centimeters or 1 liter

Using the set of metric prefixes (section 5.2), we can name multiples and submultiples of the liter. This is summarized in Table 5.4.1.

Conversions among the metric units of capacity can be performed in a manner similar to converting among metric units of length (section 5.3). Additionally, we can convert from English to metric units of capacity, and vice versa. Table 5.4.2 lists the approximate conversion factors.

Note

The liter and milliliter are the two most commonly used units of capacity. We defined liter earlier as 1000 cubic centimeters. One milliliter then is equal to 1 cubic centimeter. Thus the two units, *milliliter (ml)*, and *cubic centimeter (cc)*, are used interchangeably.

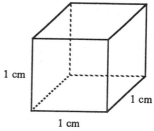

1 milliliter = 1 cubic centiliter (cc)

Example 1 (page 277, #1a,c)

Complete each of the following (parts a and c only).

Solution

a. 12 hl = _____ l

To convert from hl to l the direction is *right* and the number of positional moves is two. Thus 12 hl = 1200 l.

c. 3500 l = _____ kl

To convert from l to kl the direction is *left* and the number of positional moves is three. Therefore, 3500 l = 3.500 kl.

Table 5.4.1 Units of Capacity Measure

1 kiloliter (k*l*)*	=	1000 liters
1 hectoliter (h*l*)	=	100 liters
1 dekaliter (da*l*)	=	10 liters
1 liter (*l*)*	=	1 liter
1 deciliter (d*l*)	=	0.1 liter
1 centiliter (c*l*)*	=	0.01 liter
1 milliliter (m*l*)*	=	0.001 liter

* Units most commonly used

Table 5.4.2 Capacity English-Metric Conversions

ENGLISH TO METRIC		
When you know	Multiply by	To find
teaspoons (tsp.)	5	milliliters (m*l*)
tablespoons (tbsp.)	15	milliliters (m*l*)
fluid ounces (fl. oz.)	30	milliliters (m*l*)
cups (c.)	0.24	liters (*l*)
pints (pt.)	0.47	liters (*l*)
quarts (qt.)	0.95	liters (*l*)
gallons (gal.)	3.8	liters (*l*)
cubic feet (cu. ft.)	0.03	cubic meters (cu. m)
cubic yards (cu. yd.)	0.76	cubic meters (cu. m)
METRIC TO ENGLISH		
When you know	Multiply by	To find
milliliters (m*l*)	0.03	fluid ounces (fl. oz.)
liters (*l*)	2.1	pints (pt.)
liters (*l*)	1.06	quarts (qt.)
liters (*l*)	0.26	gallons (gal.)
cubic meters (cu. m)	35	cubic feet (cu. ft.)
cubic meters (cu. m)	1.3	cubic yards (cu. yd.)

Example 2 (page 278, #7a,e)

Which has the greatest volume in each pair (parts a and e only). (**Note:** To solve these problems we must work with equivalent units.)

Solution

a. 2 gal. or 2 k*l*

   To solve this problem we convert 2 gallons to kiloliters and then compare this measure with 2 kiloliters. Since 1 gallon is approximately equal to 3.8 liters (Table 5.4.2), 2 gallons must be approximately equal to 2×3.8, or 7.6 liters, and 7.6 liters is equal to 0.0076 kiloliters. We now have an approximate kiloliter measure for 2 gallons, namely, 0.0076 k*l*. Comparing 0.0076 k*l* with 2 k*l*, 2 k*l* is the larger measure.

e. 3 tsp. or 3 m*l*

   To solve this problem we convert 3 teaspoons to milliliters and then compare this measure with 3 milliliters. Since 1 teaspoon is approximately equal to 5 milliliters (Table 5.4.2), 3 teaspoons must be approximately equal to 3×5, or 15 milliliters. We now have an approximate milliliter measure for 3 teaspoons, namely, 15 milliliters. Comparing 15 m*l* with 3 m*l*, 15 m*l* is the larger measure. Therefore 3 tsp. is larger than 3 m*l*.

Example 3 (page 278, #9)

Find the volume of a box that is 80 cm long, 0.5 m wide, and 50 mm high. Express your answer in liters.

Solution

Recall that a liter is defined as the volume of a cube 10 cm long, 10 cm high, and 10 cm wide. To find the volume of the box in this problem (in liters) we must first convert the measurements so that they are expressed in terms of cm.

Continued on next page ...

Length: The given length is 80 cm. This measurement needs no conversion.

Width: The given width is 0.5 m. To convert 0.5 m to cm, we multiply by 100. Thus 0.5 m = 50 cm.

Height: The given height is 50 mm. To convert 50 mm to cm we divide by 10. Thus 50 mm = 5 cm.

We now find the volume by multiplying the length, width, and height.

$$V = (80\ cm) \times (50\ cm) \times (5\ cm) = 20{,}000\ cc$$

Finally, we convert 20,000 cubic centimeters to liters. Since 1 liter is equal to 1000 cubic centimeters (10 cm × 10 cm × 10 cm), 20 liters must equal 20,000 cubic centimeters. Therefore, the volume of the box is 20 liters.

SUPPLEMENTARY EXERCISE 5.4

In problems 1 through 20 convert each measure of capacity.

1. 6 cl = _____ ml
2. 4.2 l = _____ ml
3. 0.62 l = _____ ml
4. 92 dl = _____ ml
5. 38 l = _____ cl
6. 6 dl = _____ cl
7. 400 ml = _____ cl
8. 3.8 l = _____ cl
9. 16 dl = _____ l
10. 3.82 kl = _____ l
11. 900 cl = _____ l
12. 3025 ml = _____ l
13. 482 hl = _____ l
14. 1.28 cl = _____ l
15. 250 l = _____ dal
16. 309.6 l = _____ hl
17. 8000 l = _____ kl
18. 48.7 hl = _____ kl
19. 900 dal = _____ kl
20. 295 l = _____ kl

21. What is the volume of a cardboard box that measures 2 m long, 35 cm wide, and 50 cm high? Express your answer in liters.

22. How much money will Howard receive if he sells 300 *l* of wine for $2.50 per gallon?

23. Ten cubic yards of concrete are equivalent to approximately how many cubic meters?

24. Convert the following recipe ingredients to milliliters.

Lemon Filling

$2\frac{1}{2}$ tablespoons cornstarch $\frac{1}{2}$ cup water or orange juice

$\frac{3}{4}$ cup sugar   3 tablespoons lemon juice

$\frac{1}{4}$ teaspoon salt $\frac{1}{2}$ teaspoon grated lemon rind

1 tablespoon butter

5.5 MASS (WEIGHT)

In the English system of measurement we use the unit *pound* to represent the weight of an object. In the metric system we measure the mass of an object. There is an important difference between weight and mass when applied to the sciences (e.g., physics, chemistry, etc). We will not concern ourselves with this difference in this study guide.

The most commonly used unit of mass in the metric system is the **gram**. Using the set of metric prefixes (section 5.2), we can name multiples and submultiples of the gram. This is summarized in Table 5.5.1. Additionally,

- 1 metric tonne (t) = 1,000 kilograms
- 1 gram is the mass of 1 milliliter of water
- 1 kilogram is the mass of 1 liter of water
- 1 metric tonne is the mass of 1 cubic meter of water

Conversions among the metric units of mass can be performed in a manner similar to converting among metric units of length (section 5.3). We also can convert from English to metric units of mass, and vice versa. Table 5.4.2 lists the approximate conversion factors.

Table 5.5.1 Units of Mass Measure

1 kilogram (kg)*	=	1000 grams
1 hectogram (hg)	=	100 grams
1 dekagram (dag)	=	10 grams
1 gram (g)*	=	1 gram
1 decigram (dg)	=	0.1 gram
1 centigram (cg)	=	0.001 gram
1 milligram (mg)*	=	0.001 gram

* Units most commonly used

Table 5.5.2. Mass English-Metric Conversions

ENGLISH TO METRIC		
When you know	Multiply by	To find
ounces (oz.)	28	grams (g)
pounds (lb.)	0.45	kilograms (kg)
tons (T.)	0.9	tonnes (t)
METRIC TO ENGLISH		
When you know	Multiply by	To find
grams (g)	0.035	ounces (oz.)
kilograms (kg)	2.2	pounds (lb.)
tonnes (t)	1.1	tons (T.)

Example 1 (page 283, #1a,d)

Complete each of the following (parts a and d only).

Solution

a. 52 g = _____ mg

   To convert from grams to milligrams the direction is *right* and the number of positional moves is three. Thus 52 g = 52,000 mg.

d. 4.3 dg = _____ g

   To convert from decigrams to grams the direction is *left* and the number of positional moves is one. Thus 4.3 dg = 0.43 g.

Example 2 (page 283, #5a,b,f)

Convert each of the following to the indicated weight (parts a and b only).

Solution

a. 8 oz. = _____ g

   Since 1 ounce is approximately equal to 28 grams (Table 5.5.2), 8 ounces is approximately equal to $8 \times 28 = 224$ grams.

b. 100 kg = _____ lb.

   Since 1 kilogram is approximately equal to 2.2 pounds, 100 kilograms is approximately equal to $100 \times 2.2 = 220$ pounds.

f. 6000 lb. = _____ t

   To convert from pounds to tonnes we will convert from pounds to kilograms, and then from kilograms to tonnes.

   - Since 1 pound is approximately equal to 0.45 kilograms, 6000 pounds is approximately equal to $6000 \times 0.45 = 2700$ kilograms.

   - Since 1 metric tonne is equal to 1000 kilograms, 2700 kilograms is equal to 2.7 tonnes.

   Therefore, 6000 lb. $\cong$ 2.7 t.

   (**Note:** A more simple and direct approach is to divide 6000 pounds by 2000.)

Example 3 (page 283, #9)

A container is 60 centimeters long, 40 centimeters wide, and 50 centimeters high. It weighs three kilograms when it is empty. What will it weigh (in kilograms) when it is filled with cold water?

Solution

A container 60 cm by 40 cm by 50 cm has a volume of $(60 \times 40 \times 50)$ 120,000 cubic centimeters. Since 1 liter = 1000 cubic centimeters, the given container has a volume of 120 liters (divide 120,000 by 1,000). Since 1 kilogram = 1 liter, if the given container has a volume of 120 liters, then it also has a mass of 120 kilograms. The container (when empty) has a mass of 3 kilograms. Therefore the container has a total mass of 123 kilograms when it is filled with cold water.

Example 4 (page 284, #13a,e)

Which has the greater weight (parts a and e only). (**Note:** To do these problems we must work with equivalent units.)

Solution

a. 1 lb. or 1 kg

   Table 5.5.2 indicates that 1 pound is approximately equal to 0.45 kilograms. 1 kg is larger.

e. 40 oz. or 2 kg

   To solve this problem we convert ounces to grams, and then convert grams to kilograms. Since one ounce is approximately equal to 28 grams, 40 ounces is approximately equal to $40 \times 28 = 1120$ grams, which is equal to 1.12 kilograms (divide by 1000).

Continued on next page ...

As a result, 40 ounces is approximately equal to 1.12 kilograms, which is smaller than 2 kilograms. Therefore 2 kg is the larger measure.

(**Note:** An alternative solution is to note that 40 oz. = 2.5 lb. Converting kilograms to pounds, 2 kg × 2.2 lb. = 4.4 lb., which is bigger than 2.4 lb. Thus 2 kg is the larger measure.)

SUPPLEMENTARY EXERCISE 5.5

1. 12 cg = _____ mg
2. 2.675 g = _____ mg
3. 3.62 kg = _____ mg
4. 4.6 dg = _____ mg
5. 800 g = _____ cg
6. 2.3 kg = _____ cg
7. 15 g = _____ dg
8. 2.7 kg = _____ dg
9. 4 kg = _____ g
10. 600 dg = _____ g
11. 350 mg = _____ g
12. 37.5 cg = _____ g
13. 39 kg = _____ dag
14. 6300 mg = _____ dag
15. 10,000 g = _____ hg
16. 49 kg = _____ hg
17. 2,000,000 mg = _____ kg
18. 3875 g = _____ kg
19. 325,000 cg = _____ kg
20. 769.8 dag = _____ kg

21. A 160 pound man has a mass of how many kilograms?

22. An 18 ounce box of cereal has a mass of how many grams?

23. If Cheryl bought 50 kg of sugar for $20, how much did she pay per pound?

5.6 TEMPERATURE

Temperature is measured using a thermometer that is marked with a numerical scale. The scale used to measure temperature in the English system is the Fahrenheit scale, and the unit of measurement is the degree, denoted °F. The scale used to measure temperature in the metric system is the Celsius scale, and the unit of measurement is also the degree, denoted °C.

The Celsius scale is calibrated in intervals of 10 and uses two main reference temperatures: the freezing point of water (0°C) and the boiling point of water (100°C). A comparison of certain temperatures measured in both systems of measurements is shown below.

	Fahrenheit	Celsius
Boiling point of water	212°	100°
Body temperature	98.6°	37°
Room temperature	68°	20°
Freezing point of water	32°	0°

Table 5.6.1 and Table 5.6.2 summarize the conversion formulas for converting from °F and °C, respectively.

Table 5.6.1 Fahrenheit to Celsius Conversion

When you know	Subtract	Then multiply by	To find
°F	32	$\frac{5}{9}$	°C

Note: This is written as $C = \frac{5}{9}(F - 32)$

Table 5.6.2 Celsius to Fahrenheit Conversion

When you know	Multiply	Then add by	To find
°C	$\frac{9}{5}$	32	°F

Note: This is written as $F = (\frac{9}{5} \times C) + 32$

Example 1 (page 289, #1a)

Convert each Fahrenheit temperature to Celsius (part a only).

Solution

a. 104°F

Using Table 5.6.1 we first subtract 32 from 104 and then multiply this difference by $\frac{5}{9}$.

- $104 - 32 = 72$
- $\frac{5}{9} \times 72 = 40$

Therefore, 104°F = 40°C.

Note: We also could have used the formula.

$$C = \frac{5}{9}(F - 32)$$
$$C = \frac{5}{9}(104 - 32)$$
$$C = \frac{5}{9} \times (72)$$
$$C = 40$$

Example 2 (page 289, #3a)

Convert each Celsius temperature to Fahrenheit (part a only).

Solution

a. 20°C

Using Table 5.6.2 we first multiply 20 by $\frac{9}{5}$ and then add 32 to this product.

- $20 \times \frac{9}{5} = 36$
- $36 + 32 = 68$

Therefore, 20°C = 68°F.

Note: We also could have used the formula.

$$F = (\frac{9}{5} \times C) + 32$$
$$F = (\frac{9}{5} \times 20) + 32$$
$$F = 36 + 32$$
$$F = 68$$

SUPPLEMENTARY EXERCISE 5.6

In problems 1 through 16 convert each measure of temperature. (Round to the nearest degree.)

1. 80°F = _____ °C
2. 5°F = _____ °C
3. 42°F = _____ °C
4. 425°F = _____ °C
5. 212°F = _____ °C
6. 101°F = _____ °C
7. 27°F = _____ °C
8. 69°F = _____ °C
9. 10°C = _____ °F
10. 16°C = _____ °F
11. 32°C = _____ °F
12. 70°C = _____ °F
13. 2°C = _____ °F
14. 200°C = _____ °F
15. 29°C = _____ °F
16. 98°C = _____ °F

17. A baking temperature of 450°F is equal to what temperature in Celsius?

18. Normal body temperature of 98.6°F is equal to what temperature in Celsius?

19. Which is warmer: 28°C or 62°F?

20. Which is colder: 100°F or 50°C?

21. Kelly has a temperature of 39°C. Does she have a fever?

22. A temperature of 35°F in conjunction with a 15 mph wind produces a wind chill of 33°F. What are the equivalent temperatures in Celsius?

23. What would be the equivalent metric title of the film, *Fahrenheit 451*?

5.7 MISCELLANY

Since metric units of measure might one day replace English units of measure, it is helpful to know which metric units to use in place of the English units. This information is summarized in Table 5.7.1.

Table 5.7.1 Comparison of Metric and English Measurements

Unit of measure	In Place of (English)	Use (Metric)
Length	inch foot yard mile	milliliter centimeter meter kilometer
Capacity	fluid ounce pint quart gallon	milliliter liter liter liter
Weight	ounce pound ton	gram kilogram tonne
Temperature	°F	°C

5.8 CHAPTER 5 TEST

In 1 through 6 write the numeral that is associated with each prefix.

1. *deci*   2. *milli*   3. *hecto*   4. *kilo*   5. *deka*   6. *centi*

In 7 through 18 label each measure given as a measure of length, area, volume, mass, or temperature.

7. 12 *l*   8. 36 mm   9. 2.6 kg   10. 7 m^2 11. 0.3 hectare   12. 1000 cc
13. 3 tonnes   14. -3°C   15. 702 m*l* 16. 0.13 dam   17. 43 mg   18. 5 m

In 19 through 28 convert each metric measure.

19. 30 mm = _____ m   20. 5 *l* = _____ m*l*   21. 72 kg = _____ g   22. 1700 m = _____ km
23. 55 cg = _____ g   24. 6 c*l* = _____ *l*   25. 0.35 km = _____ m   26. 1.4 cm = _____ mm
27. 12 h*l* = _____ m*l* 28. 10 g = _____ mg

29. If hamburger sells for $2.89 per pound, how much would you pay for 5 kg?

30. Jim paid $1.50 per gallon for unleaded gasoline while Bill paid 40¢ per liter. Who paid more?

6. MATHEMATICAL SYSTEMS

6.1 INTRODUCTION

In a most general sense, a mathematical system can be regarded as a set of elements with at least one operation (or rule) that can be used to combine the elements. In this chapter, we will study various mathematical systems and discuss their nature and structure.

6.2 CLOCK ARITHMETIC

A mathematical system is comprised of two items: a set of elements, and one or more operations used to combine the elements. As an example of a mathematical system consider the set of whole numbers and the operation of addition.

- The set of whole numbers $\{0,1,2,3,...\}$ is the set of elements; and
- The operation used to combine the elements is addition (e.g., $5 + 8 = 13$).

Since the set of whole numbers is infinite, this mathematical system is regarded as an infinite system.

An example of a finite mathematical system is *clock arithmetic*, or the *12-hour clock system*. This system uses an ordinary clock face and only the hour hand. The set of elements of the system is $\{1,2,3,4,5,6,7,8,9,10,11,12\}$, and the operations of addition, subtraction, and multiplication are used to combine these elements.

6.2.1 Addition

The operation of addition is performed in the 12-hour clock system by moving the hour hand in a clockwise direction. For example, if the hour hand is currently at the element 6, then the expression *6 + 9* means *nine hours past six*, which places the hour hand at the element 3. Consequently, *6 + 9 = 3* in the 12-hour clock system. Using this technique for addition, an addition table (Table 6.1.1) is constructed. Notice that the sum of any two numbers in the 12-hour clock system is found in this table.

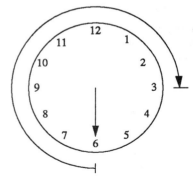

A number greater than 12 is expressed as a number in the 12-hour clock system by dividing the number by 12; the remainder is an equivalent number in this system. For example, an equivalent number for 35 in the 12-hour clock system is 11.

$$\begin{array}{r} 2 \quad \text{Quotient} \\ 12\overline{)35} \\ \underline{24} \\ 11 \quad \text{Remainder} \end{array}$$

The quotient 2 indicates that there are two complete rotations of the hour hand, beginning at 12. The remainder 11 indicates that there are eleven additional hours after the last rotation. That is,

$$\begin{array}{rcl} 35 &=& 12 + 12 + 11 \\ 35 &=& 11 \end{array}$$

Thus the hour hand is in the same position on the clock face for both 35 and 11.

Table 6.1.1   Addition in a 12-Hour Clock System

+	1	2	3	4	5	6	7	8	9	10	11	12
1	2	3	4	5	6	7	8	9	10	11	12	1
2	3	4	5	6	7	8	9	10	11	12	1	2
3	4	5	6	7	8	9	10	11	12	1	2	3
4	5	6	7	8	9	10	11	12	1	2	3	4
5	6	7	8	9	10	11	12	1	2	3	4	5
6	7	8	9	10	11	12	1	2	3	4	5	6
6	7	8	9	10	11	12	1	2	3	4	5	6
7	8	9	10	11	12	1	2	3	4	5	6	7
8	9	10	11	12	1	2	3	4	5	6	7	8
9	10	11	12	1	2	3	4	5	6	7	8	9
10	11	12	1	2	3	4	5	6	7	8	9	10
11	12	1	2	3	4	5	6	7	8	9	10	11
12	1	2	3	4	5	6	7	8	9	10	11	12

In the 12-hour clock system, where the operation is addition, the number 12 is regarded as the **identity element**. That is, any number added to 12 always yields the number itself. If the sum of two numbers yields the identity element then the numbers being added are **additive inverses** of each other. For example, 3 and 9 are additive inverses of each other since their sum is 12. Thus the inverse of 3 is 9, and the inverse of 9 is 3. It is important to note that if a system does not have an identity element then it is meaningless to speak of inverse elements.

6.2.2 Subtraction

In order to perform the operation of subtraction in the 12-hour clock system we must be sensitive to the numbers that make up the problem. To illustrate this let us assume we have the following general subtraction problem:

$$a - b = ?$$

- Case 1: (a > b)

  If a is greater than the b then we subtract as usual. For example, $8 - 5 = 3$.

- Case 2: (a < b)

  If a is less than b, then we need to express a as an equivalent 12-hour clock system number that is greater than b. To do this we add 12 or any multiple of 12 to a, since 12 is the identity element in the 12-hour clock system. Once a is greater than b we then subtract as usual. To illustrate, the subtraction problem $5 - 8 = ?$ is solved below.

$$
\begin{aligned}
&\quad 5 - 8 \\
&= (12 + 5) - 8 \\
&= 17 - 8 \\
&= 9
\end{aligned}
$$

- Case 3: (a = b)

  If a is equal to b then $a - b = 0$. Since zero is not an element in the 12-hour clock system, we add 12 to it to produce an equivalent 12-hour clock system number; this answer is 12. As a result, $a - a = 0 = 12$.

6.2.3 Multiplication

The operation of multiplication is performed in the 12-hour clock system by multiplying as usual. However, if the product is greater than 12 then the product is expressed as an equivalent 12-hour clock system number. For example, $6 \times 7 = 42$, but 42 is not an element in the 12-hour clock system. When we divide 42 by 12, we get a quotient of 3 and a remainder of 6. Thus $6 \times 7 = 6$ in the 12-hour clock system.

6.2.4 Properties of Mathematical Systems

Any mathematical system can be examined for the following properties.

1. Closure Property

    If any two elements of a set are combined under the rules of an operation, and the result is an element of the set, then the set is **closed** under that operation. In other words, the answer must be a member of the set in order for the closure property to hold.

2. Commutative Property

    Given any two elements a and b of a set and an operation $*$ to combine them, if the result is the same regardless of the order in which the elements are combined, then the set is commutative under that operation. More generally, if

    $$a * b = b * a$$

    then the commutative property holds.

3. Associative Property

    Given any three elements a, b, and c of a set and an operation $*$ to combine them, if the result is the same regardless of how the elements are grouped, then the set is associative under that operations. More generally, if

    $$(a * b) * c = a * (b * c)$$

    then the associative property holds.

4. Identity Property

    The identity property holds for a given mathematical system if the set contains an identity element. In order for such an element to exist, every a of the set when combined with the identity element e under a given operation $*$ must produce a result that is the element itself. That is, if

    $$a * e = a \quad \text{and} \quad e * a = a$$

    then the identity property holds.

5. Inverse Property

    The inverse property holds for a given mathematical system if for every element a in the system there exists an element b in the system such that when a and b are combined under a given operation $*$ their result is the identity element. That is, if

    $$a * b = e \quad \text{and} \quad b * a = e$$

    then the inverse property holds.

When a set of elements and an operation satisfy the closure, associative, identity, and inverse properties, we say that the elements form a **group** under that operation, provided that the operation is binary. If the group also satisfies the commutative property then we say the group is a **commutative group**.

Example 1 (page 307, #1d)

Evaluate 11 + 11 on a 12-hour clock.

Solution

The sum of 11 and 11 is 22. Since 22 is not an element in the 12-hour clock system, we must express it as a number in this system. We do so by dividing by 12 and focusing on the remainder. As a result, 22 ÷ 12 = 1 with a remainder of 10. Thus 11 + 11 = 10.

Example 2 (page 308, #5a)

Evaluate 5 − 7 on a 12-hour clock.

Solution

Since 5 is smaller than 7, we have a Case 2 situation. Thus we must reexpress 7 as an equivalent 12-hour clock system number. We do so by adding to it the identity element 12. As a result,

$$\begin{aligned} & 5 - 7 \\ =\ & (12 + 5) - 7 \\ =\ & 17 - 7 \\ =\ & 10 \end{aligned}$$

Hence 5 − 7 = 10.

Example 3 (page 308, #7d)

Evaluate 4 × 10 on a 12-hour clock system.

Solution

The product of 4 and 10 is 40. Since 40 is not an element in the 12-hour clock system we convert it to an equivalent number in that system. We do this by dividing 40 by 12 and focusing on the remainder. 40 ÷ 12 = 3 with a remainder of 4. Hence, 4 × 10 = 4.

Example 4 (page 308, #11a,b,d,e,f)

State the property of the 12-hour clock system that is illustrated by each of the following.

a. 7 + 6 = 1

This problem implies that two elements of the set are being combined under the operation of addition and the result is an element of the set. This illustrates the closure property of addition.

b. 7 × 5 = 5 × 7

This problem implies that the *order* in which the elements are combined under the operation of multiplication is unimportant. This illustrates the commutative property of multiplication.

d. (2 × 4) × 3 = 2 × (4 × 3)

This problem implies that *grouping* of the elements under the operation of multiplication is unimportant. (Note that the expression to the left of the equal sign shows 4 being grouped with 2. However, the expression to the right of the equal sign shows 4 being grouped with 3.) This illustrates the associative property of multiplication.

Continued on next page ...

e. $5 \times 5 = 1$

This problem illustrates two properties — the closure property of multiplication and the inverse property of multiplication. The system is closed with respect to multiplication because $5 \times 5 = 25$, which is equivalent to 1 and is an element in the 12-hour clock system. The inverse property of multiplication holds because the result of combining the given elements under the operation of multiplication yields the identity element. (Note that the element 1 is the identity element for multiplication.)

f. $12 + 7 = 7$

This problem illustrates the identity property for addition. Any number added to the identity element yields the number itself.

Example 5 (page 308, #13b,d,e)

Determine whether each statement is true or false (parts b,d, and e only).

The set of counting numbers $\{1,2,3,...\}$...

Solution

b. ... is closed with respect to subtraction.

False (Consider $3 - 4 = -1$. Negative one is not in the given set.)

d. ... is commutative with respect to division.

False (Consider $4 \div 2 \neq 2 \div 4$.)

e. ... contains an identity element for addition.

False (Zero is not an element of the given set.)

SUPPLEMENTARY EXERCISE 6.2

In problems 1 through 12 perform the indicated operations on a 12-hour clock.

1. $9 + 9$
2. $7 + (6 + 4)$
3. $(8 + 2) + 5$
4. $3 - 6$
5. $2 - (3 + 9)$
6. $10 + (6 - 3)$
7. 5×7
8. $3 \times (2 + 6)$
9. $4 + (6 \times 8)$
10. $5 \times (3 - 11)$
11. $(4 - 6) \times 8$
12. $(11 - 12) \times (3 + 8)$

In 13 through 16 state the property of the 12-hour clock system that is illustrated.

13. $9 + 3 = 12$
14. $8 \times 8 = 4$
15. $11 \times 1 = 11$
16. $2 + (5 + 7) = (5 + 7) + 2$

In 17 through 20 determine whether each statement is true or false.

The set of integers $\{...,-1,-2,-3,...,1,2,3,...\}$...

17. is closed with respect to subtraction.

18. is associative with respect to multiplication.

19. is commutative with respect to subtraction.

20. contains an identity element for addition.

6.3 MORE NEW SYSTEMS

Many systems of arithmetic can be created. In this section, Professor Setek develops two new systems. The first system, the system of *seasons*, is similar to a 4-hour clock system; each season is assigned a unique number 1, 2, 3, or 4. Specifically,

          1 = Spring   3 = Fall
          2 = Summer 4 = Winter

By relating this *season* system to the 12-hour clock system note that the identity element for addition is *Winter* (4), and the identity element for multiplication is *Spring* (1).

The second system, the system of *days of the week*, is similar to a 7-hour clock system; each day of the week is assigned a unique number 1, 2, 3, 4, 5, 6, or 7. Specifically,

          1 = Sunday   5 = Thursday
          2 = Monday   6 = Friday
          3 = Tuesday   7 = Saturday
          4 = Wednesday

Once again, by relating this system to a 7-hour clock system we find that the identity element for addition is *Saturday* (7), and the identity element for multiplication is *Sunday* (1).

The operations of addition, subtraction, and multiplication, (along with other operations), can be performed on these new mathematical systems in a similar manner to that of the 12-hour clock system. Each of these new systems also can be examined for the closure, commutative, associative, identity, and inverse properties.

Example 1 (page 313, #1)

Evaluate Summer + Fall

Solution

By replacing the elements Summer and Fall with their respective numerical values (2 and 3), the sum of Summer and Fall is 5. However, since 5 does not correspond to an element in the set we must reexpress it as an equivalent number that does correspond to an element of the set. We do this by dividing 5 by 4 and focusing on the remainder. As a result, $5 \div 4 = 1$ with a remainder of 1. Hence Summer + Fall = Spring.

Example 2 (page 313, #5)

Evaluate Spring − Summer

Solution

By replacing the elements Spring and Summer with their respective numerical values (1 and 2), we have the problem $1 - 2$. Recalling from section 6.2, this is a Case 2 problem ($a < b$), since 1 is less than 2. As a result, we add the identity element of addition (*Winter*, which has a numerical value of 4), to 1 before we perform the subtraction. Thus,

	Spring	−	Summer	
=	1	−	2	
=	(4 + 1)	−	2	
=	5	−	2	
=		3		
=		Fall		

Example 3 (page 314, #11)

Evaluate Fall × (Winter − Spring)

Solution

	Fall	×	(Winter − Spring)
=	3	×	(4 − 1)
=	3	×	(3)
=	9		

9 ÷ 4 = 2 with a remainder of 1

Therefore, Fall × (Winter − Spring) = Spring.

Example 4 (page 314, #17)

Does the *season* system form a group under the operation of multiplication?

Solution

The closure, associative, identity, and inverse properties must all hold under multiplication in order for this system to form a group under the operation of multiplication. As a result, we must investigate each of these properties.

- The closure and associative properties can be easily verified.
- The identity element is Spring, thus the identity property holds.
- In order for the inverse property to hold, each element of the set must be multiplied by another element of the set and the product must be Spring (1), which is the identity element. In other words, each element must have a *multiplicative inverse*. (**Note:** Numerical equivalents to the *season* system are placed within parentheses.)

Finding The Inverse of Spring (1)

$$1 \times 1 = 1 \quad \text{(This product is the identity element.)}$$

The inverse of Spring is Spring.

Finding The Inverse of Summer (2)

$2 \times 1 = 2$ (This product is not the identity element.)
$2 \times 2 = 4$ (This product is not the identity element.)
$2 \times 3 = 6 \equiv 2$ (This product is not the identity element.)
$2 \times 4 = 8 \equiv 4$ (This product is not the identity element.)

Note that none of the elements is the multiplicative inverse of 2. As a result, the element Summer does not have an inverse.

Finding The Inverse For Fall (3)

$3 \times 1 = 3$ (This product is not the identity element.)
$3 \times 2 = 6 \equiv 2$ (This product is not the identity element.)
$3 \times 3 = 9 \equiv 1$ (This product is the identity element.)

The inverse of Fall is Fall.

Continued on next page ...

Finding The Inverse Element For Winter (4)

$4 \times 1 = 4$ (This product is not the identity element.)
$4 \times 2 = 8 \equiv 4$ (This product is not the identity element.)
$4 \times 3 = 12 \equiv 4$ (This product is not the identity element.)
$4 \times 4 = 16 \equiv 4$ (This product is not the identity element.)

Note that none of the elements is the multiplicative inverse of 4. As a result, the element Winter does not have an inverse.

As a result of the above discussion, the *season* system is not a group under the operation of multiplication since every element of the set does not have an inverse. (**Note:** Since the element Summer did not have an inverse, we could have stopped our search at that point.)

SUPPLEMENTARY EXERCISE 6.3

In problems 1 through 15 we define a mathematical system of *time division* with each division of time listed below being assigned a unique number from 1 to 6.

$$1 = \text{day} \quad 2 = \text{week} \quad 3 = \text{month}$$
$$4 = \text{year} \quad 5 = \text{decade} \quad 6 = \text{century}$$

In 1 through 10 evaluate each expression and give your answer in terms of a *time division*.

1. week + year
2. century + decade
3. month × day
4. year × year
5. day − decade
6. month − century
7. (century + day) × month
8. decade × (year − century)
9. (year + day) − decade
10. (decade − century) + week

11. Does the time *division system* have an identity element for addition. If so, what is it?

12. What is the additive inverse of *decade*?

13. What is the multiplicative inverse of *year*?

14. Does the *time division* system form a group under the operation of addition? If not, why?

15. Does the *time division* system form a group under the operation of multiplication? If not, why?

6.4 MODULAR SYSTEMS

A mathematical system that is cyclic (e.g., the 12-hour clock system), is called a **modular system**. (Frequently the word *mod* is used in place of *modular*.) In general, a **mod m system** consists of the set of elements $\{0,1,2,3,...,m-1\}$. For example,

- The set of elements for a *mod 5* system consist of $\{0,1,2,3,4\}$
- The set of elements for a *mod 7* system consist of $\{0,1,2,3,4,5,6\}$
- The set of elements for a *mod 12* system consist of $\{0,1,2,3,4,5,6,7,8,9,10,11\}$

To express a number that is not an element of the set for a given mod system we divide the number by m; the remainder is the equivalent numerical value. Formally, we denote this equivalence as $a \equiv b \pmod{m}$, where a is the given number, b is the remainder from dividing a by m, and m is the mod. This equivalence is read *a is equivalent to b, mod m* and means that both a and b have the same remainder when they are divided by m.

The operations of addition, subtraction, and multiplication (and other operations) are performed on modular systems in the usual manner. Also, as with any mathematical system, a mod m system can be examined for the closure, commutative, associative, identity, and inverse properties.

Example 1 (page 320, #1e)

Evaluate $(3 + 2) + 4$ in mod 5.

Solution

$$\begin{aligned} & (3 + 2) + 4 \\ =\ & 5 + 4 \\ =\ & 9 \end{aligned}$$

Since 9 is not an element of the set that makes up a mod 5 system we divide 9 by 5 and look at the remainder. $9 \div 5 = 1$ with a remainder of 4. So, $(3 + 2) + 4 \equiv 4 \pmod 5$.

Example 2 (page 320, #5e)

Evaluate $3 - (2 - 4)$ in mod 5.

Solution

Working within parentheses first, note that the subtraction is a Case 2 problem where $a < b$. (See section 6.2.) This means that we must reexpress 2 in the problem $(2 - 4)$ with an equivalent mod 5 number. An equivalent mod 5 number for two is seven $(2 + 5)$.

$$\begin{aligned} & 3 - (2 - 4) \\ =\ & 3 - ((5 + 2) - 4) \\ =\ & 3 - (7 - 4) \\ =\ & 3 - 3 \\ =\ & 0 \end{aligned}$$

Hence, $3 - (2 - 4) \equiv 0 \pmod 5$.

Example 3 (page 321, #7e)

Evaluate $3 \times (4 + 2)$ in mod 5.

Solution

$$\begin{aligned} & 3 \times (4 + 2) \\ =\ & 3 \times (6) \\ =\ & 18 \end{aligned}$$

Since 18 is not a mod 5 number we reexpress it by dividing it by 5 and focusing on the remainder of the division. Hence, $3 \times (4 + 2) \equiv 3 \pmod 5$.

Example 4 (page 321, #13b,d)

Determine whether each statement is true or false (parts b and d only).

Solution

b. $22 \equiv 1 \pmod 5$

False — the statement $22 \equiv 1 \pmod 5$ implies that when 22 is divided by 5 the remainder is 1. This is not the case.

Continued on next page ...

d. $33 \equiv 5 \pmod{7}$

True — the statement $33 \equiv 5 \pmod 7$ implies that when 33 is divided by 7 the remainder is 5. This is indeed the case.

Example 5 (page 321, #17a,d)

Determine the missing number in the given problems (parts a and d only).

Solution

To solve the problems given in exercise 17 we employ a trial and error approach. This approach is outlined here.

1. Choose a number from the set of the given mod system.
2. Replace the question mark in the problem with the number selected from step 1.
3. Perform the indicated operation on the left side of the equivalence sign ($\equiv$).
4. Reexpress the result from step 3 (if necessary) as an element from the set of the given mod system.

If the answer from step 4 is equal to the number given in the problem then the number used (from step 1) is the correct answer; otherwise repeat steps 1 through 4 until a correct number is obtained, or until all of the elements of the set are tried. If a correct answer cannot be found after all the elements of the set are used then the answer is the empty set.

a. $2 \times ? \equiv 3 \pmod 7$

The elements of the set for mod 7 are $\{0,1,2,3,4,5,6\}$.

- $2 \times 0 = 0$, which is equal to 0 in mod 7
- $2 \times 1 = 2$, which is equal to 2 in mod 7
- $2 \times 2 = 4$, which is equal to 4 in mod 7
- $2 \times 3 = 6$, which is equal to 6 in mod 7
- $2 \times 4 = 8$, which is equal to 1 in mod 7
- $2 \times 5 = 10$, which is equal to 3 in mod 7 (This is what we want; the correct number is 5.)

d. $1 - ? \equiv 4 \pmod 5$

The elements of the set for mod 5 are $\{0,1,2,3,4\}$

- $1 - 0 = 1$, which is equal to 1 in mod 5
- $1 - 1 = 0$, which is equal to 0 in mod 5
- $1 - 2 = (5 + 1) - 2 = 6 - 2 = 4$, which is equal to 4 in mod 5
  (This is what we want; the correct number is 2.)

SUPPLEMENTARY EXERCISE 6.4

In problems 1 through 20 evaluate each expression in mod 8.

1. $4 + 3$
2. $5 + 6$
3. $2 - 6$
4. $1 - 0$
5. 7×3
6. 5×5
7. $(4 + 6) - 2$
8. $(5 + 2) - 7$
9. $(1 - 3) - 7$
10. $(3 - 7) - 2$
11. $(0 \times 3) + 6$
12. $(5 \times 7) + 4$
13. $(4 - 7) \times 3$
14. $(7 - 6) \times 2$
15. $5 \times (7 - 3)$
16. $6 \times (2 - 4)$
17. $4 + (3 \times 6)$
18. $3 + (5 \times 2)$
19. $(3 + 6) \times (0 - 5)$
20. $(4 - 6) + (2 - 3)$

Continued on next page ...

21. Does the mod 8 system form a group under addition? If not, why?

22. Does the mod 8 system form a group under multiplication. If not, why?

In problems 23 through 28 determine whether the given statement is true or false.

23. $0 \equiv 12 \pmod 3$ 24. $28 \equiv 6 \pmod 8$ 25. $522 \equiv 0 \pmod 2$
26. $175 \equiv 4 \pmod 9$ 27. $230 \equiv 23 \pmod{10}$ 28. $328 \equiv 16 \pmod{21}$

In problems 28 through 34 use the elements from the set of the indicated mod system to make each statement true.

29. $5 \times ? \equiv 4 \pmod 6$ 30. $? \times 8 \equiv 1 \pmod{13}$ 31. $? - 3 \equiv 5 \pmod 8$
32. $4 - ? \equiv 4 \pmod 7$ 33. $4 + ? \equiv 6 \pmod 9$ 34. $? \times 7 \equiv 3 \pmod{10}$

6.5 MATHEMATICAL SYSTEMS WITHOUT NUMBERS

The mathematical systems discussed thus far have been based on the use of numbers. The operations performed on the elements of the sets have been the typical operations of addition, subtraction, and multiplication. In this section of your textbook, examples of mathematical systems that are not based on numbers and do not employ these typical operations are presented. Such systems are usually defined by a table. To illustrate we will discuss a system described in your textbook and defined in the table below.

*	A	B	C	D
A	A	B	C	D
B	B	C	D	A
C	C	D	A	B
D	D	A	B	C

From the table the following information can be derived.

- The set of elements for the system is $\{A, B, C, D\}$. This is determined by noting the elements in the top-most row of the table (above the horizontal line), and by the elements in the left-most column of the table (to the left of the vertical line).

- The operation used to combine the elements is $*$ and is displayed in the upper left corner of the table.

- The answers that result from combining the elements under the operation are located in the central part of the table. To determine an answer when two elements are combined
  1. find the *row* headed by the first element of the problem;
  2. read across the table until we reach the *column* headed by the second element of the problem;
  3. select the element that is located at the intersection of steps 1 and 2.

This is illustrated below for the problem $C * D = ?$

		Column			
	*	A	A	C	D
	A				
Row	B				
	C				B
	D				

Thus $C * D = B$.

6.5.1 Examining Properties Using a Table

To examine a mathematical system, which has been defined by a table, for the closure, identity, inverse, and commutative properties, we can employ the following techniques. (**Note:** There are no easy techniques for determining if the associative property holds.)

<u>Closure</u>

If there are no new elements in the answer part of the table (i.e., all the answers are elements of the given set), then the system is closed.

<u>Identity</u>

An identity element exists if the central part of the table (i.e., the answer section) contains both a row and a column that matches, respectively, the row and column headings of the table. For example, in the table below an identity element exists and the identity element is the element A. Row 1 has exactly the same pattern as the row-heading, and column 1 has exactly the same pattern as the column-heading.

*	A	B	C	D
A	A	B	C	D
B	B			
C	C			
D	D			

<u>Inverse</u>

If an identity element exists then we can determine if each element of the set has an inverse. To do this we scan each row and column of the table for the identity element. If the identity element exists in both instances then the inverse of a given element is the element headed by that row or column.

<u>Commutative</u>

A system is commutative if the central part of the table is symmetrical. More specifically, when a line is drawn from the upper left corner of the table to the lower right of the table, the elements on either side of this diagonal should match.

Example 1 (page 326, #1)

Evaluate $P:Q$ for the system defined by the table below.

Solution

$P:Q$ is evaluated by finding the element located at the intersection of the row headed by P and the column headed by Q. Thus $P:Q = Q$.

:	P	Q	R	S
P	P	Q	R	Q
Q	Q	R	S	P
R	R	S	P	Q
S	S	P	Q	R

Example 2 (page 326, #13)

Which elements of the set have an inverse? Name the inverse of each of these elements. (Use the set from example 1.)

Solution

Before we can determine which elements have an inverse we must first determine if the system has an identity element. Note from the table given in Example 1 that the identity element is P. The first row of the table has exactly the same pattern has the row-heading, and the first column of the table has exactly the same pattern as the column-heading.

Note further that each row and each column of the table contains an entry of P. Thus every element has an inverse. The inverse of P is P; the inverse of Q is S; the inverse of R is R; and the inverse of S is Q.

SUPPLEMENTARY EXERCISE 6.5

The following table defines the operation # for the elements $\{T, Y, P, E\}$.

#	T	Y	P	E
T	P	T	Y	E
Y	T	Y	P	E
P	Y	P	T	E
E	E	E	X	Y

In problems 1 through 12 use the table above to evaluate each expression.

1. $Y \# T$
2. $E \# E$
3. $T \# P$
4. $P \# P$
5. $Y \# E$
6. $T \# (T \# E)$
7. $(Y \# P) \# T$
8. $(E \# P) \# Y$
9. $(Y \# Y) \# (P \# E)$

10. Is the set $\{T, Y, P, E\}$ closed with respect to the operation #? Why or why not?

11. What is the identity element (if any) of the system defined in the table?

12. Name all the elements of the set that have an inverse, and the inverse of these elements.

13. Does this system form a group under the operation #? Why or why not?

6.6 AXIOMATIC SYSTEMS

An **axiomatic system** consists of four parts, which are defined below.

- **Undefined terms** are used to form a basic vocabulary with which other terms are defined.
- **Defined terms** are necessary so as to avoid any misunderstanding in a discussion. This way the definition of a word or phrase has the same meaning to everyone.
- **Axioms** are general statements whose truth is assumed without proof. Axioms enable us to deduce the truth of other statements just as undefined terms enable us to define other terms.
- **Theorems** are statements proved by deduction.

In an axiomatic system the undefined terms, defined terms, and axioms form the basic structure of a subject. They are the tools by which other statements (i.e., theorems) are deduced. The combination of the four parts of an axiomatic system leads to the discovery and explanation of new theorems, which further develops knowledge of a subject.

Example 1 (page 332, #1)

Given the following sets of axioms, which of the models shown represent the axiomatic system? (**Note:** Capital letters represent buildings and lines represent sidewalks.)

    I. There are at least two buildings on campus.

    II. There is exactly one sidewalk between any two buildings.

    III. Not all of the buildings are on the same sidewalk.

a.
b.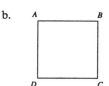

c.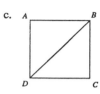
d.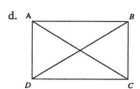

e.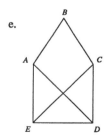
f.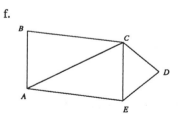

Solution

    Axiom I: All six models satisfy this axiom

    Axiom II: The following models do not satisfy this axiom:

- b — There are no sidewalks between buildings A and C, and B and D.
- c — There is no sidewalk between buildings A and C.
- e — There are no sidewalks between buildings A and C, B and D, and B and E.
- f — There are no sidewalks between buildings B and D and B and E.

    Axiom III: The remaining models a and d satisfy this axiom.

Thus, models a and d represent this axiomatic system.

(**Note:** The solutions to all the remaining odd numbered exercises in this section are, for all practical purposes, provided in the answer section of your textbook.)

SUPPLEMENTARY EXERCISE 6.6

1. Given the axioms below, prove that at least four cats must exist.

    I. There are exactly two cats in each house.

    II. For each pair of cats there is exactly one house that contains them.

    III. There is at least one house.

    IV. For each house there exists exactly one other house that has no cats in common with it.

6.7 CHAPTER 6 TEST

1. List the components of a mathematical system.
2. The statement $5 + 0 = 5$ illustrates what property?
3. List the elements of a mod 6 system.
4. What property does not have to be satisfied in order for a mathematical system to be a group?
5. Using the numbers 2, 3, and 4, write a statement that shows why the associative property of subtraction does not hold.

In problems 6 through 10 answer true or false.

6. The set of counting numbers, $\{1,2,3,...\}$, is closed with respect to addition.
7. The set of whole numbers, $\{0,1,2,3,...\}$, is commutative with respect to division.
8. The inverse of 5 (mod 7) is 3 under the operation of multiplication.
9. The equivalent of 302 in a mod 7 system is 5.
10. $4 \times 5 \equiv 5 + 4 \pmod{11}$.

In problems 11 through 15 evaluate each expression using a 12 hour clock system.

11. $9 + 8 =$ _____
12. $7 - 11 =$ _____
13. $6 \times 5 =$ _____
14. $3 \times (8 + 6) =$ _____
15. $(2 \times 9) - 8 =$ _____

In problems 16 through 18 evaluate each expression. Assume that 1 = *yes*, 2 = *no*, and 3 = *maybe*. Express all answers in terms of *yes, no,* or *maybe*.

16. yes + no = _____
17. yes − maybe = _____
18. no × (no + maybe) = _____

In problems 19 through 26 fill in the blank such that each statement will be true in the indicated mod system.

19. $2 +$ _____ $\equiv 1 \pmod 5$
20. $3 -$ _____ $\equiv 4 \pmod 5$
21. _____ $+ 4 \equiv 6 \pmod 8$
22. $4 -$ _____ $\equiv 2 \pmod 6$
23. $1 \times$ _____ $\equiv 3 \pmod 4$
24. _____ $\times 5 \equiv 4 \pmod 7$
25. $5 - (2 - 4) \equiv$ _____ $\pmod 5$
26. $(4 \times$ _____ $) - 3 \equiv 5 \pmod 6$

Continued on next page ...

Use the table below to answer problems 27 through 36.

&	?	!	%
?	:	%	!
!	?	!	%
%	!	%	:

27. ? & ! = _____
28. % & ? = _____
29. ? & ? = _____
30. % & % = _____
31. ! & % = _____
32. ? & (! & %) = _____

33. Is this system closed with respect to the operation &?

34. What is the identity element (if any) for this system?

35. What is the inverse element of the element % (if it exists)?

36. Does this system form a group under the operation &? Why or why not?

7. SYSTEMS OF NUMERATION

7.1 INTRODUCTION

In a most general sense, a system of numeration deals with the act of counting. This includes the symbols used to count, which are called **numerals**, and the rules for combining these symbols so that other numerals can be represented. In this chapter we examine many different number systems, including the systems used by the Egyptians, Greeks, Babylonians, and Chinese-Japanese.

7.2 SIMPLE GROUPING SYSTEMS

One of the oldest systems of numeration is the one used by the Egyptians. The symbols they used to represent numbers, called **hieroglyphics**, are listed in table 7.2.1. The rules they followed for combining these symbols were simple: The position of a symbol does not affect the number represented and a different symbol is used to indicate a certain group of items.

This system is an example of a simple grouping system because the values of the symbols remain constant regardless of their position. Thus the order (i.e., placement) of a symbol in a group of symbols does not affect the symbol's numerical value. As a result there are many different symbol representations for the same number. For example, to represent the number 147 in hieroglyphics we use one coiled rope, four heelbones, and seven strokes. However, the order in which these symbols are arranged is not important. As a result, the following arrangements of these symbols are three (of many) representations of the number 147.

$$e \cap \cap \cap \cap ||||||| = 147$$
$$\cap \cap \cap e |||| \cap ||| = 147$$
$$\cap || \cap || \cap || \cap | e = 147$$

Table 7.2.1 Egyptian Numerals and Their Values

Egyptian Numerals	Name	Value	Power of 10
|	Stroke	1	10^0
∩	Heelbone	10	10^1
e	Coiled Rope	100	10^2
ↄ⃯	Lotus Flower	1,000	10^3
(	Pointed Finger	10,000	10^4
⊲	Polywog	100,000	10^5
ℛ	Astonished Man	1,000,000	10^6

7.2.1 Adding and Subtracting Egyptian Numerals

The operations of addition and subtraction using hieroglyphics are straightforward.

A. To add numbers expressed as Egyptian numerals:

1. Group similar symbols together.
2. Simplify each grouping from step 1. (For example, a grouping of 12 strokes is simplified to one heelbone and two strokes.)

B. To subtract numbers expressed as Egyptian numerals:

1. Subtract similar symbols.

   (**Note:** It might be necessary to rewrite some numerals in terms of other symbols prior to a subtraction. See example 4.)

Example 1 (page 347, #1d,f)

Express the following numbers as an Egyptian numeral (parts d and f only).

Solution

d. 102

The number 102 is expressed as (100) + (2). Thus we need one coiled rope and two strokes. The order in which these symbols are arranged is not significant. As a result one representation is $e \mid \mid$.

f. 1,132

The number 1,132 is expressed as (1000) + (100) + (30) + (2). Thus we need one Lotus flower, one coiled rope, three heelbones, and two strokes. Once again, the order of these symbols is unimportant.

As a result one representation is $\overset{\ulcorner D}{\Delta} \; e \; \cap \cap \cap \mid \mid$

Example 2 (page 348, #5d,f)

Evaluate the given Egyptian numeral (parts d and f only).

Solution

d. $\mid \overset{\ulcorner D}{\Delta} \overset{\ulcorner D}{\Delta} e \, e \cap \mid \mid$

	$\mid$	$\overset{\ulcorner D}{\Delta}$	$\overset{\ulcorner D}{\Delta}$	e	e	$\cap$	$\mid$	$\mid$
=	(10,000)	+ (1000)	+ (1000)	+ (100)	+ (100)	+ (10)	+ (1)	+ (1)

= 12,212

f. $\mid \cap \mid \overset{\ulcorner D}{\Delta} e \overset{\ulcorner D}{\Delta} e$

	$\mid$	$\cap$	$\mid$	$\overset{\ulcorner D}{\Delta}$	e	$\overset{\ulcorner D}{\Delta}$	e
=	(1)	+ (10)	+ (1)	+ (1000)	+ (100)	+ (1000)	+ (100)

= 2212

Example 3 (page 348, #7c)

Add: $||||||+||||$

Solution

Grouping similar symbols yields ten strokes, which equal one heelbone. The answer is ∩.

Example 4 (page 348, #11c)

Subtract: $e ∩ | - ∩∩∩∩∩∩||$

Solution

By rewriting this problem vertically and aligning similar symbols under one another note that we subtract two strokes from one stroke, and six heelbones from one heelbone.

$$\begin{array}{r} e \quad ∩ \quad | \\ - \quad ∩∩∩∩∩ \quad || \\ \hline \end{array}$$

Since we cannot subtract two strokes from one stroke we rewrite the heelbone as ten strokes, thus yielding a total of 11 strokes.

$$\begin{array}{r} e \quad\quad\quad\quad ||||||||||| \\ - \quad ∩∩∩∩∩ \quad\quad || \\ \hline \end{array}$$

We now need to rewrite the coiled rope as ten heelbones.

$$\begin{array}{r} ∩∩∩∩∩∩∩∩∩∩ \quad ||||||||||| \\ - \quad\quad ∩∩∩∩∩ \quad\quad || \\ \hline \end{array}$$

Finally, we subtract.

$$\begin{array}{r} ∩∩∩∩∩∩∩∩∩∩ \quad ||||||||||| \\ - \quad\quad ∩∩∩∩∩ \quad\quad || \\ \hline ∩∩∩∩ \quad\quad ||||||||| \end{array}$$

Thus the answer is 4 heelbones and 9 strokes.

SUPPLEMENTARY EXERCISE 7.2

In problems 1 through 10 express each number as an Egyptian numeral.

1. 29   2. 38   3. 66   4. 101   5. 297   6. 316   7. 500   8. 1952   9. 2022   10. 11,111

In problems 11 through 20 evaluate each Egyptian numeral.

11. ∩∩∩∩||||||

12. ∩∩∩∩∩∩∩∩|||

13. e ∩|

14. $e\,e\,e\,e\,e\,e\,e$ ∩|||

15. 𓆼 𓆼 𓆼 $e\,e\,e\,e\,e$ ∩∩∩|||

16. 𓆼 𓆼 𓆼 𓆼 𓆼 $e\,e\,e\,e\,e\,e\,e$ ∩∩∩||

Continued on next page ...

SYSTEMS OF NUMERATION 147

17. $|\overset{\Gamma^D}{\Delta}\overset{\Gamma^D}{\Delta}eeeeeee\cap\cap\cap\cap\cap\cap\cap||$ 18. $|||||||||\overset{\Gamma^D}{\Delta}\overset{\Gamma^D}{\Delta}\overset{\Gamma^D}{\Delta}\overset{\Gamma^D}{\Delta}\overset{\Gamma^D}{\Delta}\overset{\Gamma^D}{\Delta}\overset{\Gamma^D}{\Delta}$

19. $\triangleleft\triangleleft e\cap\cap\cap\cap\cap\cap|||$ 20. $\triangleleft\triangleleft\triangleleft |||||\overset{\Gamma^D}{\Delta}\overset{\Gamma^D}{\Delta}\overset{\Gamma^D}{\Delta}\overset{\Gamma^D}{\Delta}\cap\cap$

In problems 21 through 25 perform the indicated operation.

21. $\cap\cap\cap\cap\cap|||||||| + e|||||$ @ 22. $\cap\cap\cap\cap||| - \cap\cap|$

23. $eeeee\cap|||||||| - \cap\cap\cap\cap\cap\cap\cap|||$

24. $|||||||||ee\cap|||||| - \overset{\Gamma^D}{\Delta}\overset{\Gamma^D}{\Delta}\overset{\Gamma^D}{\Delta}\cap\cap\cap\cap\cap||||$

25. $e\cap| + e\cap\cap|||$

7.3 MULTIPLICATIVE GROUPING SYSTEMS

A multiplicative grouping system is a system that has the following two properties.

1. Symbols are used to represent numbers that belong to a *basic* group.
2. A specific notation is used to represent numbers that are multiples of the basic group.

Two examples of such a system are the Greek and Chinese-Japanese systems of numeration.

7.3.1 The Greek System of Numeration

The Greek system of numeration uses letters to represent numbers. These letters and their respective numerical values are summarized in Table 7.3.1. In the Greek system the basic group is five. To create symbols that represent multiples of five we *cradle* other symbols within the symbol for five. This newly formed symbol represents the product of the individual symbols. That is, we multiply the respective numerical values of the individual symbols to determine the numerical value of the new symbol. For example,

$$\Gamma_\Delta = 50$$
$$\Gamma_H = 500$$
$$\Gamma_X = 5000$$
$$\Gamma_M = 50000$$

To express a number in the decimal system as an equivalent number in the Greek system, we partition the given number into groups that are represented by a Greek symbol or a multiple of a symbol. If the number can be expressed further in multiples of five, then we do so. To illustrate this, the numbers 323 and 597 are expressed in Greek numerals. This is shown on the next page.

TABLE 7.3.1 Greek Numerals and Their Values

Greek Numerals	Values
I (iota)	1
Γ (gamma)	5
Δ (delta)	10
H (eta)	100
X (chi)	1,000
M (mu)	10,000

Illustration of 323 and 597 Expressed in Greek Numerals

$$323 = 300 + 20 + 3$$
$$= (100 + 100 + 100) + (10 + 10) + (1 + 1 + 1)$$
$$= H H H \quad \Delta\Delta \quad I I I$$

$$597 = 500 + 90 + 7$$
$$= 500 + (50 + 40) + (5 + 2)$$
$$= \Gamma_H \quad \Gamma_\Delta \Delta\Delta\Delta \quad \Gamma I I$$

Note in the second illustration that 500 is represented as Γ_H rather than by H H H H H. Also, 90 is expressed as 50 + 40 so we can take advantage of the multiple of five, namely 50. By using a multiplicative grouping system such as the Greek system we use fewer symbols to express a number.

7.3.2 The Chinese-Japanese System of Numeration

The Chinese-Japanese system is another example of a multiplicative grouping system. The symbols and their respective numerical values are summarized in Table 7.3.2. Numbers are expressed in this system by writing the symbols in a vertical format (i.e., one under the other). If the symbol for 10, 100, or 1000 is placed immediately after (i.e., under) a symbol that is less in value, then multiples of 10, 100, and 1000 are created.

To illustrate, consider the symbols for ten and two. If the symbol for ten is written above the symbol for two, then the representation is 12 (10 + 2). If, however, the symbol for 10 is placed directly under the symbol for two, then the representation is 20 (2 × 10). When working with this system we must be sensitive to groupings involving multiples of 10, 100, and 1000.

Table 7.3.2 Chinese-Japanese Numerals and Their Values

Chinese-Japanese Numerals	Values	Chinese-Japanese Numerals	Values
一	1	七	7
二	2	八	8
三	3	九	9
四	4	十	10
五	5	百	100
六	6	千	1000

Example 1 (page 352, #1e)

Express 187 as a Greek numeral.

Solution

$$187 = 100 + 80 + 7$$
$$= 100 + (50 + 30) + (5 + 2)$$
$$= \text{H} \quad \Gamma_\Delta \Delta\Delta\Delta \quad \Gamma \text{II}$$

Example 2 (page 352, #5d)

Evaluate Γ_H H Γ_Δ Γ I

Solution

	Γ_H		H		Γ_Δ		Γ		I
=	(5 × 100)	+	(100)	+	(5 × 10)	+	(5)	+	(1)
=	500	+	100	+	50	+	5	+	1
=	656								

Example 3 (page 352, #7d)

Express 1776 as a Chinese-Japanese numeral.

Solution

$$1776 = 1000 + 700 + 70 + 6$$
$$= 1000 + (7 \times 100) + (7 \times 10) + 6$$
$$= 一千 \quad 七百 \quad 七十 \quad 六$$

Rewriting vertically, we have

一
千
七
百
七
十
六

Example 4 (page 352, #11b)

Evaluate 三百十五

Solution

The top two characters represent 3 × 100, and the bottom two characters represent 10 + 5. Thus the answer is 300 + 15 = 15.

SUPPLEMENTARY EXERCISE 7.3

In problems 1 through 9 express each number as a Greek or Chinese-Japanese numeral.

1. 43   2. 96   3. 212   4. 555   5. 917   6. 1,111   7. 3,355   8. 6,274   9. 11,268

In problems 10 through 15 evaluate each Greek numeral.

10. ΔIII   11. HII   12. ₣ΔΔΓIII

13. ₣XXXΗΔΔΔΔI   14. MMΡXHΔΔΔΔΓ   15. ΜMMMΡXXΗII

In problems 16 through 20 evaluate each Chinese-Japanese numeral.

16. 八十二   17. 三百三十三   18. 九百十六   19. 二千五百四十五   20. 六千二十

7.4 PLACE-VALUE SYSTEMS

7.4.1 The Babylonian System

A place-value system is a system in which the position of a symbol within a numeral determines the symbol's value. Thus each symbol of a numeral has its own **place-value**. One of the oldest place-value systems is the Babylonian system of numeration. This system used only two symbols (called *cuneiform signs*) to represent numbers. The symbol ∇ was used to represent the number one, and the symbol ◄ was used to represent the number 10. These two symbols were used to represent numbers from 1 to 60. The Babylonians also used a place-value system; the symbol for ten was always placed to the left of the symbol for one.

To represent numbers greater than 60 the Babylonians used a place-value system (called a *sexagesimal system*), which was based on powers of 60. Each position is 60 times greater than the position to the immediate right.

$$..., \quad 216{,}000 \quad 3600 \quad 60 \quad 1$$
$$60^3 \quad 60^2 \quad 60^1 \quad 60^0$$

The following procedure is used to express a number using Babylonian symbols.

Expressing a Number in the Babylonian System

1. Determine the largest base 60 place-value that is contained in the given number. (For example, the largest base 60 place-value that is contained in the number 438 is 60. The reason for this is that the next largest base 60 place-value is 3600, which is not at all contained in the number 438. Similarly, the largest base 60 place-value that is contained in the number 9,283 is 3600.)

2. Divide the given number by the base 60 place-value determined from step one. The quotient is the number of groups (i.e., multiples) of this place-value that are contained within the given number.

3. If the remainder from the division of step two is greater than 60, then repeat steps one and two; otherwise express each group with the appropriate arrangement of the cuneiform symbols.

To illustrate this procedure let us express the number 2693 using Babylonian symbols.

- The largest base 60 place-value contained in the number 2963 is 60.

- When we divide 2963 by 60 we obtain a quotient of 49 and a remainder of 3.

- As a result of the division the number 2963 is represented as a base 60 number as

$$(49 \times 60) + (3 \times 1)$$

To represent this using Babylonian symbols we must remember to place the symbol that represents ten to the left of the symbol for one. (This implies multiplication.) Thus, 2,963 is expressed using Babylonian symbols as

≪≪∇∇∇∇∇   ∇∇∇
≪≪∇∇∇∇

49 *sixties*   3 *ones*

7.4.2 The Hindu-Arabic System

A second place-value system is the Hindu-Arabic system, which is the system of numeration we use today. This system uses the symbols 0, 1, 2, 3, 4, 5, 6, 7, 8, 9 to represent numbers and is based on powers of ten. Place-value is arranged in such a manner that each position is ten times greater than the position to the immediate right.

$$..., \quad 1000 \quad 100 \quad 10 \quad 1$$
$$10^3 \quad 10^2 \quad 10^1 \quad 10^0$$

Numbers are written in terms of powers of 10 by combining multiples of these powers. For example, the number 369 is expressed in powers of 10 as

$$
\begin{aligned}
369 &= 300 + 60 + 9 \\
&= (3 \times 100) + (6 \times 10) + (9 \times 1) \\
&= (3 \times 10^2) + (6 \times 10^1) + (9 \times 10^0)
\end{aligned}
$$

Numbers written in this form are said to be in **expanded notation**.

Example 1 (page 358, #1e)

Express the number 349 using Babylonian symbols.

Solution

- The largest base 60 place-value contained in 349 is 60.
- When 349 is divided by 60 the quotient is 5 and the remainder is 49. Thus there are 5 *sixties* and 49 *ones* that make up 349.
- Using the proper arrangement of the cuneiform symbols we have

$$\begin{array}{cc} \triangledown\triangledown\triangledown & \blacktriangleleft\blacktriangleleft\triangledown\triangledown\triangledown\triangledown\triangledown \\ \triangledown\triangledown & \blacktriangleleft\blacktriangleleft\triangledown\triangledown\triangledown\triangledown \end{array}$$

5 *sixties*   9 *ones*

Example 2 (page 358, #5b)

Evaluate

Solution

$$\begin{array}{cc} \triangledown\triangledown & \blacktriangleleft\blacktriangleleft\triangledown\triangledown \\ 2\ sixties & 22\ ones \end{array}$$

$$= \quad 120 \quad + \quad 22$$
$$= \quad 142$$

Example 3 (page 358, #7b,e)

Write the following numbers in expanded notation (parts b and e only).

Solution

b. 378

$$\begin{aligned} 378 &= 300 + 70 + 8 \\ &= (3 \times 100) + (7 \times 10) + (8 \times 1) \\ &= (3 \times 10^2) + (7 \times 10^1) + (8 \times 10^0) \end{aligned}$$

e. *Ten thousand four hundred one*

	Ten thousand	four hundred	one
=	10000	400	1
=	(1×10000) +	(4×100) +	(1×1)
=	(1×10^4) +	(4×10^2) +	(1×10^0)

Example 4 (page 358, #11b,f)

Write the following numbers in base 10 notation (parts b and f only).

Solution

b. *Two thousand three hundred eleven*

	Two thousand		three hundred		eleven
=	2000		300		11
=	2000	+	300	+	11
=	2311				

f. $(4 \times 10^4) + (3 \times 10^2) + (1 \times 10^0)$

	(4×10^4)	+	(3×10^2)	+	(1×10^0)
=	(4×10000)	+	(3×100)	+	(1×1)
=	(40000)	+	(300)	+	(1)
=	40301				

SUPPLEMENTARY EXERCISE 7.4

In problems 1 through 8 write each number as a Babylonian numeral.

1. 21   2. 69   3. 88   4. 121   5. 316   6. 529   7. 1,164   8. 3,817

In problems 9 through 16 evaluate each Babylonian numeral.

9. ◄∇∇∇∇   10. ◄◄◄◄◄∇∇   11. ∇   ◄◄◄∇∇∇   12. ∇∇∇   ◄◄◄◄◄ ∇∇∇∇

13. ∇∇∇∇∇∇∇   ◄∇∇∇∇∇∇∇∇   14. ◄∇∇∇∇∇   ∇∇∇∇∇
15. ◄∇∇∇∇∇∇   ◄◄◄◄   16. ∇   ∇∇∇∇∇   ◄◄∇∇∇∇

In problems 17 through 24, write each number in expanded notation.

17. 825   18. 3,283   19. 6,408   20. 42,054

21. Four hundred four
22. Twenty-five thousand, eight hundred fourteen
23. Thirty thousand, seven hundred
24. One hundred one thousand, six

In problems 25 through 30 write each number in base 10 notation.

25. Two hundred sixteen
26. Nine hundred forty-seven thousand, six hundred eighteen
27. Seven hundred million, six thousand, one hundred twenty-five
28. $(6 \times 10^3) + (5 \times 10^2) + (3 \times 10^1) + (2 \times 10^0)$
29. $(8 \times 10^4) + (3 \times 10^2) + (7 \times 10^0)$
30. $(2 \times 10^5) + (3 \times 10^4) + (4 \times 10^3) + (5 \times 10^2) + (6 \times 10^1)$

7.5 NUMERATION IN BASES OTHER THAN 10

Any counting number b that is greater than one can be used as the base for a place value number system similar to the decimal number system (where the base b is equal to 10). Such systems of numeration use b symbols and consist of the counting numbers

$$0, 1, 2, 3, ..., b - 1$$

For example, the base ten system of numeration consists of the ten symbols 0, 1, 2, 3, 4, 5, 6, 7, 8, and 9. These symbols are commonly called the **digits** of the system. The position of each digit determines the numerical value associated with it. For instance, in the base ten system each position has a numerical value ten times greater than that of the position to its immediate right. As a result we can represent each position as a power of ten.

$$... \quad 10^4 \quad 10^3 \quad 10^2 \quad 10^1 \quad 10^0$$

In general, for any base b, each position has a numerical value that is b times greater than that of the position to its immediate right. Thus each position is represented as a power of the base b.

$$... \quad b^4 \quad b^3 \quad b^2 \quad b^1 \quad b^0$$

7.5.1 Converting a Base b Number ($b \neq 10$) to a Base 10 Number

To convert a number expressed in a base other than 10 to a base 10 number:

1. Write the number in expanded notation. (That is, express the product of each digit of the number and its corresponding place value as a sum.)

2. Perform the multiplication and add the products.

7.5.2 Converting a Base 10 Number to a Base b Number ($b \neq 10$)

To convert a number from base 10 to another base:

1. Divide the number, and each succeeding quotient, by the base b until a zero quotient is obtained. Record the remainder of the successive divisions.

2. Collect the sequence of remainders in the reverse order from which they were obtained. This number is the base b representation of the given number.

Example 1 (page 365, #5c)

Change 231_{five} to base 10 notation.

Solution

Since the given number is in base 5, the place values are in powers of 5. As a result we have,

$$\begin{align} 231_{\text{five}} &= (2 \times 5^2) + (3 \times 5^1) + (1 \times 5^0) \\ &= (2 \times 25) + (3 \times 5) + (1 + 1) \\ &= 50 + 15 + 1 \\ &= 66_{\text{ten}} \end{align}$$

SYSTEMS OF NUMERATION   155

Example 2 (page 365, #7e)

Change 121 to base 5 notation.

Solution

To change 121 to base 5 notation, divide 121 (and each subsequent quotient) by 5 until a zero quotient is obtained. Record all remainders.

Divisions	Quotients	Remainders
121 ÷ 5	24	1
24 ÷ 5	4	4
4 ÷ 5	0	4

Collecting the remainders in reverse order we get 441. Thus, $121 = 441_5$.

Example 3 (page 365, #11e)

Change 243_{twelve} to base 10 notation.

Solution

A base of 12 implies that the place values are in powers of 12. As a result we have,

$$
\begin{aligned}
243_{twelve} &= (2 \times 12^2) + (4 \times 12^1) + (3 \times 3^0) \\
&= (2 \times 144) + (4 \times 12) + (3 \times 1) \\
&= 288 + 48 + 3 \\
&= 339_{ten}
\end{aligned}
$$

Example 4 (page 365, #13d)

Change 137 to base 12 notation.

Solution

To express 137 in base 12 divide 137 (and each subsequent quotient) by 12 until a zero quotient is obtained. Record all remainders.

Divisions	Quotients	Remainders
137 ÷ 12	11	5
11 ÷ 12	0	11 (E)

Collecting the remainders in reverse order we get E5. Thus $137_{ten} = E5_{twelve}$.

Example 5 (page 365, #17d)

Change $TE5_{twelve}$ to base 10 notation.

Solution

$$
\begin{aligned}
TE5_{twelve} &= (T \times 12^2) + (E \times 12^1) + (5 \times 12^0) \\
&= (10 \times 12^2) + (11 \times 12^1) + (5 \times 12^0) \\
&= (10 \times 144) + (11 \times 12) + (5 \times 1) \\
&= 1440 + 132 + 5 \\
&= 1{,}577_{ten}
\end{aligned}
$$

SUPPLEMENTARY EXERCISE 7.5

In problems 1 through 16 express each number in base 10.

1. 22_{five} 2. 11_{five} 3. 102_{five} 4. 314_{five} 5. 341_{five} 6. 413_{five} 7. 2403_{five}
8. 4424_{five} 9. 89_{twelve} 10. 36_{twelve} 11. $5E_{twelve}$ 12. $T9_{twelve}$ 13. 507_{twelve} 14. $3TE_{twelve}$
15. $T0E_{twelve}$ 16. $109E_{twelve}$

In problems 17 through 24 change each number to base 5 and base 12.

17. 8   18. 31   19. 83   20. 119   21. 367   22. 928   23. 2,005   24. 5,103

7.6 BASE 5 ARITHMETIC

The operations of addition, subtraction, multiplication, and division can be extended to numbers expressed in terms of base 5 (as well as any other base). To perform these operations we proceed in a manner similar to that when working with base 10 numbers. One note of caution, though. In base 5 only the digits 0, 1, 2, 3, and 4 are used. Note also that the base subscript is now written using a numeral. For example, 17_{five} is now expressed as 17_5.

Example 1 (page 377, #1d)

Add: $123_5 + 124_5$

Solution

- Add the *ones* column. The sum of 3 and 4 is 7. However, since we are working in base 5, $7_{10} = 12_5$. Therefore, write 2 and carry 1.

$$
\begin{array}{r}
1 \\
1\ 2\ 3_5 \\
+\ 1\ 2\ 4_5 \\
\hline
2_5
\end{array}
$$

- Next add the *fives* column. The sum of 1, 2, and 2 is 5. However, 5 is not a base five digit. Converting to base five: $5_{10} = 10_5$. Thus write 0 and carry 1.

$$
\begin{array}{r}
1\ 1 \\
1\ 2\ 3_5 \\
+\ 1\ 2\ 4_5 \\
\hline
0\ 2_5
\end{array}
$$

- Finally, we add the *twenty-fives* column.

$$
\begin{array}{r}
1\ 1 \\
1\ 2\ 3_5 \\
+\ 1\ 2\ 4_5 \\
\hline
3\ 0\ 2_5
\end{array}
$$

Thus, the sum is 302_5.

SYSTEMS OF NUMERATION

Example 2 (page 377, #5e)

Subtract: $4211_5 - 1232_5$

Solution

- To subtract the *ones* column borrow from the *fives* column.

$$\begin{array}{rrrr} & & 0 & 6 \\ 4 & 2 & \not{1} & \not{1}_5 \\ -1 & 2 & 3 & 2_5 \\ \hline & & & 4_5 \end{array}$$

- To subtract the *fives* column borrow from the *twenty-fives* column.

$$\begin{array}{rrrr} & & 5 & \\ & 1 & \not{0} & 6 \\ 4 & \not{2} & \not{1} & \not{1}_5 \\ -1 & 2 & 3 & 2_5 \\ \hline & & 2 & 4_5 \end{array}$$

- To subtract the *twenty-fives* column borrow from the *125s* column.

$$\begin{array}{rrrr} & 6 & 5 & \\ 3 & 1 & \not{0} & 6 \\ \not{4} & \not{2} & \not{1} & \not{1}_5 \\ -1 & 2 & 3 & 2_5 \\ \hline & 4 & 2 & 4_5 \end{array}$$

- Finally, subtract the *125s* column.

$$\begin{array}{rrrr} & 6 & 5 & \\ 3 & 1 & \not{0} & 6 \\ \not{4} & \not{2} & \not{1} & \not{1}_5 \\ -1 & 2 & 3 & 2_5 \\ \hline 2 & 4 & 2 & 4_5 \end{array}$$

Thus the answer is 2424_5.

Example 3 (page 377, #7e)

Multiply: $324_5 \times 23_5$

Solution

- First multiply 324_5 by the 3 of the number 23_5. (That is, each digit of the first number is multiplied by the *ones* digit of the second number.) Note that each product obtained is converted to base 5.

$$\begin{array}{rrrr} & 1 & 2 & \\ & 3 & 2 & 4_5 \\ \times & & 2 & 3_5 \\ \hline 2 & 0 & 3 & 2_5 \end{array}$$

Continued on next page ...

- Next multiply 324_5 by the 2 of the number 23_5. (That is, each digit of the first number is multiplied by the *fives* digit of the second number.) Note that each product obtained is converted to base 5.

$$\begin{array}{rrrrr} & & 1 & 1 & \\ & & 3 & 2 & 4_5 \\ \times & & & 2 & 3_5 \\ \hline & 2 & 0 & 3 & 2_5 \\ 1 & 2 & 0 & 3_5 & \end{array}$$

- Finally, add the partial products.

$$\begin{array}{rrrrr} & & 1 & 1 & \\ & & 3 & 2 & 4_5 \\ \times & & & 2 & 3_5 \\ \hline & 2 & 0 & 3 & 2_5 \\ 1 & 2 & 0 & 3_5 & \\ \hline 1 & 4 & 1 & 1 & 2_5 \end{array}$$

The product is 14112_5.

Example 4 (page 377, #11e)

Divide: $243_5 \div 11_5$

Solution

Perhaps the easiest way to perform a division problem of this nature (i.e., when the divisor is a two-digit number) is to convert the problem to a base 10 problem and divide as usual. Once this is done, we then convert the quotient and any subsequent remainder back to the given base.

- Convert the divisor and dividend to base 10.

$$11_5 = 6_{10} \quad \text{and} \quad 243_5 = 73_{10}$$

- Perform the equivalent base 10 division.

$$73 \div 6 = 12 \text{ with a remainder of } 1$$

- Convert the quotient and remainder to an equivalent base 5 number.

$$12_{10} = 22_5 \quad \text{and} \quad 1_{10} = 1_5$$

The quotient is 22_5 and the remainder is 1_5.

SUPPLEMENTARY EXERCISE 7.6

In problems 1 through 20 perform the indicated operation.

1. $31_5 + 23_5$
2. $24_5 + 43_5$
3. $132_5 + 211_5$
4. $341_5 + 212_5$
5. $4124_5 + 4022_4$
6. $10_5 - 2_5$
7. $41_5 - 32_5$
8. $140_5 - 24_5$
9. $301_5 - 213_5$
10. $4213_5 - 204_5$
11. $124_5 \times 4_5$
12. $243_5 \times 11_5$
13. $302_5 \times 24_5$
14. $432_5 \times 34_5$
15. $213_5 \times 231_5$
16. $41_5 \div 3_5$
17. $243_5 \div 2_5$
18. $123_5 \div 24_5$
19. $410_5 \div 32_5$
20. $2034_5 \div 41_5$

7.7 BINARY NOTATION AND OTHER BASES

Conversions to other bases from base 10 and conversions to base 10 from other bases can be accomplished by using the procedures presented in section 7.5. Also, the operations of addition, subtraction, and multiplication can be performed on numbers expressed in terms of other bases in a manner similar to that discussed in section 7.6.

Example 1 (page 385, #1e)

Change 1011_2 to base 10 notation.

Solution

$$
\begin{aligned}
1011_2 &= (1 \times 2^3) + (0 \times 2^2) + (1 \times 2^1) + (1 \times 2^0) \\
&= (1 \times 8) + (0 \times 4) + (1 \times 2) + (1 \times 1) \\
&= 8 + 0 + 2 + 1 \\
&= 11_{10}
\end{aligned}
$$

Example 2 (page 385, #5d)

Change 21 to binary notation.

Solution

Divisions	Quotients	Remainders
21 ÷ 2	10	1
10 ÷ 2	5	0
5 ÷ 2	2	1
2 ÷ 2	1	0
1 ÷ 2	0	1

Thus, $21 = 10101_2$.

Example 3 (page 385, #7f)

Add: $110_2 + 110_2$

Solution

To find this sum add as usual except all partial sums must be converted to base 2 prior to continuing with the addition. The completed problem is shown here.

$$
\begin{array}{r}
1 \\
1\ 1\ 0_2 \\
+\ 1\ 1\ 0_2 \\
\hline
1\ 1\ 0\ 0_2
\end{array}
$$

Example 4 (page 386, #11d)

Subtract: $1010_2 - 101_2$

Solution

To find this difference subtract as usual except all partial differences must be converted to base 2 prior to continuing with the subtraction. The completed problem is shown here.

$$
\begin{array}{cccc}
0 & 2 & 0 & 2 \\
\cancel{1} & \cancel{0} & \cancel{1} & \cancel{0} \\
  & 1 & 0 & 1_2 \\
\hline
  & 1 & 0 & 1_2
\end{array}
$$

Example 5 (page 386, #13d)

Multiply: $110_2 \times 11_2$

Solution

$$
\begin{array}{cccc}
  & 1 & 1 & 0_2 \\
\times & & 1 & 1_2 \\
\hline
  & 1 & 1 & 0_2 \\
1 & 1 & 0_2 & \\
\hline
1 & 0 & 0 & 1 & 0_2
\end{array}
$$

Example 6 (page 386, #17)

Convert 434_5 to base 4.

Solution

- First convert 434_5 to base 10.

$$
\begin{aligned}
434_5 &= (4 \times 5^2) + (3 \times 5^1) + (4 \times 5^0) \\
&= (4 \times 25) + (3 \times 5) + (4 \times 1) \\
&= 100 + 15 + 4 \\
&= 119
\end{aligned}
$$

- Now convert 119 to base 4.

Divisions	Quotients	Remainder
$119 \div 4$	29	3
$29 \div 4$	7	1
$7 \div 4$	1	3
$1 \div 4$	0	1

Collecting the remainders in reverse order we get $434_5 = 1313_4$.

Example 7 (page 386, #21c)

Determine whether the statement $1011_2 = 21_5$ is true or false.

Solution

To solve this problem convert each number to base 10 and then compare the base 10 equivalents.

- Convert 1011_2 to a base ten number.

$$
\begin{aligned}
1011_2 &= (1 \times 2^3) + (0 \times 2^2) + (1 \times 2^1) + (1 \times 2^0) \\
&= (1 \times 8) + (0 \times 4) + (1 \times 2) + (1 \times 1) \\
&= 8 + 0 + 2 + 1 \\
&= 11_{10}
\end{aligned}
$$

- Convert 21_5 to a base ten number.

$$
\begin{aligned}
21_5 &= (2 \times 5^1) + (1 \times 5^0) \\
&= (2 \times 5) + (1 \times 1) \\
&= 10 + 1 \\
&= 11_{10}
\end{aligned}
$$

Since $11_{10} = 11_{10}$, the statement is true.

SUPPLEMENTARY EXERCISE 7.7

In problems 1 through 8 change to base 10 notation.

1. 1001_2 2. 1100_2 3. 1111_2 4. 1110_2 5. 10010_2 6. 10111_2 7. 11101_2 8. 10101_2

In problems 9 through 16 change the given base 10 number to binary notation.

9. 19 10. 20 11. 24 12. 30 13. 43 14. 51 15. 62 16. 109

In problems 17 through 28 perform the indicated operation.

17. $100_2 + 111_2$ 18. $1101_2 + 110_2$ 19. $1101_2 + 1011_2$ 20. $1110_2 + 1111_2$
21. $100_2 - 10_2$ 22. $110_2 - 11_2$ 23. $1001_2 - 110_2$ 24. $1100_2 - 101_2$
25. $110_2 \times 100_2$ 26. $111_2 \times 110_2$ 27. $1111_2 \times 101_2$ 28. $1011_2 \times 1101_2$

29. Convert 523_6 to base 3. 30. Convert 2614_7 to base 9.
31. Convert 3023_4 to base 5. 32. Convert 101101_2 to base 8.

33. Convert 1277_8 to base 16. (**Note:** Let A = 10, B = 11, C = 12, ..., F = 15.)

7.8 CHAPTER 7 TEST

In problems 1 through 10 indicate whether the statement is true or false.

1. The Babylonian system of numeration is a base 20 system.
2. The Chinese-Japanese system of numeration allows multiples of 10, 100, and 1000 to be created.
3. In the Egyptian system of numeration the order in which the symbols are placed is not important.
4. The Greek system of numeration is an example of an additive (i.e. simple grouping) system.
5. In base 5 the digits that can be used to create numbers are 1, 2, 3, 4, and 5.
6. In expanded notation 309 is equal to $(3 \times 10^3) + (9 \times 10^1)$.
7. The 4 in the number 246_8 represents four *eights*.
8. The numeral 3 represents the same number in both base 6 and base 9.
9. $222_3 > 11100_2$.
10. The number TE_{12} is equal to 10×11, or 110.

In problems 11 through 14 convert the given number to base ten.

11. 48_9 12. 205_6 13. 1100101_2 14. $5TE_{12}$

In problems 15 through 20 convert the given number to the indicated base.

15. $83_{10} = \underline{}_4$ 16. $21_{10} = \underline{}_{12}$ 17. $66_{10} = \underline{}_3$
18. $105_{10} = \underline{}_8$ 19. $621_7 = \underline{}_6$ 20. $104_5 = \underline{}_2$

In problems 21 through 29 perform the indicated operation. Express all answers in the given base.

21. $214_6 + 503_6$ 22. $1022_3 + 211_3$ 23. $825_9 - 456_9$
24. $4037_8 - 2573_8$ 25. $101101_2 \times 11010_2$ 26. $3012_4 \times 213_4$
27. $5156_7 \times 2044_7$ 28. $1043_5 \div 33_5$ 29. $4204_5 \div 21_5$

8. SETS OF NUMBERS AND THEIR STRUCTURE

8.1 INTRODUCTION

In this chapter we examine various sets of numbers. Included in the discussion are the sets of natural numbers, whole numbers, integers, rational numbers, irrational numbers, and real numbers. The operations of addition, subtraction, multiplication, and division are applied to these sets of numbers.

8.2 NATURAL NUMBERS - PRIMES AND COMPOSITES

One classification of numbers frequently used is the set of counting numbers, or more formally, the set of **natural numbers**. The set of natural numbers consists of the elements $\{1,2,3,...\}$.

Any natural number can be expressed as the product of two or more natural numbers. The numbers that are multiplied together to form a product are called **factors** of the product. For example, in the equation $2 \times 4 = 8$, the 2 and 4 are factors of the product 8.

To determine if a number n is a factor of a given number, we divide the given number by n. If the division yields a zero remainder then n is a factor of the given number. (Thus factors are also thought of as *divisors* of a product.)

If a natural number is greater than one and has only two factors, namely one and the number itself, then the number is called a **prime number**. The smallest prime number is two. Two is the only prime number that is even.

If a natural number is greater than two and is not prime, then the number is called a **composite number**. A composite number has factors other than one and itself.

A natural number is classified in one of three ways.

- The number could be 1
- The number could be prime
- The number could be composite

8.2.1 Determining If a Natural Number is Prime

Prime numbers are very useful in certain areas of mathematics and it is important that we be able to determine if a natural number is prime. To do so we use the following procedure. (**Note:** This procedure is different than that discussed in your textbook. It is presented here as an alternative approach.)

To Determine If A Natural Number Is Prime

1. Divide the given number, first by 2, then by 3, 4, 5, etc., until either (a) the quotient is less than the divisor, or (b) a remainder of zero is obtained (whichever occurs first).

2. If a remainder of zero is obtained, then the given number is composite; otherwise if every division example produces a non zero remainder, then the number is prime.

To illustrate this procedure, the numbers 35 and 47 are checked to determine whether or not they are prime.

Determine If 35 is Prime or Composite

Divisions	Quotients	Remainders	
35 ÷ 2	17	1	
35 ÷ 3	11	2	
35 ÷ 4	8	3	
35 ÷ 5	7	0	**Stop**

The division was stopped becasue the remainder is zero. Thus 35 is composite.

Determine If 47 is Prime or Composite

Divisions	Quotients	Remainders	
47 ÷ 2	23	1	
47 ÷ 3	15	2	
47 ÷ 4	11	3	
47 ÷ 5	9	2	
47 ÷ 6	7	5	
47 ÷ 7	6	5	**Stop**

The division was stopped because the quotient 6 is less than the divisor 7. Since every division produced a non-zero remainder the number 47 is prime.

According to the Fundamental Theorem of Arithmetic, every composite number can be expressed as a product of its prime factors. This factoring process is called **prime factorization**, and except for the order of the factors, the resulting prime factors are unique. One method that is used to determine the prime factors of a composite number is to first factor the given number into any two easily recognizable factors. Each subsequent factor is then further factored until all the factors are prime. An example of this method of finding the prime factors of a number is given below for the number 180. A tree diagram is used to illustrate the procedure.

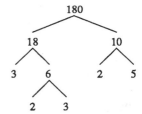

Two factors of 180 are 10 and 18. Working with the factor 10, note that it contains factors of 2 and 5, both of which are prime. Focusing on the factor 18, note that it contains factors of 2 and 9. Since 2 is prime it does not contain any additional prime factors. The factor 9, however, does contain factors of 3 and 3, both of which are prime. Consequently, the prime factors of 180 are 2, 2, 3, 3, and 5. Thus $180 = 2 \times 2 \times 3 \times 3 \times 5$.

Note that the final prime factors of a composite number are unique, but the intermediate factoring results are not. For example, we could have begun the prime factorization of 180 with the factors 9 and 20, or 3 and 60. Also, in the example, we could have chosen 3 and 6 as the factors of 18, rather than 2 and 9.

8.2.2 Divisibility Rules

To assist you in determining the prime factors of a number, the following divisibility rules for 2, 3, 4, 5, 6, 8, 9, 10, and 11 are provided.

SETS OF NUMBERS AND THEIR STRUCTURE   165

Divisibility Rules for 2, 3, 4, 5, 6, 8, 9, 10, and 11

- **2:** A number is divisible by 2 if it is an even number (i.e., it ends in 0, 2, 4, 6, or 8).
- **3:** A number is divisible by 3 if the sum of its digits is divisible by 3. (e.g., 876 is divisible by 3 since $8 + 7 + 6 = 21$, which is divisible by 3.)
- **4:** A number is divisible by 4 if its two right-most digits form a number that is divisible by 4. (e.g., 724 is divisible by 4 since 24 is divisible by 4.)
- **5:** A number is divisible by 5 if it is a multiple of 5 (i.e., it ends in either 0 or 5).
- **6:** A number is divisible by 6 if it is divisible by both 2 and 3.
- **8:** A number is divisible by 8 if its three right-most digits form a number that is divisible by 8.
- **9:** A number is divisible by 9 if the sum of its digits is divisible by 9.
- **10:** A number is divisible by 10 if it ends in 0.
- **11:** A number is divisible by 11 if the sum of the digits in the 1's, 100's, 10,000's, etc. position minus the sum of of the digits in the 10's, 1,000's, 100,000's, etc. position is either 0 or divisible by 11.

Example 1 (page 401, #3a)

Test the number 2,688 for divisibility by 2, 3, 4, 5, 6, 8, 9, 10, and 11.

Solution

- **2:** 2,688 is an even number. Thus it is divisible by 2.
- **3:** The sum of the digits of 2,688 is equal to $2 + 6 + 8 + 8$, which is 24. Since 24 is divisible by 3, the number 2,688 is divisible by 3.
- **4:** The two right most digits of 2,688 are 8 and 8, which form the number 88. Since 88 is divisible by 4, the number 2,688 is divisible by 4.
- **5:** 2,688 is not a multiple of 5 (it does not end in a 0 or 5). Thus it is not divisible by 5.
- **6:** Since 2,688 is divisible by both 2 and 3, it is also divisible by 6
- **8:** The three right most digits of 2,688 are 6, 8, and 8, which form the number 688. Since 688 is divisible by 8, the number 2,688 is divisible by 8.
- **9:** The sum of the digits that make up 2,688 is 24, which is not divisible by 9. Thus 2,688 is not divisible by 9.
- **10:** Since 2,688 is not a multiple of 10 (it does not end in zero), it is not divisible by 10.
- **11:** To determine whether or not 2,688 is divisible by 11, we consider the digits from right to left, beginning with the 1's position.
    - a. Add the *odd-numbered* digits; that is, add the digit in the 1's position to the digit in the 100's position: 2,6̲88̲. For this example we add 6 and 8, which is 14.
    - b. Add the *even-numbered* digits; that is, add the digit in the 10's position to the digit in the 1,000's position: 2̲,68̲8. For this example we add 2 and 8, which is 10.
    - c. The difference between the sum of the *even-numbered* digits and the sum of the *odd-numbered* digits is $14 - 10 = 4$. Since this difference is not 0 or divisible by 11, the number 2,688 is not divisible by 11.

As a result of the above discussion, 2,688 is divisible by 2, 3, 4, 6, and 8.

Example 2 (page 401, #7d)

Determine whether 243 is prime or composite.

Solution

The sequence of divisions is shown below.

Divisions	Quotients	Remainders
243 ÷ 2	121	1
243 ÷ 3	81	0

Stop. (The remainder is zero.)

Since the remainder is zero 243 is composite.

Example 3 (page 401, #9e)

Determine the prime factors of 625.

Solution

A tree diagram is used to generate the prime factors of 625.

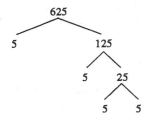

Thus, $625 = 5 \times 5 \times 5 \times 5$.

SUPPLEMENTARY EXERCISE 8.2

In problems 1 through 5 test each number for divisibility by 2,3,4,5,8,9,10, and 11.

1. 628   2. 3,917   3. 21,648   4. 109,520   5. 6,314,829

In problems 6 through 20 determine whether each number is prime or composite.

6. 21   7. 48   8. 39   9. 63   10. 29   11. 87   12. 120   13. 129
14. 147   15. 183   16. 199   17. 207   18. 293   19. 667   20. 781

In problems 21 through 35 determine the prime factors of each number.

21. 24   22. 27   23. 28   24. 39   25. 44   26. 48   27. 120   28. 128
29. 147   30. 183   31. 207   32. 221   33. 460   34. 500   35. 999

SETS OF NUMBERS AND THEIR STRUCTURE   167

8.3 GREATEST COMMON DIVISOR AND LEAST COMMON MULTIPLE

Two concepts that utilize prime factorization are the greatest common divisor (GCD), which is also referred to as the greatest common factor (GCF), and the least common multiple (LCM).

8.3.1 Greatest Common Factor

Common factors are factors that are common to two or more given numbers. For example, the factors of 12 are 1,2,3,4,6, and 12, and the factors of 16 are 1,2,4,8, and 16. The factors common to both 12 and 16 are 1, 2, and 4. (The common factors are denoted with an underscore in the illustration below.)

$$\textbf{Factors of 12:} \quad \underline{1}, \underline{2}, 3, \underline{4}, 6, 12$$

$$\textbf{Factors of 16:} \quad \underline{1}, \underline{2}, \underline{4}, 8, 16$$

In the example above note that 4 is the greatest (i.e., largest) common factor.

One method that is used to find the greatest common factor of two or more numbers uses the prime factors of the numbers. This method is outlined below.

<u>To Find The Greatest Common Factor of Two Or More Numbers</u>

1. Find the prime factors of each number.
2. Select those prime factors that are common and multiply them. This product is the GCF of the given numbers. (**Note:** If there are no prime factors in common, then the GCF is 1 and the numbers are said to **relatively prime**.)

This procedure is demonstrated in Example 1.

GCFs are used to reduce fractions. A fraction is considered to be reduced to lowest terms when its numerator and denominator are relatively prime. The following procedure is used to accomplish this.

<u>To Reduce A Fraction To Lowest Terms</u>

1. Factor the numerator and denominator of the fraction into their respective prime factors.
2. Cancel any factors that are common to both the numerator and denominator.
3. Multiply the remaining factors of the numerator. This product is the numerator of the reduced fraction.
4. Multiply the remaining factors of the denominator. This product is the denominator of the reduced fraction.

This procedure is illustrated in Example 2.

8.3.2 Least Common Multiple

Multiples of a number are derived by *counting* in multiples of the number. For example,

- multiples of 2 are 2, 4, 6, 8, 10, ...
- multiples of 3 are 3, 6, 9, 12, 15, ...
- multiples of 4 are 4, 8, 12, 16, 24, ...
- and so forth.

Common multiples of two or more numbers are those multiples that are common to the given numbers. For example, the common multiples of 2 and 3 are 6, 12, 18, ... (The common multiples shown are denoted with an underscore in the illustration below.)

Multiples of 2: 2, 4, 6, 8, 10, 12, 14, 16, 18, 20, ...

Multiples of 3: 3, 6, 9, 12, 15, 18, 21, ...

Note that for any two or more numbers there are an infinite number of common multiples. It should be clear from the illustration above that 6 is the least (i.e., smallest) common multiple of 2 and 3. The least common multiple (LCM) is the smallest number that is a multiple of each of the given numbers.

Although we can employ prime factorization to find the LCM of two or more numbers, we choose not to do so here. Rather, suffice it to say that to find the LCM of two or more numbers begin listing the multiples of each given number until you encounter the smallest number that is common to all the sets of multiples. This procedure is identical to what was just illustrated. (See also Example 3.) LCMs are used in adding and subtracting unlike fractions (i.e., fractions that do not have the same denominators). This is illustrated in Examples 4 and 5.

Example 1 (page 408, #1d)

Find the GCF of 52 and 78.

Solution

- Find the prime factors of each number.

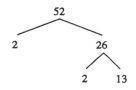 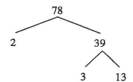

- The common prime factors are 2 and 13.

$$52 = \underline{2} \times 2 \times \underline{13}$$
$$78 = \underline{2} \times 3 \times \underline{13}$$

The GCF is $2 \times 13 = 26$.

Example 2 (page 408, #3a)

Reduce $\dfrac{30}{36}$ to lowest terms.

Solution

- Factor the numerator and denominator into prime factors.

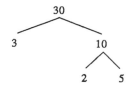 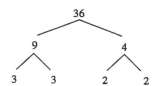

Continued on next page ...

SETS OF NUMBERS AND THEIR STRUCTURE 169

Thus $30 = 2 \times 3 \times 5$ and $36 = 2 \times 2 \times 3 \times 3$.

- Cancel common factors and multiply the remaining factors of the numerator and denominator, respectively.

$$\frac{30}{36} = \frac{2 \times 3 \times 5}{2 \times 2 \times 3 \times 3} = \frac{2^1 \times 3^1 \times 5}{2_1 \times 2 \times 3_1 \times 3} = \frac{1 \times 1 \times 5}{1 \times 2 \times 1 \times 3} = \frac{5}{6}$$

Thus $\frac{30}{36}$ reduced to lowest terms is $\frac{5}{6}$.

Example 3 (page 409, #5c)

Find the LCM of 15 and 24.

Solution

List the multiples of each number and select the first (i.e., the smallest) number that is common to both 15 and 24.

Multiples of 15: 15, 30, 45, 60, 75, 90, 105, 120, 135, ...

Multiples of 24: 24, 48, 72, 96, 120, 144, ...

The LCM of 15 and 24 is 120.

Example 4 (page 409, #7b)

Add: $\frac{5}{9} + \frac{1}{12}$

Solution

- Since the fractions are unlike we express them as like fractions. To do this we find a common denominator by finding the LCM of the denominators 9 and 12.

Multiples of 9: 9, 18, 27, 36, ...

Multiples of 12: 12, 24, 36, ...

The LCM is 36.

- Rewrite each fraction as an equivalent fraction with a denominator of 36.

$$\frac{5}{9} = \frac{20}{36} \quad \text{and} \quad \frac{1}{12} = \frac{3}{36}$$

- Now add the like fractions by adding their numerators and placing the sum over the common denominator.

$$\frac{20}{36} + \frac{3}{36} = \frac{23}{36}$$

Thus $\frac{5}{9} + \frac{1}{12} = \frac{23}{36}$.

170   STUDENT STUDY GUIDE

Example 5 (page 409, #7e)

Subtract: $\dfrac{10}{11} - \dfrac{4}{5}$

Solution

- Since the fractions are unlike we rewrite them as like fractions. Note that the denominators 11 and 5 are both prime numbers. As a result the LCM is their product. Thus the LCM of 11 and 5 is 55.

$$\dfrac{10}{11} = \dfrac{50}{55} \quad \text{and} \quad \dfrac{4}{5} = \dfrac{44}{55}$$

- Now subtract.

$$\dfrac{50}{55} - \dfrac{44}{55} = \dfrac{6}{55}$$

Thus, $\dfrac{10}{11} - \dfrac{4}{5} = \dfrac{6}{55}$.

SUPPLEMENTARY EXERCISE 8.3

In problems 1 through 12 find the GCF of the numbers given.

1. 9, 6
2. 4, 8
3. 3, 9
4. 8, 6
5. 7, 4
6. 24, 64
7. 45, 18
8. 40, 50
9. 130, 15
10. 500, 220
11. 20, 15, 30
12. 58, 40, 27

In problems 13 through 24 find the LCM of the numbers given.

13. 4, 6
14. 7, 8
15. 9, 15
16. 8, 14
17. 5, 18
18. 20, 16
19. 42, 18
20. 50, 30
21. 100, 30
22. 75, 220
23. 8, 10, 12
24. 18, 22, 30

In problems 25 through 36 reduce each fraction to lowest terms.

25. $\dfrac{5}{15}$
26. $\dfrac{6}{18}$
27. $\dfrac{8}{32}$
28. $\dfrac{12}{20}$
29. $\dfrac{14}{16}$
30. $\dfrac{50}{85}$

31. $\dfrac{14}{49}$
32. $\dfrac{16}{30}$
33. $\dfrac{15}{36}$
34. $\dfrac{57}{123}$
35. $\dfrac{72}{248}$
36. $\dfrac{75}{125}$

In problems 37 through 48 perform the indicated operation. Reduce all answers to lowest terms.

37. $\dfrac{1}{8} + \dfrac{1}{2}$
38. $\dfrac{5}{16} + \dfrac{3}{4}$
39. $\dfrac{3}{5} + \dfrac{1}{6}$
40. $\dfrac{3}{4} + \dfrac{3}{5}$

41. $\dfrac{7}{18} + \dfrac{3}{4}$
42. $\dfrac{7}{15} + \dfrac{7}{9}$
43. $\dfrac{1}{3} - \dfrac{1}{6}$
44. $\dfrac{10}{12} - \dfrac{1}{3}$

45. $\dfrac{7}{8} - \dfrac{3}{5}$
46. $\dfrac{3}{4} - \dfrac{2}{3}$
47. $\dfrac{7}{10} - \dfrac{4}{15}$
48. $\dfrac{7}{12} - \dfrac{7}{18}$

8.4 INTEGERS

The set of natural numbers was defined previously as {1,2,3,...}. If we include the element 0 in this set we have the set of whole numbers {0,1,2,3,...}. Both sets of numbers are represented by a number line as shown below. (**Note:** The arrowhead of the number line implies that the set of numbers continues without end.)

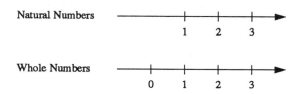

If we extend the number line for the set of whole numbers to the left of zero, another set of numbers, called the negative numbers, is formed. Negative numbers are written with a negative sign (−) placed before them. The numbers to the right of zero are referred to as positive numbers. Positive numbers are written either with a positive sign (+) placed before them, or with no sign at all. The number zero is neither positive nor negative.

Numbers that are used to name the positive natural numbers, the negative natural numbers, and zero are called **integers**. The set of integers also can be graphed on a number line. This is shown below.

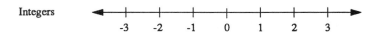

Thus the set of integers is {...,−3,−2,−1,0,1,2,3,...}. Observe that numbers increasing in numerical value are placed to the right on a number line, whereas numbers decreasing in numerical value are placed to the left. Also, a sign attached to a number indicates position to the left or right of zero. A positive sign (+), or the absence of a sign, implies a position to the right of zero. A negative sign (−) implies a position to the left of zero.

8.4.1 Addition of Integers

When adding two nonzero integers there are three distinct possibilities that must be noted.

1. Both integers are positive
2. Both integers are negative
3. One integer is positive and the other is negative

In the first two cases if the two integers being added both have the same sign (i.e., $(+a) + (+b)$, or $(-a) + (-b)$), then we do the following.

[1] Add the numbers without regard to their signs.

[2] The sign of the answer is the same sign as the sign of the given numbers.

In the third case, where one integer is positive and the other is negative, (i.e., $(+a) + (-b)$, or $(-a) + (+b)$), we must consider the distance each integer is from zero. With this in mind we do the following.

[1] Subtract the shorter distance from the longer distance.

[2] The sign of the answer is the same sign as that of the integer that represents the longer distance.

Addition of integers is demonstrated in Examples 1 and 2.

8.4.2 Subtraction of Integers

Subtracting an integer is equivalent to adding the integer's opposite value. That is,

$$a - (+b) = a + (-b)$$
$$\text{and} \quad a - (-b) = a + (+b)$$

Thus it is possible to change a subtraction problem to an equivalent addition problem by

1. changing the subtraction sign to an addition sign; and
2. reversing the sign of the second integer.

Once this has been done we employ the rules of addition presented earlier in the discussion. Subtraction of integers is demonstrated in Example 3.

8.4.3 Multiplication of Integers

When multiplying two nonzero integers there are three distinct possibilities that must be noted.

1. Both integers are positive
2. Both integers are negative
3. One integer is positive and the other is negative

In the first case the product of two positive integers is positive. In the second case the product of two negative integers is positive. Finally, in the third case, the product of a positive integer and a negative integer is negative. More formally,

$$(+a) \times (+b) = +(ab)$$
$$(-a) \times (-b) = +(ab)$$
$$(+a) \times (-b) = -(ab)$$
$$(-a) \times (+b) = -(ab)$$

Thus to multiply integers we multiply the numbers without regard to their sign. The sign of the product is then determined by examining the signs of the factors and using the scheme above. Multiplication of integers is demonstrated in Example 4.

Example 1 (page 421, #1a)

Evaluate $2 + (-3)$

Solution

- Using a number line we first start at 0 and move 2 units in a positive direction.

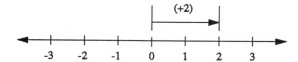

- From this point we move 3 units in a negative direction.

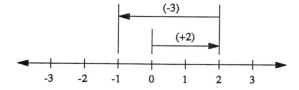

Continued on next page ...

• The sum is the number at which we end. Thus 2 + (−3) = −1.

Note that this problem also could be evaluated using the rules presented earlier in the discussion. Since the two integers have different signs (i.e., one is positive and the other is negative) we do the following:

[1] Determine which integer represents the greater distance from zero. In this example, −3 is farther from zero than 2.

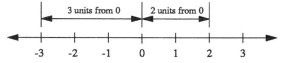

[2] Subtract the shorter distance from the longer distance (3 − 2 = 1).

[3] The sign of the answer is negative since it is the negative integer that represents the greater distance from zero. Thus 2 + (−3) = −1.

Example 2 (page 421, #1d)

Evaluate −3 + 5

Solution

Since one integer is positive and the other is negative, we do the following.

- 5 represents the greater distance from zero.
- 5 − 3 = 2
- The answer is positive.

Thus −3 + 5 = +2.

Example 3 (page 421, #3a)

Evaluate 3 − 5

Solution

Change the subtraction problem to an equivalent addition problem. This is done by changing the subtraction sign to an addition sign and reversing the sign of the second integer.

$$\begin{aligned} & \ 3 \ - \ 5 \\ &= \ 3 \ - \ (+5) \\ &= \ 3 \ + \ (-5) \end{aligned}$$

Now add the integers by following the rules for addition. Since one integer is positive and the other is negative, do the following:

- 5 represents the greater distance from zero.
- 5 − 3 = 2
- The answer is negative.

Thus 3 − 5 = −2.

Example 4 (page 421, #7a)

Evaluate $(-4) \times (-5)$

Solution

Multiplying the integers without regard to their sign we get $4 \times 5 = 20$. Since both integers are negative the product is positive. Thus $(-4) \times (-5) = +20$.

Example 5 (page 421, #7c)

Evaluate $(-4) \times (-6) \times (-1)$

Solution

$$
\begin{aligned}
& (-4) \times (-6) \times (-1) \\
= \ & (+24) \times (-1) \\
= \ & -24
\end{aligned}
$$

Example 6 (page 421, #9c)

Evaluate $16 - 2(3 + 4 \div 2)$.

Solution

Evaluate the expression within parentheses first. (**Note:** The correct order of operations is followed.)

$$
\begin{aligned}
& 16 - 2(3 + 4 \div 2) \\
= \ & 16 - 2(3 + 2) \\
= \ & 16 - 2(5) \\
= \ & 16 - 10 \\
= \ & 6
\end{aligned}
$$

Example 7 (page 422, #11d)

Replace the ? with =, >, or < to make the sentence $-3 + 2 \ ? \ 2 - (-3)$ **true**.

Solution

First evaluate the expression on each side of the question mark.

$$-3 + 2 = -1 \quad \text{and} \quad 2 - (-3) = 5$$

Now compare results. Clearly, -1 is smaller than 5. Thus ? should be replaced with **<**.

SUPPLEMENTARY EXERCISE 8.4

In problems 1 through 30 evaluate the given expression.

1. $4 + 2$
2. $-1 + 4$
3. $-8 + (-2)$
4. $-8 + (+2)$
5. $6 - 7$
6. $4 - (-8)$
7. $-3 - (+4)$
8. $0 - (-4)$
9. $-6 - (-10)$
10. $0 + (--3)$
11. $-4 - (-4)$
12. $0 - 9$

Continued on next page ...

13. (–1) × 7
14. (–3) × (–7)
15. 4 × (–6)
16. (–8) × (+3)
17. (–7) × 4
18. –3 × 0
19. (5) × (+4)
20. (–1) × (–1)
21. –4 + (–12) + 10
22. 3 + (–18) – (–6)
23. –20 – (–8) – 41
24. (–5) × (3) × (–2)
25. (–8) × (–3) × (–5)
26. 7 × (–5) × (–2) × 5
27. (–7 + 4) × (–2)
28. (4 – (–6)) × (–7)
29. (3 + (–2)) × (5 – 3)
30. (–2 – (–7)) × (6 + (–2))
31. 4 × (–3) + 2
32. –5 × (–6 + 4)
33. 9 + (–3) × 2
34. –2 × (–2 + (–2))
35. 8 – 5 × (–4 + 2)
36. –2 × 3 + 5 × (–2)

In problems 37 through 42 replace each question mark with =, <, or > to make the sentence true.

37. –5 ? –3
38. 3 ? –3
39. –1 + (–2) ? 5 × (–1)
40. –3 + (–3) ? (–7 – (–2))
41. (–8) × (–2) ? –8 + (–2)
42. 5 + (–4) – 3 ? 5 – (–4) + 3

8.5 RATIONAL NUMBERS

Any number that written in the form $\frac{a}{b}$ where both a and b are integers, and $b \neq 0$, is called a **rational number**. As a result of this definition we conclude that all natural numbers, whole numbers, and integers can be classified as rational numbers since they can be expressed as a quotient of two integers. For example,

- The set of natural numbers $\{1,2,3,...\}$ can be expressed as rational numbers.

$$\left\{ \frac{1}{1}, \frac{2}{1}, \frac{3}{1}, \cdots \right\}$$

- The set of whole numbers $\{0,1,2,3,...\}$ can be expressed as rational numbers.

$$\left\{ \frac{0}{1}, \frac{1}{1}, \frac{2}{1}, \frac{3}{1}, \cdots \right\}$$

- The set of integers $\{...,-3,-2,-1,0,1,2,3,...\}$ can be expressed as rational numbers.

$$\left\{ \frac{-3}{1}, \frac{-2}{1}, \frac{-1}{1}, \frac{0}{1}, \frac{1}{1}, \frac{2}{1}, \frac{3}{1}, \cdots \right\}$$

8.5.1 Addition of Rational Numbers

A general rule for adding any two rational numbers is as follows:

If $\frac{a}{b}$ and $\frac{c}{d}$ are rational numbers,

then $\frac{a}{b} + \frac{c}{d} = \frac{(a \times d) + (b \times c)}{(b \times d)}$.

8.5.2 Subtraction of Rational Numbers

Negative fractions are written by placing a negative sign in front of the fraction. Since a fraction is negative when either its numerator or denominator (but not both) is negative, the following are equivalent.

$$-\frac{a}{b} = \frac{-a}{b} = \frac{a}{-b}$$

Consider now the general subtraction problem, $\frac{a}{b} - \frac{c}{d}$. In our earlier work we subtracted integers by adding the opposite value of the second integer to the first integer. Since integers are rational numbers this procedure is extended to subtracting rational numbers. (See next page.)

$$\frac{a}{b} - \frac{c}{d}$$

$$= \frac{a}{b} + \frac{(-c)}{d}$$

$$= \frac{(a \times d) + (b \times (-c))}{(b \times d)}$$

8.5.3 Multiplication of Rational Numbers

A general rule for multiplying two rational numbers is as follows.

If $\frac{a}{b}$ and $\frac{c}{d}$ are rational numbers,

$$\text{then } \frac{a}{b} \times \frac{c}{d} = \frac{a \times c}{b \times d}$$

It is helpful to first cancel out factors common to both the numerator and denominator. In any event, the multiplication of rational numbers involves multiplying the numerators to get the numerator of the answer, and multiplying the denominators to get the denominator of the answer.

8.5.4 Division of Rational Numbers

A general rule for dividing two rational numbers is to multiply the first fraction by the *reciprocal* of the second fraction. (**Note:** A reciprocal is formed by interchanging the numerator and denominator.) In general,

$$\frac{a}{b} \div \frac{c}{d} = \frac{a}{b} \times \frac{d}{c} \text{ (where } b \neq 0, c \neq 0, \text{ and } d \neq 0\text{)}$$

In the event the division problem is expressed in the form of a *complex fraction*,

$$\frac{\frac{a}{b}}{\frac{c}{d}}$$

the procedure we follow is not any different.

$$\frac{\frac{a}{b}}{\frac{c}{d}} = \frac{a}{b} \div \frac{c}{d} = \frac{a}{b} \times \frac{d}{c} \text{ (where } b \neq 0, c \neq 0, \text{ and } d \neq 0\text{)}$$

If the numerator or denominator (or both) of a complex fraction contains more than one fraction or whole number, first perform the indicated operations in the numerator and denominator, and then simplify the resulting complex fraction. In other words, express the numerator and denominator of a complex fraction as single fractions before evaluating the complex fraction. An example of this is presented in Example 6.

Example 1 (page 431, #3a)

Evaluate: $\frac{4}{5} + \frac{1}{7}$

Solution

Use the rule for addition addition of rational numbers.

$$\frac{4}{5} + \frac{1}{7} = \frac{(4 \times 7) + (5 \times 1)}{(5 \times 7)} = \frac{28 + 5}{35} = \frac{33}{35}$$

Example 2 (page 431, #3e)

Evaluate: $\dfrac{13}{16} - \dfrac{4}{5}$

Solution

Use the rule for subtraction of rational numbers.

$$\dfrac{13}{16} - \dfrac{4}{5}$$
$$= \dfrac{13}{16} + \left(\dfrac{-4}{5}\right)$$
$$= \dfrac{(13 \times 5) + (16 \times (-4))}{(16 \times 5)}$$
$$= \dfrac{65 - 64}{80}$$
$$= \dfrac{1}{80}$$

Example 3 (page 431, #5a)

Evaluate $\dfrac{4}{5} \times \dfrac{2}{7}$

Solution

Use the rule for multiplication of rational numbers.

$$\dfrac{4}{5} \times \dfrac{2}{7} = \dfrac{4 \times 2}{5 \times 7} = \dfrac{8}{35}$$

Example 4 (page 431, #5e)

Evaluate $\dfrac{3}{11} \div \dfrac{4}{9}$

Solution

Use the rule for division of rational numbers.

$$\dfrac{3}{11} \div \dfrac{4}{9}$$
$$= \dfrac{3}{11} \times \dfrac{9}{4}$$
$$= \dfrac{27}{44}$$

Example 5 (page 431, #7c)

Simplify: $\dfrac{\frac{2}{5}}{3}$

Solution

Use the rule for simplifying complex fractions.

$$\dfrac{\frac{2}{5}}{3} = \dfrac{\frac{2}{5}}{\frac{3}{1}} \qquad (\textit{Express 3 as a fraction.})$$

$$= \frac{2}{5} \div \frac{3}{1}$$

$$= \frac{2}{5} \times \frac{1}{3}$$

$$= \frac{2}{15}$$

Example 6 (page 431, #7d)

Simplify: $\dfrac{1 + \frac{1}{2}}{2 - \frac{1}{3}}$

Solution

Express the numerator and denominator of the complex fraction as single fractions.

	Numerator		Denominator
	$1 + \frac{1}{2}$		$2 - \frac{1}{3}$
=	$\frac{1}{1} + \frac{1}{2}$	=	$\frac{2}{1} - \frac{1}{3}$
=	$\frac{(1 \times 2) + (1 \times 1)}{1 \times 2}$	=	$\frac{2}{1} + \left[\frac{-1}{3}\right]$
=	$\frac{2 + 1}{2}$	=	$\frac{(2 \times 3) + (1 \times (-1))}{(1 \times 3)}$
=	$\frac{3}{2}$	=	$\frac{6 - 1}{3}$
		=	$\frac{5}{3}$

Continued on next page ...

Simplify the complex fraction.

$$\frac{1+\frac{1}{2}}{2-\frac{1}{3}} = \frac{\frac{3}{2}}{\frac{5}{3}} = \frac{3}{2} \div \frac{5}{3}$$

$$= \frac{3}{2} \times \frac{3}{5}$$

$$= \frac{9}{10}$$

Example 7 (page 432, #13a)

Determine if the statement $\frac{4}{7} > \frac{2}{3}$ is true or false

Solution

To compare two rational numbers do the following:

[1] Rewrite the fractions so that they have a common denominator. (Recall the LCM concept from section 8.3.)

Multiples of 7   7, 14, $\underline{21}$, ...

Multiples of 3:   3, 6, 9, 12, 15, 18, $\underline{21}$, ...

The LCM of 7 and 3 is 21. Therefore,

$$\frac{4}{7} = \frac{12}{21} \quad \text{and} \quad \frac{2}{3} = \frac{14}{21}$$

[2] Compare the numerators of the like fractions. Since $12 < 14$, $\frac{12}{21} < \frac{14}{21}$.

As a result, $\frac{4}{7} < \frac{2}{3}$. Hence the given statement is false.

SUPPLEMENTARY EXERCISE 8.5

In problems 1 through 36 perform the indicated operation.

1. $\frac{3}{8} + \frac{1}{2}$
2. $\frac{2}{9} + \frac{2}{3}$
3. $\frac{5}{16} + \frac{3}{4}$
4. $\frac{2}{3} + \frac{2}{5}$
5. $\frac{1}{2} + \frac{3}{7}$
6. $\frac{2}{3} + \frac{1}{4}$
7. $\frac{7}{12} + \frac{2}{9}$
8. $\frac{8}{15} + \frac{4}{9}$
9. $\frac{5}{18} + \frac{5}{12}$
10. $\frac{7}{8} - \frac{3}{4}$
11. $\frac{2}{3} - \frac{5}{9}$
12. $\frac{1}{6} - \frac{1}{24}$
13. $\frac{5}{7} - \frac{2}{3}$
14. $\frac{4}{5} - \frac{5}{7}$
15. $\frac{5}{8} - \frac{1}{3}$
16. $\frac{5}{8} - \frac{5}{12}$
17. $\frac{5}{6} - \frac{3}{10}$
18. $\frac{2}{9} - \frac{1}{6}$
19. $\frac{3}{4} \times \frac{5}{7}$
20. $\frac{2}{3} \times \frac{4}{5}$
21. $\frac{7}{8} \times \frac{3}{4}$
22. $\frac{5}{6} \times \frac{10}{11}$
23. $\frac{2}{5} \times \frac{1}{6}$
24. $\frac{6}{15} \times \frac{5}{9}$

Continued on next page ...

25. $3 \times \dfrac{1}{2}$ 26. $\dfrac{3}{4} \times 7$ 27. $6 \times \dfrac{3}{8}$ 28. $\dfrac{5}{6} \div \dfrac{2}{5}$ 29. $\dfrac{3}{7} \div \dfrac{5}{9}$ 30. $\dfrac{3}{4} \div \dfrac{4}{5}$

31. $\dfrac{5}{8} \div \dfrac{2}{3}$ 32. $\dfrac{7}{8} \div \dfrac{7}{12}$ 33. $\dfrac{9}{10} \div \dfrac{4}{5}$ 34. $\dfrac{2}{3} \div 8$ 35. $\dfrac{6}{7} \div 9$ 36. $5 \div \dfrac{1}{8}$

In problems 37 through 45 simplify each complex fraction.

37. $\dfrac{\frac{7}{8}}{\frac{2}{3}}$ 38. $\dfrac{\frac{5}{7}}{\frac{4}{9}}$ 39. $\dfrac{\frac{6}{11}}{2}$ 40. $\dfrac{3 - \frac{1}{5}}{2 + \frac{1}{8}}$ 41. $\dfrac{5 + \frac{1}{7}}{4 - \frac{5}{6}}$

42. $\dfrac{\frac{3}{8} + 2}{1 - \frac{5}{9}}$ 43. $\dfrac{\frac{1}{7} + \frac{2}{3}}{\frac{3}{8}}$ 44. $\dfrac{\frac{6}{9}}{\frac{3}{5} - \frac{1}{4}}$ 45. $\dfrac{\frac{2}{7} + \frac{3}{8}}{\frac{1}{5} - \frac{1}{9}}$

In problems 46 through 54 compare each pair of fractions by placing <, > or = between them.

46. $\dfrac{2}{3}, \dfrac{4}{5}$ 47. $\dfrac{7}{8}, \dfrac{3}{5}$ 48. $\dfrac{7}{9}, \dfrac{2}{3}$ 49. $\dfrac{4}{5}, \dfrac{5}{7}$ 50. $\dfrac{3}{4}, \dfrac{5}{6}$

51. $\dfrac{2}{3}, \dfrac{1}{4}$ 52. $\dfrac{3}{5}, \dfrac{6}{11}$ 53. $\dfrac{7}{12}, \dfrac{5}{8}$ 54. $\dfrac{3}{14}, \dfrac{2}{9}$

8.6 RATIONAL NUMBERS AND DECIMALS

Any fraction in the form $\dfrac{a}{b}$ can be expressed as a decimal by dividing the denominator into the numerator. This division process, however, leads to three types of decimals: terminating, repeating nonterminating, and nonrepeating nonterminating.

- A **terminating decimal** results if the division process terminates (i.e., a zero remainder is obtained).

- A **repeating nonterminating decimal** (also referred to as a *repeating decimal*) results if the division process does not terminate, but after a certain point begins to repeat itself. (Repeating decimals are written with a horizontal bar placed over those digits of the quotient that repeat.)

- A **nonrepeating, nonterminating decimal** results if the division process does not terminate and does not repeat. (This will be the topic of section 8.7.)

Every rational number can be represented by either (one or the other but not both) a terminating decimal or a repeating decimal. Conversely, every decimal that either terminates or repeats can be represented as a rational number. It is not difficult to express a rational number as either a terminating or repeating decimal. Simply divide the denominator into the numerator. Expressing a terminating decimal as a rational number is not difficult either. Simply write the decimal digit(s) as the numerator of the fraction and write the place value of the decimal as the denominator of the fraction. To express a repeating decimal as a rational number, however, requires a more involved process. To illustrate this process we express the repeating decimal $0.0132\overline{424}$ as a rational number.

Step 1

Let x represent the given decimal.

$$x = 0.01324\overline{24}$$

Continued on next page ...

Step 2

Multiply the equation from step 1 by a power of 10 that represents the *length* of the block of digits that repeat. In this example, the repeating block consists of the digits 2 and 4. Since these digits occupy two decimal places, the length of the repeating block is *hundredths*. Thus we multiply the equation from step 1 by 100.

$$100x = 1.3242\overline{24}$$

Step 3

Subtract the equation in step 1 from the equation in step 2.

$$\begin{array}{rcl} 100x & = & 1.3242\overline{24} \\ - \quad x & = & 0.0132\overline{24} \\ \hline 99x & = & 1.311 \end{array}$$

Step 4

Using the equation derived from step 3 divide the dumber on the right side of the equal sign by the number on the left side of the equal sign.

$$x = \frac{1.311}{99}$$

Step 5

If the fraction derived from step 4 contains integers in both the numerator and denominator then this is the fractional representation of the given decimal. If this is not the case, then multiply both the numerator and denominator by an appropriate power of ten so that they will contain integers.

$$\frac{1.311}{99} \times \frac{1000}{1000} = \frac{1311}{99000}$$

Thus $0.0132\overline{24} = \frac{1311}{99000} = \frac{437}{33000}$.

Example 1 (page 441, #1b)

Express $\frac{5}{16}$ as a decimal.

Solution

To convert $\frac{5}{16}$ to a decimal divide the denominator into the numerator.

```
         0.3125
    16 ⟌ 5.0000
         48
         ──
         20
         16
         ──
          40
          32
          ──
           80
           80
           ──
            0   Stop
```

Thus $\frac{5}{16}$ is equal to the terminating decimal 0.3125.

Example 2 (page 441, #1f)

Express $\dfrac{15}{37}$ as a decimal.

Solution

$$
\begin{array}{r}
0.405405 \\
37\,\overline{\smash{)}\,15.00000} \\
\underline{148} \\
200 \\
\underline{185} \\
150 \\
\underline{148} \\
200 \\
\underline{185} \\
15
\end{array}
$$

Note that the division process repeats. Thus $\dfrac{15}{37}$ is equal to the repeating decimal $0.\overline{405}$.

Example 3 (page 441, #5a)

Express 0.125 as a rational number.

Solution

Since 0.125 is a terminating decimal, we write the decimal digits (125) over the place value of the decimal (*thousandths*). Thus, $0.125 = \dfrac{125}{1000}$ or $\dfrac{1}{8}$ in reduced form.

Example 4 (page 441, #5e)

Express $6.2\overline{81}$ as a rational number.

Solution

- Let $x = 6.2\overline{81}$
- Since there are two repeating digits (8 and 1) multiply the equation from step 1 by 100.

$$
\begin{array}{rrl}
 & 100x & =\ 628.18\overline{181} \\
- & x & =\ 6.28\overline{181} \\
\hline
 & 99x & =\ 621.9
\end{array}
$$

- Divide the number on the right side by the number on the left side.

$$x = \dfrac{621.9}{99}$$

- Multiply the fraction by ten so that both the numerator and denominator are integers.

$$\dfrac{621.9}{99} \times \dfrac{10}{10} = \dfrac{6219}{990}$$

Thus $6.2\overline{81} = \dfrac{6219}{990}$ or $\dfrac{691}{110}$ in reduced form.

Example 5 (page 441, #9e)

Find a rational number between $\frac{3}{4}$ and $\frac{9}{11}$.

Solution

- Add the two fractions.

$$\frac{3}{4} + \frac{9}{11} = \frac{(3 \times 11) + (4 \times 9)}{(4 \times 11)} = \frac{33 + 36}{44} = \frac{69}{44}$$

- Multiply by any positive fraction less than 1. (We choose to use $\frac{1}{2}$.)

$$\frac{1}{2} \times \frac{69}{44} = \frac{69}{88}$$

Thus a rational number between $\frac{3}{4}$ and $\frac{9}{11}$ is $\frac{69}{88}$.

(Note: $\frac{3}{4} = 0.75$, $\frac{9}{11} = 0.\overline{81}$, and $\frac{69}{88} = 0.784090$.)

SUPPLEMENTARY EXERCISE 8.6

In problems 1 through 8 express each fraction as a decimal.

1. $\frac{3}{5}$ 2. $\frac{19}{20}$ 3. $\frac{9}{16}$ 4. $\frac{7}{28}$ 5. $\frac{43}{99}$ 6. $\frac{8}{37}$ 7. $\frac{17}{22}$ 8. $\frac{17}{21}$

In problems 9 through 16 express each decimal as a rational number. Reduce all answers to lowest terms.

9. 0.85 10. 0.009 11. 0.016 12. 0.20 13. $0.\overline{28}$ 14. $0.\overline{111}$ 15. $1.\overline{2}$ 16. $1.58\overline{3}$

In problems 17 through 20 find a rational number between the given pair of rational numbers.

17. $\frac{4}{9}$, $\frac{7}{9}$ 18. $\frac{2}{3}$, $\frac{3}{5}$ 19. $\frac{7}{8}$, $\frac{3}{4}$ 20. $\frac{5}{6}$, $\frac{3}{7}$

8.7 IRRATIONAL NUMBERS AND THE SET OF REAL NUMBERS

In the previous section it was pointed out that every rational number can be represented by either a terminating or a repeating decimal. We learned that every decimal that either terminates or repeats can be represented as a rational number. This statement implies that if a given decimal is neither terminating nor repeating then it cannot represent a rational number. A decimal of this form (i.e., a nonterminating, nonrepeating decimal) represents an **irrational number**. The sets of rational numbers and irrational numbers are disjoint (their intersection is empty). The union of these two sets however make up a new set, which we call the set of **real numbers**. Thus a real number is either rational or irrational, but not both. The following tree diagram depicts the structure of the real number system and its subsets.

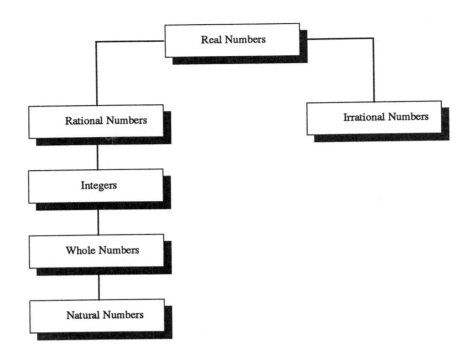

Example 1 (page 445, #1a)

Classify $\frac{1}{3}$ as rational or irrational.

Solution

The fraction $\frac{1}{3}$ is equal to the repeating decimal, $0.\overline{3}$. Thus it is a rational number.

Example 2 (page 445, #1b)

Classify −2 as a rational or irrational number.

Solution

−2 is an integer and all integers are rational numbers.

Example 3 (page 445, #1f)

Classify $\sqrt{4}$ as rational or irrational.

Solution

Since 4 is a *perfect square* the square root of all perfect squares is a rational number.

SETS OF NUMBERS AND THEIR STRUCTURE

Example 4 (page 445, #9)

Let R = the set of real numbers, I = the set of integers, Y = the set of rational numbers, and Z = the set of irrational numbers. Determine whether each statement is true or false.

Solution

a. $R \cap Y = I$

False; the intersection of reals and rationals is rationals.

b. $I \cup Y = Z$

False; the union of integers and rationals is rationals since integers are a subset of rationals.

c. $Y \subset R$

True; all of the rational numbers are contained in the set of real numbers.

d. $R \subset Z$

False; the set of real numbers can be thought of as the universal set, which cannot be contained in the set of irrational numbers.

e. $Y \cap Z = R$

False; the intersection of rational numbers and irrational numbers is empty.

f. $Y \cup Z = R$

True; by definition.

SUPPLEMENTARY EXERCISE 8.7

In problems 1 through 8 classify each number as rational or irrational.

1. $\dfrac{2}{3}$ 2. 0.142 3. $\sqrt{9}$ 4. $\dfrac{\sqrt{2}}{3}$ 5. $\sqrt{5}+1$ 6. $\dfrac{3}{\sqrt{4}}$ 7. $\dfrac{\sqrt{3}}{4\sqrt{3}}$ 8. $3.2014014\overline{014}$

In problems 9 through 15 complete the following chart by placing an X in the column(s) for each number.

	Number	Natural Number	Whole Number	Integer	Rational Number	Irrational Number	Real Number
9.	-5						
10.	-3.2						
11.	0						
12.	$\sqrt{16}$						
13.	$0.\overline{43}$						
14.	$3\sqrt{2}$						
15.	$\dfrac{5}{8}$						

8.8 SCIENTIFIC NOTATION (OPTIONAL)

Scientific notation is a method that is used to express very large or very small numbers conveniently. A number expressed in scientific notation is written as the product of a number between 1 and 10 (or -1 and -10) and a power of ten. For example, all of the numbers below are written in scientific notation.

$$5000 = 5.0 \times 10^3$$
$$8{,}000{,}000 = 8.0 \times 10^6$$
$$0.00003 = 3.0 \times 10^{-5}$$
$$3{,}026{,}000 = 3.026 \times 10^6$$
$$0.005206 = 5.206 \times 10^{-3}$$

In general, a number expressed in scientific notation is of the form $N \times 10^m$ where $1 \leq |N| < 10$ and m is an integer that indicates the number of places the decimal point of the original number was moved. The following general procedure is used to express a number in scientific notation.

To Express A Number In Scientific Notation

1. Viewing the number from left to right place a decimal point after the first nonzero digit you encounter. The *starting number*, N, is now a number between 1 and 10 (or -1 and -10).

2. Count the number of digits that are to the right of this newly placed decimal point. This count becomes the exponent for the power of ten; that is, it is the numerical value for m in the general form above. (**Note:** m is positive if the count is to the right; otherwise it is negative.)

3. Express the starting number, N, using only the *significant digits* of the original number. (**Note:** A significant digit is a nonzero digit, or any digit zero that is between two nonzero digits. For example, in the number 29,000,000, only the 2 and 9 are significant. In the number 29,006,000, the two zeros between the 9 and 6 are significant in addition to the nonzero digits 2, 9, and 6.)

When multiplying or dividing numbers expressed in scientific notation, observe the following rules of exponents:

1. If b is any real number and m and n are natural numbers, then

$$b^m \times b^n = b^{m+n}$$

2. If b is any real number ($b \neq 0$) and m and n are natural numbers, then

$$b^m \div b^n = b^{m-n}$$

Example 1 (page 450, #3)

Write 235,000 in scientific notation.

Solution

The starting number, N, is 2.35, and the exponent, m, is 5 since there are five digits to the right of the newly placed decimal. Therefore, $235{,}000 = 2.35 \times 10^5$ in scientific notation.

Example 2 (page 450, #11)

Express 0.0002345 in scientific notation.

Solution

$N = 2.345$, and $m = -4$, since we divide 2.345 by 10000. (**Note:** Dividing by 10000 is equivalent to multiplying by 10^{-4}.) Therefore, $0.0002345 = 2.345 \times 10^{-4}$ in scientific notation.

SETS OF NUMBERS AND THEIR STRUCTURE 187

Example 3 (page 451, #25)

Express 4.859×10^7 in standard form.

Solution

Multiplication by 10^7 implies that the decimal point is to be moved 7 places to the right. Therefore, $4.859 \times 10^7 = 48590000 = 48{,}590{,}000$.

Example 5 (page 451, #33)

Evaluate $60{,}500 \times 0.0034$ using scientific notation.

Solution

- Express each factor in scientific notation.

$$60{,}500 = 6.05 \times 10^4 \quad \text{and} \quad 0.0034 = 3.4 \times 10^{-3}$$

- Multiply the N of the first expression by the N of the second expression, and add their respective exponents.

$$\begin{aligned}
& 6.05 \times 10^4 \quad \times \quad 3.4 \times 10^{-3} \\
=\ & (6.05)(3.4) \quad \times \quad \left[10^4\right]\left[10^{-3}\right] \\
=\ & (20.57) \quad \times \quad \left[10^{4+(-3)}\right] \\
=\ & (20.57) \quad \times \quad \left[10^1\right]
\end{aligned}$$

The answer is 20.57×10^1, or 205.7.

Example 6 (page 451, #43)

Evaluate $\dfrac{0.0204}{0.00012}$ using scientific notation.

Solution

- Express the numerator and denominator in scientific notation.

$$0.0204 = 2.04 \times 10^{-2} \quad \text{and} \quad 0.00012 = 1.2 \times 10^{-4}$$

- Divide the N of the first expression by the N of the second expression, and add their respective exponents.

$$\begin{aligned}
& \dfrac{2.04 \times 10^{-2}}{1.2 \times 10^{-4}} \\
=\ & \left[\dfrac{2.04}{1.2}\right] \quad \times \quad \left[\dfrac{10^{-2}}{10^{-4}}\right] \\
=\ & (1.7) \quad \times \quad \left[10^{-2-(-4)}\right] \\
=\ & (1.7) \quad \times \quad \left[10^2\right]
\end{aligned}$$

Thus the answer is 1.7×10^2, or 170.

SUPPLEMENTARY EXERCISE 8.8

In problems 1 through 8 express the given number in scientific notation.

1. 3,000,000,000
2. 12,100,000
3. 403,000
4. 207,001,000,000
5. 0.00000006
6. 0.102
7. 0.0000204
8. −0.001102

In problems 9 through 12 express the given number in standard form.

9. 4.6×10^8
10. 6.01×10^{-3}
11. 99×10^3
12. -1.701×10^4

In problems 13 through 21 evaluate the given expression using scientific notation. You may leave your answers in scientific form.

13. $(4.2 \times 10^2)(7 \times 10^7)$
14. $(6.3 \times 10^{-4})(3.6 \times 10^{-1})$
15. $(1.2 \times 10^{-6})(5 \times 10^2)$

16. $\dfrac{7.2 \times 10^2}{1.5 \times 10^5}$
17. $\dfrac{7.2 \times 10^{-2}}{1.5 \times 10^5}$
18. $\dfrac{7.2 \times 10^{-2}}{1.5 \times 10^{-5}}$

19. $\dfrac{.00067 \times 20000000}{0.0004}$
20. $\dfrac{3900000}{0.013 \times 0.0000001}$
21. $\dfrac{.000021 \times 210000}{30000 \times 0.00003}$

8.9 CHAPTER 8 TEST

In problems 1 through 10 answer true or false.

1. All prime numbers are odd.
2. The number zero is considered to be positive.
3. A number without a sign in front of it is considered to be positive.
4. When adding a positive integer and a negative integer the sum is always negative.
5. The product of two negative integers is positive.
6. All integers can be expressed as rational numbers.
7. The number zero cannot be expressed as rational number.
8. All rational numbers can be expressed as a terminating decimal or a repeating decimal.
9. Some integers can be expressed as irrational numbers.
10. All irrational numbers are real numbers.

In problems 11 through 14 determine whether each number is prime or composite.

11. 101
12. 91
13. 147
14. 449

15. Find the GCF of 52 and 78.
16. Find the GCF of 72 and 90.
17. Find the LCM of 30 and 21.
18. Find the LCM of 72 and 150.

Continued on next page ...

In problems 19 through 27 evaluate the given problem.

19. $\dfrac{4}{5} + \dfrac{7}{8}$ 20. $\dfrac{5}{7} - \dfrac{3}{5}$ 21. $\dfrac{2}{9} \times \dfrac{6}{7}$ 22. $\dfrac{3}{4} \div \dfrac{5}{6}$ 23. $(-4) + (-5)$

24. $(-3 + 8) - (-2)$ 25. $2 \times (-6)$ 26. $(-3 - 4) \times (-5)$ 27. $(4 - 9) \times (2 - (-3))$

28. Express $\dfrac{3}{5}$ as a decimal.   29. Express $\dfrac{9}{16}$ as a decimal.

30. Express $0.\overline{15}$ as a rational number.   31. Express $3.\overline{143}$ as a rational number.

In problems 32 through 37 classify each number as rational or irrational.

32. 3.14 33. $\sqrt{8}$ 34. $2.\overline{14}$ 35. $5 - \sqrt{4}$ 36. $\dfrac{2\sqrt{3}}{\sqrt{3}}$ 37. $\dfrac{5}{\sqrt{2}}$

In problems 38 through 40, evaluate using scientific notation.

38. $(460000000)(0.0000092)$ 39. $\dfrac{0.00000042}{700000}$ 40. $\dfrac{300000 \times 0.0015}{2500 \times 0.000002}$

9. AN INTRODUCTION TO ALGEBRA

9.1 INTRODUCTION

Algebra is a branch of mathematics that generalizes the facts of arithmetic. To provide for this generalization, letters are used to represent numbers. These letters are called **variables**. In this chapter we discuss various basic properties of algebra and show the relationship of algebra to topics discussed previously in the textbook.

9.2 OPEN SENTENCES AND THEIR GRAPHS

A mathematical sentence that shows equality is referred to as an **equation**. Equations can be true, false, or open. For example:

- The numerical equation, $3 + 2 = 5$, is a true sentence since $3 + 2$ is equal to 5
- The numerical equation, $3 + 2 = 10$, is a false sentence since $3 + 2$ does not equal 10.
- The algebraic equation, $x + 6 = 10$, is an open sentence; its truth value is unknown until the variable x is replaced with a specific numerical value.

In the equation, $x + 6 = 10$, if the variable x is replaced with 4, then the equation is true. The number 4 is the flsolution to the equation. The **solution set** of an open sentence consists of the set of elements that make the sentence true when they are substituted for the variable(s). (Note that if any number other than 4 is substituted for x in the equation $x + 6 = 10$, the equation will be false.)

A mathematical sentence does not always have to show equality. Sometimes it shows inequality. An inequality is obtained by replacing the equal sign in an equation with an inequality symbol. The symbols of inequality are summarized in table 9.1.1.

When the solution set of an equation or inequality is determined, it is usually graphed. To graph a solution set means to locate the point(s) on a number line that corresponds to the element(s) that make up the solution set. This is illustrated in Examples 4, 5, and 6.

Table 9.1.1 Inequality Symbols

Symbol	Read as	Meaning
<	*less than*	Indicates all numbers that are smaller in numerical value than a given number
≤	*less than or equal to*	Indicates all numbers that are smaller than or equal to the numerical value of a given number
>	*greater than*	Indicates all numbers that are larger in numerical value than a given number
≥	*greater than or equal to*	Indicates all numbers that are larger than or equal to the numerical value of a given number

Example 1 (page 463, #5)

Find the solution set for the equation $5 - 2 = 4 + x$ where the numerical value for x must be a whole number.

Solution

First evaluate the left side of the equation.

$$\begin{aligned} 5 - 2 &= 4 + x \\ 3 &= 4 + x \end{aligned}$$

The only number that makes this equation true is the integer -1. However, x must be a whole number. Since there are no whole numbers that make the equation true, the solution set is the empty set, $\emptyset$.

Example 2 (page 464, #11)

Find the solution set for $x + 2 < 7$, where the numerical value for x must be a natural number.

Solution

The set of natural numbers is $\{1,2,3,...\}$. If x is replaced with 1, 2, 3, or 4, the inequality is true; any other number makes the inequality false.

$$\begin{aligned} 1 + 2 < 7 &\rightarrow 3 < 7 \text{ (true)} \\ 2 + 2 < 7 &\rightarrow 4 < 7 \text{ (true)} \\ 3 + 2 < 7 &\rightarrow 5 < 7 \text{ (true)} \\ 4 + 2 < 7 &\rightarrow 6 < 7 \text{ (true)} \\ 5 + 2 < 7 &\rightarrow 7 < 7 \text{ (false; 7 is not less than 7)} \\ 6 + 2 < 7 &\rightarrow 8 < 7 \text{ (false; 8 is not less than 7)} \\ \text{etc.} \end{aligned}$$

Example 3 (page 464, #17)

Find the solution set for $0 \leq x < 5$ where the numerical value(s) for x must be an integer.

Solution

The set of integers is $\{...,-3,-2,-1,0,1,2,3,...\}$. The inequality $0 \leq x < 5$ means that x is a number between 0 and 5, and includes 0 but cannot equal 5. The integers that satisfy these requirements are $\{0,1,2,3,4\}$.

Example 4 (page 464, #21)

Graph the solution set for $x + 3 = 4$. The replacement set is the set of real numbers.

Solution

The only solution to this equation is 1. The graph of the solution set is shown below.

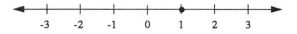

Example 5 (page 464, #29)

Graph the solution set for $x + 2 \geq 0$. The replacement set is the set of real numbers.

Solution

This inequality is true when x is replaced by any real number that is either equal to -2 or greater than -2. The graph of this solution set includes all real numbers that lie to the right of -2, and the point -2 itself. Note that we show the inclusion of the point -2 with a closed circle.

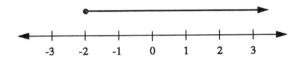

Example 6 (page 464, #33)

Graph the solution set for $-2 < x \leq 1$. The replacement set is the set of real numbers.

Solution

This inequality is true when x is replaced by any real number that is between -2 and 1, including 1 but not -2. Note that we show the exclusion of the point -2 by an open circle.

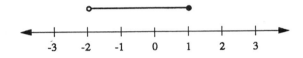

SUPPLEMENTARY EXERCISE 9.2

In 1 through 10 find the solution set for each open sentence. The replacement set is the set of integers.

1. $x + 6 = 3$ 2. $x - 8 = -2$ 3. $9 + x = 13$ 4. $x - 12 = 13$ 5. $x - 27 = -8$
6. $18 = x + 3$ 7. $14 = x - 8$ 8. $15 = 16 + x$ 9. $-14 = -18 + x$ 10. $-10 = -11 + x$

In 11 through 20 find and graph the solution set for each inequality. Unless otherwise noted, the replacement set is the set of real numbers.

11. $x > 4$ 12. $x \leq -2$ 13. $x - 8 \leq 3$ 14. $x - 4 > 3$ 15. $x + 8 \leq 7$ 16. $x - 3 > 1$

17. $0 \leq x < 2$ (x is an integer) 18. $-5 \leq x \leq 3$ (x is a whole number) 19. $-2 < x \leq 5$ 20. $1 \leq x \leq 4$

9.3 ALGEBRAIC NOTATION

In algebra the product of two or more factors is written without a multiplication symbol. For example,

- $2 \times x$ is written as $2x$
- $3 \times y$ is written as $3y$
- $x \times y$ is written as xy
- $3 \times x \times y$ is written as $3xy$

and so forth.

Whenever a variable is written without a numerical *coefficient*, the coefficient is understood to be 1. Thus x means *one times x*. If a given algebraic expression contains terms that have the same variable and the same exponent for each corresponding variable, then the terms are called **like terms**. For example, $5x$ and $2x$ are like terms. However, $5x$ and $2y$ are **unlike terms**. Expressions that contain like terms are simplified by *combining* the like terms. This is done by adding or subtracting like terms. For example,

- $5x + 2x$ is simplified to $(5 + 2)x = 7x$
- $5x - 2x$ is simplified to $(5 - 2)x = 3x$
- $-3y + 2y$ is simplified to $(-3 + 2)y = -y$

and so on.

The multiplication of a single term also is simplified. For example,

- $3(2x)$ is simplified to $(3 \times 2)x = 6x$
- $5(3y)$ is simplified to $(5 \times 3)y = 15y$

In the event a multiplication involves more than one term, the **distributive property of multiplication** is used. In general this property states that for any numbers a, b, and c

$$a(b + c) = a(b) + a(c)$$

As an example, consider the problem of simplifying the expression $5x + 3(2x + 4)$. In this expression, note that there are like terms ($5x$ and $2x$) involving the variable x. Before these like terms are combined the parentheses must be removed. To do this, we use the distributive property.

- Remove the parentheses using the distributive property.

$$5x + 3(2x + 4)$$
$$= 5x + 3(2x) + 3(4)$$

- Simplify the products.

$$\begin{aligned} &\ 5x\ +\ 3(2x)\ +\ 3(4) \\ &=\ 5x\ +\ 6x\ +\ 12 \end{aligned}$$

- Combine the like terms.

$$\begin{aligned} &\ 5x + 6x\ +\ 12 \\ &=\ 11x\ +\ 12 \end{aligned}$$

As a result, the expression $5x + 3(2x + 4)$ is simplified to the expression $11x + 12$. Note that $11x$ and 12 cannot be combined. The reason is that $11x$ is an x term (it is a term associated with the variable x), whereas 12 is a number (i.e., constant) term. They are unlike terms.

Example 1 (page 466, #17)

Simplify $2y + 2 + 3y$.

Solution

- First collect the like terms. To do this identify all like terms and group them together.

$$\begin{aligned} &\ 2y + 2 + 3y \\ &=\ 2y + 3y + 2 \end{aligned}$$

Continued on next page ...

- Now combine the like terms.

$$= \begin{array}{ccc} 2y + 3y & + & 2 \\ 5y & + & 2 \end{array}$$

The simplified expression is $5y + 2$.

Example 2 (page 466, #25)

Simplify $4(3x) + 2$.

Solution

$$\begin{array}{rcc} & 4(3x) & + \quad 2 \\ = & (4 \times 3)x & + \quad 2 \\ = & 12x & + \quad 2 \end{array}$$

Example 3 (page 466, #33)

Simplify $4x + 3x + 5 - 3$.

Solution

$$\begin{array}{rcc} & 4x + 3x & + \quad 5 - 3 \\ = & 7x & + \quad 2 \end{array}$$

Example 4 (page 466, #43)

Simplify $4x - 2 - x + 3(x + 1)$.

Solution

$$4x - 2 - x + 3(x + 1)$$

$$\begin{array}{rccccccc} = & 4x & - & 2 & - & x & + & 3(x) & + & 3(1) \\ = & 4x & - & 2 & - & x & + & 3x & + & 3 \\ = & 4x & - & x & + & 3x & - & 2 & + & 3 \\ = & 3x & & & + & 3x & - & 2 & + & 3 \\ = & & & 6x & & & + & & & 1 \\ = & 6x + 1 \end{array}$$

SUPPLEMENTARY EXERCISE 9.3

In problems 1 through 30 simplify each algebraic expression.

1. $4x + 2x$
2. $5y + 3y$
3. $7x - 3x$
4. $5x - x$
5. $5(2x)$
6. $3(3y)$
7. $3x + 3$
8. $8y + 2$
9. $\dfrac{12x}{6}$
10. $6x + 3 + 5x$
11. $2y + 5 - y$
12. $5 + 3x + 4$
13. $2x - 5 + 4$
14. $3 + 2y - 1 + y$
15. $x + 1 + 2x + 2$
16. $4y - 3 + 2 - y$
17. $5 + 3x - 8 - x$
18. $3x - 4 - 3x + 2$
19. $2(3x) - 4$
20. $5(2x) - 5 + x$

Continued on next page ...

21. $2x - 4 + 3(2x)$
22. $3(2x - 4) + 5$
23. $2(5x + 2) - 6$
24. $4 + 3(x - 1) + 2$
25. $2y + 5(3y - 6)$
26. $3(6 - 2y) + 8y$
27. $3x + 2(3 - 2x)$
28. $2x - 4 + x + 3(2x - 1)$
29. $5(2x + 5) - 3x + 4$
30. $2(3x - 1) + 5(x - 4)$

9.4 MORE OPEN SENTENCES

The overall objective when solving an algebraic equation for a specified variable is to isolate the variable from the other terms in the equation. To do so it is necessary to transform the equation, using correct mathematical procedure, to an equation of the form

$$x = \text{expression} \quad \text{or} \quad \text{expression} = x$$

where x is the variable under consideration.

The types of equations that are given can be transformed into this form by using *inverse* operations. When solving these equations it is imperative that the equality of the equation be maintained at all times. This is accomplished by performing exactly the same operation to both sides of the equation. The equality of an equation is preserved if (1) the same quantity is added to both sides of the equation, (2) the same quantity is subtracted from both sides of the equation, (3) both sides of the equation ar multiplied by the same quantity, and (4) both sides of the equation are divided by the same nonzero quantity.

The following procedure is given as a guide to help you solve simple algebraic equations in one variable, and is used in the examples that follow.

To Solve A Simple Algebraic Equation In One Variable

1. Remove any grouping symbols by applying the distributive property of multiplication.

2. Independently working with each side of the equation, combine all like terms.

3. Isolate the variable by

    a. *undoing* any additions or subtractions first, and then

    b. *undoing* any multiplications or divisions.

4. Check the solution obtained by substituting it into the original equation for the variable and evaluating the equation. If the result is true, then the solution is correct; otherwise there is an error somewhere in the process.

Example 1 (page 471, #3)

Solve $6 = x - 2$.

Solution

Since there are no grouping symbols to remove or like terms to combine, isolate the variable by adding 2 to both sides. (**Note:** By adding 2 to both sides, we *undo* the subtraction on the right side of the equation.)

$$\begin{aligned} 6 &= x - 2 \\ 6 + 2 &= x - 2 + 2 \\ 8 &= x \end{aligned}$$

The solution is 8. (The check is left for the reader.)

Example 2 (page 471, #7)

Solve $2y - 2 = 10$.

Solution

Since there are no grouping symbols to remove or like terms to combine, isolate the variable.

- Add 2 to both sides to *undo* the subtraction.

$$\begin{aligned} 2y - 2 &= 10 \\ 2y - 2 + 2 &= 10 + 2 \\ 2y &= 12 \end{aligned}$$

- Now divide both sides of the equation by 2 to *undo* the multiplication.

$$\begin{aligned} 2y &= 12 \\ \frac{2y}{2} &= \frac{12}{2} \\ y &= 6 \end{aligned}$$

The solution is 6. (The check is left to the reader.)

Example 3 (page #13)

Solve $2x + 2 = x + 3$.

Solution

Since there are no grouping symbols to remove or like terms to combine isolate the variable.

- Subtract x from both sides of the equation.

$$\begin{aligned} 2x + 2 &= x + 3 \\ 2x - x + 2 &= x - x + 3 \\ x + 2 &= 3 \end{aligned}$$

- Subtract 2 from both sides of the equation.

$$\begin{aligned} x + 2 &= 3 \\ x + 2 - 2 &= 3 - 2 \\ x &= 1 \end{aligned}$$

The solution is 1. (The check is left to the reader.)

Example 4 (page 472, #21)

Solve $\frac{x}{4} + 1 = 15$.

Solution

Since there are no grouping symbols to remove or like terms to combine isolate the variable.

- *Undo* the addition.

$$\begin{aligned} \frac{x}{4} + 1 &= 15 \\ \frac{x}{4} + 1 - 1 &= 15 - 1 \\ \frac{x}{4} &= 14 \end{aligned}$$

Continued on next page ...

- *Undo* the multiplication.

$$\frac{x}{4} = 14$$

$$(4)\frac{x}{4} = (4)14$$

$$x = 56$$

The solution is 56. (The check is left to the reader.)

Example 5 (page 472, #31)

Solve $2(3x + 1) = 8$.

Solution

- Remove the parentheses by applying the distributive property.

$$2(3x + 1) = 8$$
$$2(3x) + 2(1) = 8$$
$$6x + 2 = 8$$

- Isolate the variable by subtracting 2 from both sides of the equation and then dividing both sides of the equation by 6.

$$6x + 2 = 8$$
$$6x + 2 - 2 = 8 - 2$$
$$6x = 6$$
$$\frac{6x}{6} = \frac{6}{6}$$
$$x = 1$$

The solution is 1. (The check is left to the reader.)

Example 6 (page 472, #37)

Solve $x + 2(3x + 1) = 9$

Solution

- Remove the parentheses.

$$x + 2(3x + 1) = 9$$
$$x + 6x + 2 = 9$$

- Combine the like terms on the left side of the equation.

$$x + 6x + 2 = 9$$
$$7x + 2 = 9$$

- Isolate the variable.

$$7x + 2 = 9$$
$$7x + 2 - 2 = 9 - 2$$
$$7x = 7$$
$$\frac{7x}{7} = \frac{7}{7}$$
$$x = 1$$

The solution is 1. (The check is left to the reader.)

SUPPLEMENTARY EXERCISE 9.4

In problems 1 through 25 solve the given equation.

1. $x + 9 = 3$
2. $19 = x - 5$
3. $6x = 12$
4. $7x = -28$
5. $14 = \dfrac{x}{2}$
6. $\dfrac{x}{3} = 4$
7. $3x + 5 = 32$
8. $9 + 6x = 21$
9. $\dfrac{x}{5} + 6 = 9$
10. $\dfrac{x}{8} - 6 = -3$
11. $5 + \dfrac{x}{2} = 10$
12. $\dfrac{3x}{7} + 4 = 7$
13. $2x + x = 30$
14. $7x - 5x = 12$
15. $2x + 5x + 6 = 83$
16. $14x - 3 = 4x - 13$
17. $5 - 6x = 5 + 6x$
18. $7x = 3x - 8$
19. $2x + 16 - 8x = 4 + 3x - 4x - 8$
20. $3(x - 4) = 3$
21. $2(2x + 1) = 10$
22. $5(4 + 3x) = 50$
23. $2 + 3(x - 4) = 17$
24. $3x - 6 + 2(x - 8) = 3$
25. $2x + 2(3x - 8) = 4(x - 5)$

9.5 PROBLEM SOLVING

In your textbook, Professor Setek presents a list of suggestions that can be used as an aid in solving word problems. This list is reproduced here. As a further aid to solving word problems, a translation chart is presented in table 9.5.1.

<u>Suggestions For Solving Word Problems</u>

1. Read the problem carefully.
2. Reread the problem carefully.
3. If possible, draw a diagram to assist in interpreting the given information.
4. Translate the English phrases into mathematical phrases and choose a variable for the unknown quantity.
5. Write the equation using all of the above information.
6. Solve the equation.
7. Check the solution to determine whether it satisfies the original problem (not the equation).

Example 1 (page 480, #1)

Three more than a certain number is 10. Find the number.

Solution

• Translate the statement into a mathematical equation.

Three	*more than*	*a certain number*	*is*	*10*
3	+	x	=	10

• Solve the equation.

$$3 + x = 10$$
$$x = 10 - 3$$
$$x = 7$$

The number is 7. (The check is left to the reader.)

Table 9.5.1 Translation Chart for Solving Word Problems

The English word or phrase	Means	Examples	
		Arithmetic	Algebra
Plus The sum of More than Increased by Added to	Addition	$4 + 2$	$x + y$
Minus Decreased by Reduced by Less Take away (2) subtracted from (4) (2) less than (4)	Subtraction	$4 - 2$	$x - y$
Multiply Times Product of Multiplied by Of	Multiplication	$(4)(2)$	xy
Twice Double Twice as much	Multiply by 2	$(2)(4)$	$2x$
Quotient of Ratio of (4) divided by (2) (2) divides (4)	Division	$\dfrac{4}{2}$	$\dfrac{x}{y}$
A certain number A variable An unknown	Use a letter to represent the unknown quantity	not applicable	$x, y,$ or z
Is equal to Equals Is Equal to Are Were The sum is The difference is The same as The result is Leaves Makes Was The product is The quotient is Yields	Equality (=)		

Example 2 (page 480, #5)

The sum of two consecutive integers is 15. Find the numbers.

Solution

The numbers we are looking for are consecutive. This implies that the second integer has a numerical value that is one larger than the first integer.

$$x = \text{the first integer}$$
$$x + 1 = \text{the second integer}$$

Now translate and solve the resulting equation.

$$\text{The sum of two consecutive integers is } 15$$

$$(x) + (x + 1) = 15$$

$$(x) + (x + 1) = 15$$
$$x + x + 1 = 15$$
$$2x + 1 = 15$$
$$2x = 15 - 1$$
$$2x = 14$$
$$x = \frac{14}{2}$$
$$x = 7$$

The numbers are 7 and 8. (The check is left for the reader.)

Example 3 (page 480, #7)

Julia is 7 years younger than Lewis and the sum of their ages is 53. How old is each?

Solution

$$\text{let } x = \text{Lewis' age}$$
$$\text{then } x - 7 = \text{Julia's age}$$

$$\text{the sum of their ages is } 53$$

$$(x) + (x - 7) = 53$$
$$x + x - 7 = 53$$
$$2x - 7 = 53$$
$$2x = 53 + 7$$
$$2x = 60$$
$$x = \frac{60}{2}$$
$$x = 30$$

Lewis is 30 years old and Julia is 23 years old. (The check is left to the reader.)

AN INTRODUCTION TO ALGEBRA

Example 4 (page 480, #13)

The length of a rectangle is 2 meters longer than twice its width. If the perimeter of the rectangle is 100 meters, find the dimensions of the rectangle.

Solution

Since the length is given in terms of the width we represent the width with a variable.

$$\text{let } x = \text{the width of the rectangle}$$

We can now express the length in terms of the same variable.

length is	2 meters longer than	twice its width
=	2 +	$2x$

Therefore x = width and $2 + 2x$ = length.

The perimeter of a rectangle is equal to the sum of its four sides, which is 100 meters (as given in the problem). This leads to the following equation, which is solved below.

$$\begin{aligned}
\text{width} + \text{width} + \text{length} + \text{length} &= 100 \\
(x) + (x) + (2 + 2x) + (2 + 2x) &= 100 \\
x + x + 2 + 2x + 2 + 2x &= 100 \\
x + x + 2x + 2x + 2 + 2 &= 100 \\
6x + 4 &= 100 \\
6x &= 100 - 4 \\
6x &= 96 \\
x &= \frac{96}{6} \\
x &= 16
\end{aligned}$$

The width is 16 meters and the length is 34 meters. (The check is left to the reader.)

Example 5 (page 481, #19)

David has $2.25 in dimes and quarters in his pocket. If he has twice as many dimes as quarters, how many of each type of coin does he have?

Solution

Let x = the number of quarters. Since there are twice as many dimes as quarters the number of dimes is $2x$.

$$\begin{aligned}
x &= \text{number of quarters} \\
2x &= \text{number of dimes}
\end{aligned}$$

Since a quarter is worth 25¢ or ($0.25), the product of the number of quarters (x) and 0.25 is equal to total amount of money in quarters. Similarly, since a dime is worth 10¢ or ($0.10), the product of the number of dimes ($2x$) and 0.10 is equal to the total amount of money in dimes.

$$\begin{aligned}
(0.25)(x) &= \text{total monetary value in quarters} \\
(0.10)(2x) &= \text{total monetary value in dimes}
\end{aligned}$$

Continued on next page ...

Finally, we know that the total amount of money in dimes and quarters is $2.25, as given in the problem. With this last piece of information we now have an equation to solve.

$$(0.25)(x) + (0.10)(2x) = 2.25$$
$$0.25x + 0.20x = 2.25$$
$$0.45x = 2.25$$
$$x = \frac{2.25}{0.45}$$
$$x = 5$$

There are 5 quarters ($1.25) and 10 dimes ($1.00).

Example 6 (page #27)

A collection of 30 quarters and dimes amounts to $5.55. How many of each kind of coin are there?

Solution

We are told there is a total of 30 coins involving only 2 different types of coins (dimes and quarters). If we let x equal the number of quarters, then the number of dimes is the difference between 30 and the number of quarters.

$$x = \text{number of quarters}$$
$$30 - x = \text{number of dimes}$$

The total monetary value of the coins can be found as we did in example 5 above.

$$(0.25)(x) = \text{total monetary value in quarters}$$
$$(0.10)(30 - x) = \text{total monetary value in dimes}$$

Finally, we know that the total amount of money in dimes and quarters is $5.55, as given in the problem. With this last piece of information we now have an equation to solve.

$$(0.25)(x) + (0.10)(20 - x) = 5.55$$
$$0.25x + (0.10)(30) - (0.10)(x) = 5.55$$
$$0.25x + 3 - 0.10x = 5.55$$
$$0.25x - 0.10x + 3 = 5.55$$
$$0.15x + 3 = 5.55$$
$$0.15x = 5.55 - 3$$
$$0.15x = 2.55$$
$$x = \frac{2.55}{0.15}$$
$$x = 17$$

There are 17 quarters ($4.25) and 13 dimes ($1.30).

AN INTRODUCTION TO ALGEBRA

SUPPLEMENTARY EXERCISE 9.5

In 1 through 10 use algebra to solve the given problems.

1. A number increased by 18 is equal to 22. Find the number.
2. Three times a number less four is 20. Find the number.
3. The sum of twice a number and seven yields a difference of four times the number and five. Find the number.
4. Twenty-eight is the difference of four times a number and 16. Find the number.
5. One number is four times a second number. If their difference is 18, find the numbers.
6. Given two numbers, the second number is eight more than the first. If three times the first number is increased by five times the second number, the sum is 80. Find the two numbers.
7. Kelly is two years older than Kristin. If twice Kristin's age is increased by five times Kelly's age the sum is 73. Find the ages of the two girls.
8. A pair of pants costs $22 more than a shirt. If two pairs of pants and three shirts cost $134, find the cost of each.
9. The length of a tennis court for doubles play is six feet longer than twice its width. If its perimeter is 228 feet, find the dimensions.
10. Elizabeth has three times as many quarters as dimes, which total $6.80. How many of each coin does she have?

9.6 LINEAR EQUATIONS IN TWO VARIABLES

Any equation of the form $Ax + By = C$, where A, B, and C are real numbers is called a **linear equation**. Linear equations take as their solution **ordered pairs**. To find an ordered pair that satisfies a given a linear equation we apply the following procedure.

To Find An Ordered Pair For a Given Linear Equation

1. Arbitrarily select a number for either x or y. (Typically we choose a number for x.)
2. Substitute the number chosen into the equation for the specified variable and solve the equation for the remaining variable.

Example 1 (page 486, #5)

Find three solutions for the equation $2x + y = -6$.

Solution

Since we can select any number for x (or y) we choose to let $x = 0, 4,$ and -2. We now substitute these numbers for x into the given equation (separately) and find the corresponding numbers for y.

If $x = 0$ then	If $x = 4$ then	If $x = -2$ then
$2x + y = -6$	$2x + y = -6$	$2x + y = -6$
$2(0) + y = -6$	$2(4) + y = -6$	$2(-2) + y = -6$
$0 + y = -6$	$8 + y = -6$	$-4 + y = -6$
$y = -6$	$y = -6 - 8$	$y = -6 + 4$
	$y = -14$	$y = -2$

The ordered pairs are $(0, -6)$, $(4, -14)$, and $(-2, -2)$.

Example 2 (page 486, #17)

Find three solutions for the equation $2x - 3y = -11$.

Solution

We choose to let $x = 0, 5,$ and -1.

If $x = 0$ then			If $x = 5$ then			If $x = -1$ then		
$2x - 3y$	$=$	-11	$2x - 3y$	$=$	-11	$2x - 3y$	$=$	-11
$2(0) - 3y$	$=$	-11	$2(5) - 3y$	$=$	-11	$2(-1) - 3y$	$=$	-11
$0 - 3y$	$=$	-11	$10 - 3y$	$=$	-11	$-2 - 3y$	$=$	-11
$-3y$	$=$	-11	$-3y$	$=$	$-11 - 10$	$-3y$	$=$	$-11 + 2$
y	$=$	$\frac{-11}{-3}$	$-3y$	$=$	-21	$-3y$	$=$	-9
y	$=$	$\frac{11}{3}$	y	$=$	$\frac{-21}{-3}$	y	$=$	$\frac{-9}{-3}$
			y	$=$	7	y	$=$	3

The ordered pairs are $(0, \frac{11}{3})$, $(5, 7)$, and $(-1, -3)$.

SUPPLEMENTARY EXERCISE 9.6

In problems 1 through 8 determine if the given ordered pair satisfies the given equation. Answer yes or no.

1. $x - y = 6$ $(3, -9)$ 2. $5x - y = -4$ $(-2, -6)$ 3. $3x - y = 8$ $(0, -8)$ 4. $6x + y = 5$ $(-1, -1)$
5. $2x - y = 8$ $(0, 8)$ 6. $3x + y = -4$ $(2, -10)$ 7. $4x - 3y = 16$ $(1, -4)$ 8. $x - 5y = 6$ $(-4, 2)$

In 9 through 20 find the corresponding numerical value for y for the given value of x in each equation.

9. $x + y = 3$ $(x = 6)$ 10. $2x + y = -6$ $(x = -8)$ 11. $5x - y = 3$ $(x = 2)$
12. $x - 2y = -4$ $(x = 5)$ 13. $x + 3y = 12$ $(x = -3)$ 14. $5x - y = -8$ $(x = 2)$
15. $4x - 3y = 12$ $(x = 0)$ 16. $2x + 3y = 7$ $(x = -3)$ 17. $-2x + 3y = -2$ $(x = 4)$
18. $5x - 3y = 0$ $(x = -1)$ 19. $-x - y = -1$ $(x = 2)$ 20. $-3x - 2y = -2$ $(x = -2)$

9.7 GRAPHING EQUATIONS

The Cartesian coordinate system consists of two lines, one vertical and one horizontal, that cross at their respective zero points. The vertical line is called the **y-axis**, the horizontal line is called the **x-axis**, and the point of intersection is called the **origin**.

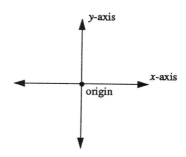

The coordinates of a point graphed in the Cartesian coordinate system are represented by the ordered pair (x, y), where x represents movement left or right, and y represents movement up or down. If x or y (or both) have a numerical value of zero, then this indicates no movement in the respective direction. As a result, the origin is represented by the ordered pair $(0, 0)$.

To Locate Or Graph A Point That Corresponds To An Ordered Pair (x, y)

1. Begin at the origin and move x units along the x-axis.

    a. If x is positive, move right.

    b. If x is negative, move left.

    c. If x is zero, then there is no horizontal movement.

2. From the location arrived at as a result of step 1 above, move y units along the y-axis.

    a. If y is positive, move up.

    b. If y is negative, move down.

    c. If y is zero, then there is no vertical movement.

The graph of an equation of the form $Ax + By = C$, where both A and B are not zero, is a straight line. To graph a straight line we plot the ordered pairs that satisfy the equation. Typically, we choose three ordered pairs, plot them as points in the Cartesian coordinate system, and draw a line through them.

When we graph a linear equation, it is often convenient to find the points where the graph crosses each axis. The point at which a line crosses an axis is called an **intercept**. The x intercept is the point where the line crosses the x-axis. This point is found by substituting zero for y in the equation and solving for x. The y intercept is the point where the line crosses the y-axis. This point is found by substituting zero for x in the equation and solving for y.

A **system** of two linear equations is solved by graphing each equation, independently, on the same set of axes. If the two lines intersect at exactly one point, then the solution is the ordered pair that represents this point of intersection. If there are no points of intersection, then there is no solution and the system is said to be **inconsistent**. (Graphically, we have parallel lines.) If the graph of the two equations results in the same line, then there are an infinite number of solutions and the system is said to be **dependent**.

Example 1 (page 497, #9)

Graph $2x + 3y = -6$.

Solution

First select three points that satisfy the equation. It is most convenient to choose the x and y intercepts as two of these points. The third point is arbitrary.

x intercept	y intercept	Let $x = 3$
If $y = 0$ then	If $x = 0$ then	If $x = 3$ then
$2x + 3y = -6$	$2x + 3y = -6$	$2x + 3y = -6$
$2x + 3(0) = -6$	$2(0) + 3y = -6$	$2(3) + 3y = -6$
$2x + 0 = -6$	$0 + 3y = -6$	$6 + 3y = -6$
$2x = -6$	$3y = -6$	$3y = -6 + -6$
$x = \dfrac{-6}{2}$	$y = \dfrac{-6}{3}$	$3y = -12$
$x = -3$	$y = -2$	$y = \dfrac{-12}{3}$
		$y = -4$

The ordered pairs are $(-3, 0)$, $(0, -2)$, and $(3, -4)$

Continued on next page ...

Plotting these points and drawing a line through them produces the graph of $2x + 3y = -6$.

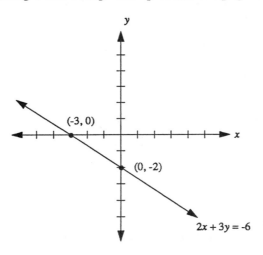

Example 2 (page 497, #23)

Graph the equations $2x + y = -3$ and $-x + 2y = 4$ on the same set of axes, and determine their solution (if any).

Solution

- Find three points that satisfy each equation.

   a. Three points that satisfy the equation $2x + y = -3$ are $(0, -3)$, $(-1, -1)$, and $(3, -9)$.

   b. Three points that satisfy the equation $-x + 2y = 4$ are $(0, 2)$, $(-4, 0)$, and $(4, 4)$.

- Graph the two equations on the same set of axes and determine their point of intersection.

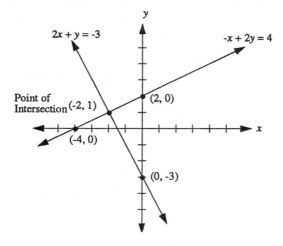

The solution is the point $(-2, 1)$. This means that $(-2, 1)$ satisfies both equations.

AN INTRODUCTION TO ALGEBRA 207

SUPPLEMENTARY EXERCISE 9.7

In problems 1 through 8 plot each ordered pair given.

1. (−3, 2)   2. (−2, −3)   3. (4, 0)   4. (0, −3)   5. (−3, 0)   6. (2, 4)   7. (5, −1)   8. (0, 2)

In problems 9 through 17 graph each equation.

9. $x + y = 3$ 10. $x + y = -2$ 11. $x - y = 4$ 12. $x - y = -5$ 13. $2x - y = 3$
14. $x - 2y = 6$ 15. $2x + 3y = 12$ 16. $5x - 3y = 15$ 17. $-3x - 2y = 18$

In problems 18 through 20, solve each system of equations graphically.

18. $3x - 2y = 4$ and $5x + 6y = -12$
19. $2x + y = 4$ and $2x + y = -5$
20. $5x - 4y = -10$ and $x + 3y = 17$

9.8 THE SLOPE OF A LINE (OPTIONAL)

In the figure at the right, notice that as you travel from point A to point B along line segment AB (denoted as $\overline{AB}$) you go up a vertical distance of 10 feet when you travel 30 feet. This 10 feet vertical change is called the *rise*, and the 30 feet horizontal change is called the *run*. The ratio of rise to run is a measure of a line's steepness and is called the *slope* of a line. In the figure at the right, the slope of $\overline{AB}$ is equal to $\frac{10 \text{ feet}}{30 \text{ feet}} = \frac{1}{3}$ foot. A slope of $\frac{1}{3}$ means that the line rises 1 foot in each run of 3 feet, or equivalently, the line rises $\frac{1}{3}$ foot in each run of 1 foot. The greater the magnitude of the slope, the steeper the line.

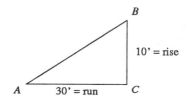

The slope of a line is calculated by finding the change in vertical distance (i.e., *the change in y*) and dividing it by the change in horizontal distance (i.e., *the change in x*). Vertical and horizontal changes are found by selecting two points on a line and finding the difference between the y-values and the difference between the x-values, respectively. For example, if line AB (denoted $\overleftrightarrow{AB}$) contained the points (1, 2) and (4, 7), then the change in y is $7 - 2 = 5$, and the change in x is $4 - 1 = 3$. Thus the slope of $\overleftrightarrow{AB}$ is equal to $\frac{5}{3}$. This means that there is a rise of 5 units for every run of 3 units. In general, if we let m represent slope, and (x_1, y_1) and (x_2, y_2) be any two points on a line, then the slope of the line is given by the formula,

$$m = \frac{rise}{run} = \frac{y_2 - y_1}{x_2 - x_1}$$

It makes no difference which two points are selected when determining the slope of a line.

9.8.1 The Slope of a Horizontal Line

The slope of any horizontal line is always zero. To illustrate, consider $\overleftrightarrow{AB}$, which contains the points $A(1, -1)$ and $B(3, -1)$, in the figure at the right. If we let $x_1 = 1$, $y_1 = -1$, $x_2 = 3$, and $y_2 = -1$, then

$$m = \frac{y_2 - y_1}{x_2 - x_1} = \frac{-1-(-1)}{3-1} = \frac{0}{2} = 0$$

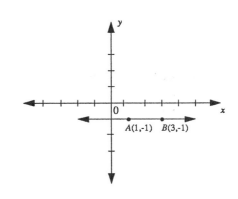

9.8.2 The Slope of a Vertical Line

The slope of any vertical line is always undefined. To illustrate, consider $\overleftrightarrow{AC}$, which contains the points $A(1, -1)$ and $C(1, 2)$, in the figure at the right. If we let $x_1 = 1$, $y_1 = -1$, $x_2 = 1$, and $y_2 = 2$, then

$$m = \frac{y_2 - y_1}{x_2 - x_1} = \frac{2-(-1)}{1-1} = \frac{3}{0}, \text{ which is undefined.}$$

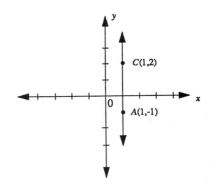

9.8.3 The Slope of Parallel Lines

If two lines are vertical, then the lines must be parallel to the y-axis. Since the slope of a vertical line is undefined, it follows then that the slope of all lines parallel to the y-axis is undefined. Consequently, it is of no interest to discuss the slope of vertical parallel lines.

On the other hand, if two lines are horizontal, then the lines must be parallel to the x-axis. Note that the slopes of these lines are equal; each line has a slope of zero. In general, it can be proven mathematically that if two nonvertical lines are parallel then their slopes are equal. The proof, though, requires a working knowledge of similar triangles (Chapter 10). As a result, the statement is only demonstrated.

Consider the figure at the right, which shows the graph of $\overleftrightarrow{AB}$ containing the points $A(-1, 3)$ and $B(2, 6)$, and $\overleftrightarrow{CD}$ containing the points $C(0, 0)$ and $D(4, 4)$. It is also given that $\overleftrightarrow{AB}$ is parallel to $\overleftrightarrow{CD}$. Let m_1 = the slope of $\overleftrightarrow{AB}$ and m_2 = the slope of $\overleftrightarrow{CD}$. The slopes of these lines are calculated below.

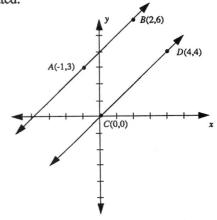

$$m_1 = \frac{y_2 - y_1}{x_2 - x_1} = \frac{6-3}{2-(-1)} = \frac{3}{3} = 1$$

$$m_2 = \frac{y_2 - y_1}{x_2 - x_1} = \frac{4-0}{4-0} = \frac{4}{4} = 1$$

Note that $m_1 = m_2$. The two lines, which were given to be parallel, have equal slopes.

9.8.4 The Slope of Perpendicular Lines

If two lines (neither of which is vertical) are perpendicular, then it can be proven mathematically that the product of their slopes is -1. That is, the slope of one line is the negative reciprocal of the slope of the second line. Using mathematical notation, if $l_1 \perp l_2$ then $m_2 = \frac{-1}{m_1}$. (Note: l_1 is read as *line 1*, m_1 is read as *the slope of line 1*, and $\perp$ is read as *is perpendicular to*.) To prove this statement, a working knowledge of the Pythagorean Theorem (Chapter 10) is required. As a result, the statement is only demonstrated.

Consider the figure at the right, which is the same figure as that given in the previous subsection, but now includes the graph of $\overleftrightarrow{EF}$ containing the points $E(3, 1)$ and $F(-2, 6)$. It is also given that $\overleftrightarrow{AB}$ is perpendicular to $\overleftrightarrow{EF}$. Since $\overleftrightarrow{AB}$ and $\overleftrightarrow{EF}$ are nonvertical lines that are perpendicular to each other, they should have negative reciprocal slopes. In the previous illustration, it was found that the slope of $\overleftrightarrow{AB}$ was 1. Thus the slope of $\overleftrightarrow{EF}$ should be -1. Let m_3 = the slope of $\overleftrightarrow{EF}$. The slope of $\overleftrightarrow{EF}$ is calculated below.

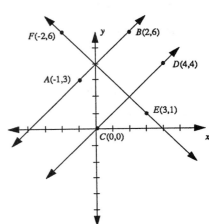

AN INTRODUCTION TO ALGEBRA

$$m_3 = \frac{y_2 - y_1}{x_2 - x_1} = \frac{6-1}{-2 - (-3)} = \frac{5}{-5} = -1$$

Since $1 \times (-1) = -1$, the slopes of $\overleftrightarrow{AB}$ and $\overleftrightarrow{EF}$ are negative reciprocals.

The converse of this statement is also true. That is, if two lines have negative reciprocal slopes, then the lines are perpendicular. Consequently, *two nonvertical lines are perpendicular if and only if their slopes are negative reciprocals*. It should be noted that this statement does not apply to vertical or horizontal lines. The slope of vertical lines is undefined (you cannot take the negative reciprocal of something that is not defined), and the slope of horizontal lines is zero (zero does not have a negative reciprocal). Any vertical line is perpendicular to any horizontal line, however.

Example 1 (page 504, #3)

Find the slope of the line containing the points $E(1, -2)$ and $F(2, 3)$.

Solution

Let $x_1 = 1$, $y_1 = -2$, $x_2 = 2$, and $y_2 = 3$.

$$m = \frac{y_2 - y_1}{x_2 - x_1} = \frac{3-(-2)}{2-1} = \frac{5}{1} = 5$$

Example 2 (page 504, #7)

Find the slope of the line containing $M(-2, 5)$ and $N(-2, 3)$.

Solution

Let $x_1 = -2$, $y_1 = 5$, $x_2 = -2$, and $y_2 = 3$.

$$m = \frac{y_2 - y_1}{x_2 - x_1} = \frac{3-5}{-2-(-2)} = \frac{-2}{0}, \text{ which is undefined}$$

Example 3 (page 504, #9)

Find the slope of the line containing $Q(3, -3)$ and $R(-3, -3)$.

Solution

Let $x_1 = 3$, $y_1 = -3$, $x_2 = -3$, and $y_2 = -3$.

$$m = \frac{y_2 - y_1}{x_2 - x_1} = \frac{-3-(-3)}{-3-3} = \frac{0}{-6} = 0$$

Example 4 (page 504, #15c)

Find the slope of the line that is perpendicular to a line whose slope is $\frac{3}{4}$.

Solution

Perpendicular lines must have negative reciprocal slopes. Since $\frac{3}{4} \times (-\frac{4}{3}) = -1$, the slope is $\frac{-4}{3}$.

210 STUDENT STUDY GUIDE

Example 5 (page 505, #17)

Show that the line that passes through $A(1, 3)$ and $B(5, 6)$ is parallel to the line that passes through the points $C(5, 1)$ and $D(9, 4)$.

Solution

- The slope of $\overleftrightarrow{AB}$ is equal to
$$\frac{6 - 3}{5 - 1} = \frac{3}{4}$$

- The slope of $\overleftrightarrow{CD}$ is equal to
$$\frac{4 - 1}{9 - 5} = \frac{3}{4}$$

Since both lines have the same slope they must be parallel.

SUPPLEMENTARY EXERCISE 9.8

In problems 1 through 12 find the slope of the line containing the indicated pairs of points.

1. $A(1, 6)$ and $B(0, -3)$
2. $A(-4, 5)$ and $B(-7, -8)$
3. $A(2, -1)$ and $B(2, 6)$
4. $A(-9, 3)$ and $B(2, 3)$
5. $A(10, -4)$ and $B(5, 6)$
6. $A(8, 0)$ and $B(11, -4)$
7. $A(0, 0)$ and $B(4, 2)$
8. $A(3, 2)$ and $B(0, 5)$
9. $A(-1, -1)$ and $B(-2, -2)$
10. $A(5, 3)$ and $B(5, -5)$
11. $A(4, -3)$ and $B(-2, 6)$
12. $A(-6, -3)$ and $B(2, -5)$

9.9 THE EQUATION OF A STRAIGHT LINE (OPTIONAL)

In this section our goal is to write the equation of a line given some specific characteristics about the line. These characteristics include the slope of the line and specific points that make up the line. Two formulas that are used to write the equation of a line are the point-slope formula and the slope intercept formula. Each formula is discussed in the subsections below.

9.9.1 The Point-Slope Formula

The graph at the right shows a line that contains the points $P(x, y)$ and $P_1(x_1, y_1)$. Note that $x \neq x_1$. Since the slope of a line is the ratio of the change in y to the change in x, the slope of the line in the graph at the right is

$$m = \frac{y - y_1}{x - x_1}$$

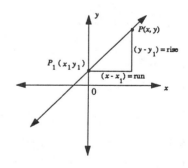

Since $x \neq x_1$, it follows that $x - x_1 \neq 0$ and hence the denominator of the fraction above is not zero. If we multiply both sides of the equation by the expression $(x - x_1)$ we get

$$y - y_1 = m(x - x_1)$$

This resulting equation is called the *point-slope formula* and is used to find the equation of a line given the line's slope and a point on the line. In this formula m represents the slope of the line, and x_1 and y_1 represent the respective x and y values of the point that lies on the line. The use of the formula is demonstrated in Examples 1 and 2 below.

9.9.2 The Slope-Intercept Formula

The graph at the right shows a line that contains the points $P(x, y)$ and $A(0, b)$. Note that point A is the y-intercept (see section 9.7), and $x \neq 0$. The slope of this line is

$$m = \frac{y - b}{x - 0} = \frac{y - b}{x}$$

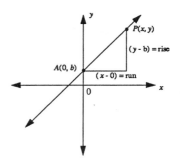

Since $x \neq 0$, it follows that the denominator of the fraction above is not zero. This equation is solved for y by multiplying both sides of the equation by x and then adding b to each side.

$$\begin{aligned} m &= \frac{y-b}{x} &&\text{(Multiply both sides by } x.) \\ mx &= y - b &&\text{(Add } b \text{ to both sides.)} \\ mx + b &= y &&\text{(Rearrange the equation.)} \\ y &= mx + b \end{aligned}$$

This resulting equation, $y = mx + b$, is called the *slope-intercept formula* and is used to find the equation of a line given the line's slope, m, and it's y-intercept $(0, b)$. The use of the formula is demonstrated in Examples 3 and 4 below. (**Note:** The slope-intercept formula also can be derived from the point-slope formula by using the point $(0, b)$ and the slope m.)

The slope-intercept form can be used to graph a line. Consider the line whose equation is $y = \frac{-3}{4}x + 5$. Since the equation is in the form $y = mx + b$, we can identify the slope, m, and the y-intercept, $(0, b)$: $m = \frac{-3}{4}$ and the y-intercept is $(0, 5)$. To sketch the graph of this line we first plot the point $(0, 5)$. (Refer to the figure at the right.) Using the fact that slope is equal to $\frac{rise}{run}$, beginning at $(0, 5)$ we move three units down (-3) and four units right (4). This leaves us at the point $(4, 2)$. We sketch this point and draw the line that contains these two points.

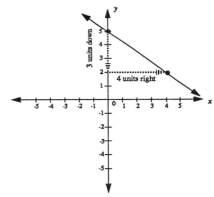

Example 1 (page 516, #5)

Write the equation of the line that passes through $E(-3, 0)$ and has slope $m = \frac{2}{3}$.

Solution

Since a point on the line and the slope are given, use the point-slope formula.

$$\begin{aligned} y - y_1 &= m(x - x_1) \\ y - 0 &= \frac{2}{3}(x - (-3)) \\ y &= \frac{2}{3}(x + 3) \\ y &= \frac{2}{3}x + 2 \end{aligned}$$

Example 2 (page 516, #15)

Write the equation of the line that passes through $E(-2, -3)$ and $F(-4, -5)$.

Solution

In order to write the equation of a line using either of the slope formulas, the slope must be known. Since the slope is not given in this example, we calculate it. Once the slope is determined, we then use the point-slope formula to write the equation of the line.

- Find the slope of $\overleftrightarrow{EF}$.

$$m = \frac{y_2 - y_1}{x_2 - x_1} = \frac{-5 - (-3)}{-4 - (-2)} = \frac{-2}{-2} = 1.$$

- Use point $E(-2, -3)$ and $m = 1$ to write the equation of $\overleftrightarrow{EF}$.

$$\begin{aligned} y - y_1 &= m(x - x_1) \\ y - (-3) &= 1(x - (-2)) \\ y + 3 &= x + 2 \\ y &= x - 1 \end{aligned}$$

Example 3 (page 516, #23)

Find the slope and y-intercept of the line whose equation is $y = 2x - 1$ and then graph the line.

Solution

Since $y = 2x - 1$ is of the form $y = mx + b$, the slope and y-intercept are readily identified: $m = 2$ and the y-intercept is $(0, -1)$.

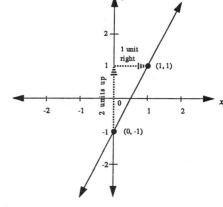

To sketch the graph of this line, first plot the point $(0, -1)$. Using the fact that slope is equal to $\frac{rise}{run}$, beginning at $(0, -1)$ move two units up (2) and one unit right (1). This leaves us at the point $(1, 1)$. Sketch the point $(1, 1)$ and draw the line that contains these two points.

Example 4 (page 517, #29)

Find the slope and y-intercept of the line whose equation is $4x - 3y + 3 = 0$ and graph the line.

Solution

The given equation is not in slope-intercept form. Using the tools of algebra, this equation is transformed into slope-intercept form. Once the equation is expressed in the proper form, the slope and y-intercept are identified, and a sketch of the graph of the line is obtained.

Continued on next page ...

- Rewrite the equation is slope-intercept form.

$$4x - 3y + 3 = 0$$
$$4x + 3 = 3y$$
$$3y = 4x + 3$$
$$y = \frac{4}{3}x + 1$$

$m = \frac{4}{3}$ and the y-intercept is $(0, 1)$.

- Sketch the graph of the line.

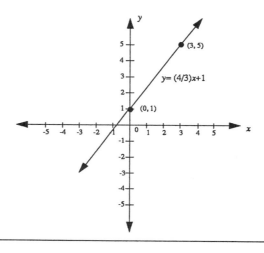

Example 5 (page 517, #33)

Find the equation of the line passing through $(1, 1)$ and parallel to the line whose equation is $2x + 3y = 1$.

Solution

The slope of the line whose equation we seek is equal to the slope of the given line ($2x + 3y = 1$) because nonvertical parallel lines have equal slopes. Once this slope is identified, the point-slope formula is used to write the equation of the line that is parallel to the given line and passes through the point $(1, 1)$.

- Find the slope of the given line.

$$2x + 3y = 1$$
$$3y = -2x + 1$$
$$y = \frac{-2}{3}x + \frac{1}{3}$$

The slope of the line whose equation we seek is $\frac{-2}{3}$.

Continued on next page ...

- Use the point-slope form to find the equation of the line.

$$y - y_1 = m(x - x_1)$$
$$y - 1 = \frac{-2}{3}(x - 1)$$
$$y - 1 = \frac{-2x}{3} + \frac{2}{3}$$
$$y = \frac{-2x}{3} + \frac{2}{3} + \frac{3}{3}$$
$$y = \frac{-2}{3}x + \frac{5}{3}$$

Example 6 (page 517, #35)

Find the equation of the line passing through $(-2, 0)$ and perpendicular to the line whose equation is $2x + 3y = 1$.

Solution

Note that the given equation is the same equation given in the previous example. Thus the slope is $m = \frac{-2}{3}$. In this problem, however, the line whose equation we seek is perpendicular to the given line. If two lines, neither of which are vertical, are perpendicular, then their slopes are negative reciprocals. Since $\frac{-2}{3} \times \frac{3}{2} = -1$, the negative reciprocal of $\frac{-2}{3}$ is $\frac{3}{2}$. Therefore, $\frac{3}{2}$ is the slope of the line whose equation we seek. We now use the point-slope form to find the equation of the line.

$$y - y_1 = m(x - x_1)$$
$$y - 0 = \frac{3}{2}(x - (-2))$$
$$y = \frac{3}{2}(x + 2)$$
$$y = \frac{3}{2}x + \frac{6}{2}$$
$$y = \frac{3}{2}x + 3$$

SUPPLEMENTARY EXERCISE 9.9

For problems 1 through 9 write the equation of the line that passes through the indicated point and has the indicated slope. Express your answers in slope-intercept form.

1. $A(-1, 5), m = \frac{3}{4}$
2. $B(-6, -4), m = 1$
3. $C(0, 9), m = \frac{-1}{2}$
4. $D(2, 0), m = -3$
5. $E(3, -8), m = \frac{4}{9}$
6. $F(2, 3), m = 2$
7. $G(-1, -3), m = -4$
8. $H(3, 0), m = \frac{1}{2}$
9. $I(-2, 4), m = \frac{-5}{4}$

Continued on next page ...

For problems 10 through 18, write the equation of the line that passes through the indicated points. Express your answers in slope-intercept form.

10. $A\,(7, 10)$ and $B\,(1, 5)$
11. $C\,(-3, -1)$ and $D\,(4, -3)$
12. $E\,(0, -4)$ and $F\,(8, 0)$
13. $G\,(-5, 6)$ and $H\,(-3, 10)$
14. $I\,(5, 8)$ and $J\,(7, 2)$
15. $K\,(2, 4)$ and $L\,(5, 1)$
16. $M\,(0, 0)$ and $N\,(2, -3)$
17. $O\,(-2, -3)$ and $P\,(2, -1)$
18. $Q\,(-5, -2)$ and $R\,(-3, -1)$

In problems 19 through 24 find the slope and y-intercept of the line whose equation is given and the sketch the graph of the line.

19. $y = 2x + 1$
20. $y = x$
21. $y = -\frac{2}{3}x + 4$
22. $x - y = 3$
23. $2x + 3y = 8$
24. $3x - 4y = 16$

In problems 25 through 30 write an equation of a line that is:

25. parallel to the line whose equation is $y = 2x + 3$ and whose y-intercept is $(0, 1)$.
26. parallel to the line whose equation is $x - y = 2$ and has the same y-intercept as $3x + y = -2$.
27. parallel to the line whose equation is $3y - 2x = 2$ and passes through the point $(-3, -1)$.
28. perpendicular to the line whose equation is $y = \frac{3}{5}x + 2$ and whose y-intercept is $(0, -3)$.
29. perpendicular to the line whose equation is $y = -3x + 4$ and passes through the origin.
30. perpendicular to the line whose equation is $4x - y = 5$ and passes through the point $(6, -1)$.

9.10 GRAPHING $y = ax^2 + bx + c$

An equation of the form $y = ax^2 + bx + c$ is called a **quadratic equation**. Quadratic equations are graphed by generating ordered pairs that satisfy the given equation. The graph of a quadratic equation is called a **parabola**, which is a u-shaped curve.

To Graph a Quadratic Equation of the Form $y = ax^2 + bx + c$

1. Determine the coordinates of the vertex (i.e., the turning point) of the curve.

    a. Find the x coordinate by using the equation $x = \frac{-b}{2a}$.

    b. Substitute this numerical value for x into the given equation and solve for y.

    The resulting ordered pair is the coordinate of the vertex.

2. Choose two or three numbers for x on either side of the vertex and find the corresponding numerical values for y.

3. Plot on the same set of axes all the points found.

4. Trace the curve through these points.

Example 1 (page 521, #1)

Graph $y = x^2 - 1$

Solution

Note that the equation $y = x^2 - 1$ is of the form $y = ax^2 + bx + c$, where $a = 1$, $b = 0$, and $c = -1$.

- Find the coordinates of the vertex.

$$
\begin{array}{l|l}
\text{Find } x & \text{Find } y \\
\text{Let } x = \dfrac{-b}{2a} & \text{If } x = 0 \\
x = \dfrac{-0}{2(1)} & \text{then } y = x^2 - 1 \\
x = \dfrac{0}{2} & y = (0)^2 - 1 \\
x = 0 & y = 0 - 1 \\
& y = -1
\end{array}
$$

The vertex is $(0, -1)$.

- Choose three points on each side of the vertex, plot these points and trace the curve through them.

x	y	
-3	8	
-2	3	
-1	0	
0	-1	* Vertex
1	0	
2	3	
3	8	

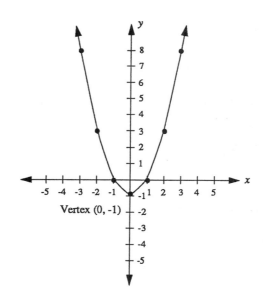

Example 2 (page 521, #11)

Graph $y = -x^2 + 6x - 9$.

Solution

The equation $y = -x^2 + 6x - 9$ is a quadratic equation with $a = -1$, $b = 6$, and $c = -9$. Note that the curve opens down since a is negative.

- Find the coordinates of the vertex.

$$\begin{array}{ll} \text{Find } x & \text{Find } y \\ \text{Let } x = \dfrac{-b}{2a} & \text{If } x = 3 \\ x = \dfrac{-6}{2(-1)} & \text{then } y = -x^2 + 6x - 9 \\ x = \dfrac{-6}{-2} & y = -(3)^2 + 6(3) - 9 \\ x = 3 & y = -9 + 18 - 9 \\ & y = 0 \end{array}$$

The vertex is (3, 0).

- Choose three points on each side of the vertex, plot these points and trace the curve through them.

x	y	
0	-9	
1	-4	
2	-1	
3	0	* Vertex
4	-1	
5	-4	
6	-9	

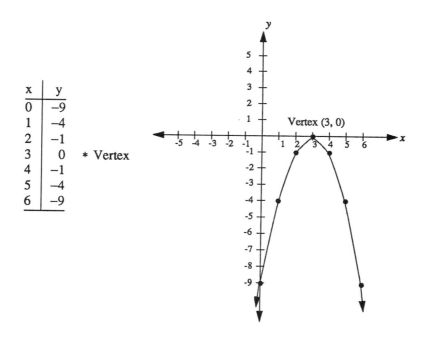

218 STUDENT STUDY GUIDE

SUPPLEMENTARY EXERCISE 9.10

In problems 1 through 9 graph each equation.

1. $y = x^2 + 1$
2. $y = x^2 - 5$
3. $y = x^2 - 2x$
4. $y = x^2 + 6x$
5. $y = x^2 - 4x - 5$
6. $y = x^2 - 6x + 5$
7. $y = -x^2 - 6x - 5$
2. $y = -x^2 + 2x - 3$
9. $y = x^2 + 2x + 2$

9.11 INEQUALITIES IN TWO VARIABLES

Linear inequalities in two variables are solved graphically.

<u>To Solve a Linear Inequality in Two Variables</u>

1. Treat the inequality as an equation and graph this equation. The graph that results is a line that is called the *boundary line*. The boundary line is graphed using

    a. a broken (i.e., dashed) line if the inequality involves a less than (<) or a greater than (>) sign;

    b. a solid line if the inequality involves a less than or equal to sign (≤), or a greater than or equal to sign (≥).

2. Choose a test point either above or below the boundary line. Be careful not to select a point that lies on the boundary line.

3. Substitute the x and y coordinates of the test point into the inequality and evaluate the inequality. If the result is a true statement then shade the region that contains the test point; otherwise shade the region that does not contain the test point.

Example 1 (page 528, #1)

Graph $x + y > 5$.

Solution

- Graph the boundary line whose equation is $x + y = 5$. The line is dashed since the inequality does not involve an equality.

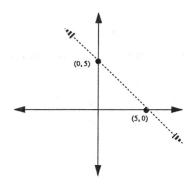

Continued on next page ...

- A test point of (0, 0), which is below the boundary line, is used to evaluate the inequality.

$$x + y > 5$$
$$(0) + (0) > 5$$
$$0 > 5 \quad \text{(False)}$$

- Since the test point yields a false statement the region below the boundary line is incorrect. Thus the region above the boundary line gets shaded.

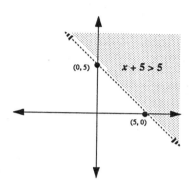

Example 2 (page 528, #13)

Graph $-3x + 5y \leq 15$.

Solution

- Graph the boundary line whose equation is $-3x + 5y = 15$. Note that the line is solid.

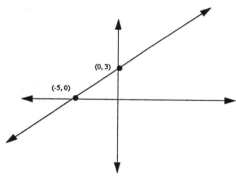

- Select a test point (we choose (0, 0)), and evaluate the inequality.

$$-3x + 5y \leq 15$$
$$-3(0) + 5(0) \leq 15$$
$$0 + 0 \leq 15$$
$$0 \leq 15 \quad \text{(True)}$$

Continued on next page ...

- Since the test point yields a true statement, the region that contains the test point, namely the region below the boundary line, is the correct region.

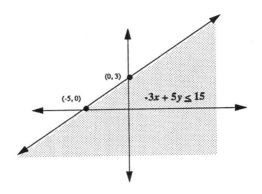

SUPPLEMENTARY EXERCISE 9.11

In problems 1 through 9 graph each inequality.

1. $x + y > 3$ 2. $x - y < 2$ 3. $3x - y \geq 0$ 4. $2x + y \leq -5$ 5. $2x - 3y > -12$
6. $5x - 4y \geq 20$ 7. $-x + y > 8$ 8. $-2x - 3y \geq -6$ 9. $-3x + 4y < 24$

9.12 LINEAR PROGRAMMING

The basic goal of a linear programming problem is to either maximize or minimize some entity, subject to certain restrictions. The quantity for which an optimal solution is desired is represented by a mathematical equation and is called the **objective function**. The restrictions of the problem are represented by linear inequalities and are called the **constraints** of the problem. When graphing these constraints on the same set of axes a region known as the **feasible region** is produced.

The underlying theory of all linear programming problems states that if there is a unique solution that maximizes or minimizes an objective function, then the solution must occur at one of the corners of the feasible region. Thus each corner is considered a **feasible solution** to the problem. The corner that will produce a maximum or minimum value is called the **optimal feasible solution**.

To Solve a Linear Programming Problem

1. Determine the objective function and constraints and translate them into mathematical statements. (Recall that the objective function is an equation and the constraints are inequalities.)

2. Graph each inequality on the same set of axes and determine the feasible region. This will be the region formed by the intersection of the graphs of the inequalities.

3. Find the corner points of the feasible region and substitute them into the objective function.

    a. If the objective function is to be maximized then the optimal solution occurs at the corner point that yields the largest numerical value.

    b. If the objective function is to be minimized then the optimal solution is the corner point that yields the smallest numerical value.

    (**Note:** It is possible that the feasible region could be empty. If this is indeed the case then there is no optimal solution.)

Example 1 (page 534, #3)

Find the maximum value of $P = 4x + 3y$ under the following conditions.
$x \geq 0, \ y \geq 0, \ 2x + y \leq 12, \ x + y \leq 7$.

Solution

- The objective function is $P = 4x + 3y$
- The constraints are $x \geq 0, y \geq 0, 2x + y \leq 12,$ and $x + y \leq 7$
- Graphing the constraints produces the following feasible region and corner points of $(0,0), (0,7), (6,0),$ and $(5,2)$.

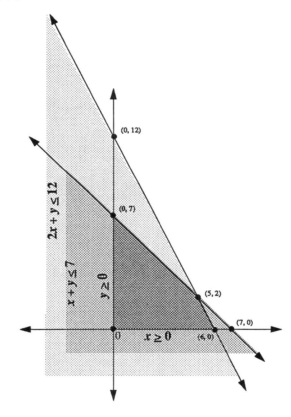

- Substituting the x and y values for each corner point into the objective function yields a maximum of 26 when x is equal to 5 and y is equal to 2. (The reader should confirm this.)

Example 2 (page 534, #7)

Frank Sloane raises pheasants and partridges and has room for at most 100 birds. It costs him $2 to raise a pheasant and $3 to raise a partridge, and he has $240 to cover these costs. If he can make a profit of $7 on each pheasant and $8 on each partridge, how many of each bird should he raise in order to maximize his profit?

Continued on next page ...

Solution

Let x = the number of partridges Frank should raise
and y = the number of partridges he should raise

- From the problem it is known that Frank has room for at most 100 birds. That is, the total number of pheasants and partridges cannot exceed 100. This is a constraint. Thus,

$$x + y \leq 100$$

- Since it costs Frank \$2 to raise each pheasant and \$3 to raise each partridge,

$2x$ = the amount of money needed to raise pheasants
and $3y$ = the amount of money needed to raise partridges

- Frank has \$240 to cover his expenses for raising these birds. This implies that the total amount of money he will need to raise pheasants and partridges cannot exceed \$240. This is another constraint.

$$2x + 3y \leq 240$$

- Frank would like to make a profit of \$7 on each pheasant and \$8 on each partridge. This is the objective function.

$$P = 7x + 8y$$

- Graphing the constraints produces the following feasible region and corner points of (0, 0), (0, 80), (100, 0), and (60, 40). ($x \geq 0$ and $y \geq 0$ are built into the problem.)

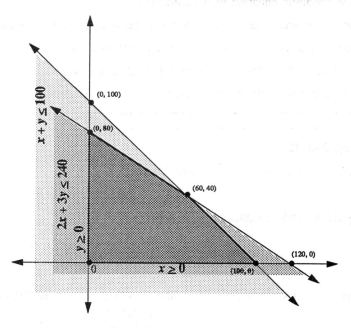

- Substituting the x and y values for the corner points into the objective function yields a maximum profit of \$740 when x (the number of pheasants) is equal to 60, and y (the number of partridges) is equal to 40. (The reader should confirm this.)

SUPPLEMENTARY EXERCISE 9.12

1. Find the maximum value of $P = 2x + 5y$, subject to the following conditions:
   $2x + y \leq 16, \quad x + y \leq 10, \quad x \geq 0, \quad y \geq 0$.

2. Find the maximum value of $P = 4x + 7y$, subject to the following conditions:
   $x + 2y \leq 10, \quad 2x + y \leq 8, \quad x \geq 0, \quad y \geq 0$.

3. Find the maximum value of $P = 5x + 3y$, subject to the following conditions:
   $3x + 10y \leq 150, \quad 3x + 2y \leq 54, \quad x \geq 0, \quad y \geq 0$.

9.13 QUADRATIC EQUATIONS (OPTIONAL)

A **quadratic equation**, or a **second degree equation** in one variable, is an equation that can be written in the **standard form** $ax^2 + bx + c = 0$, where a, b, and c are real numbers and $a \neq 0$. Any quadratic equation that is expressed in standard form can be solved using the quadratic formula. That is, if $ax^2 + bx + c = 0$, then

$$x = \frac{-b \pm \sqrt{b^2 - 4ac}}{2a}$$

Note that this equation yields at most two solutions or **roots**. If we let r_1 and r_2 represent the solutions of a quadratic equation, then

$$r_1 = \frac{-b + \sqrt{b^2 - 4ac}}{2a} \quad \text{and} \quad r_2 = \frac{-b - \sqrt{b^2 - 4ac}}{2a}$$

The following procedure is given as a guide to solving quadratic equations by the quadratic formula.

To Solve a Quadratic Equation by the Quadratic Formula

1. Express the quadratic equation in standard form.
2. Identify the numerical values of a, b, and c by comparing with $ax^2 + bx + c = 0$.
3. Substitute the values of a, b, and c in the quadratic formula.
4. Solve the equation for x by evaluating the formula from step 3.
5. Check the solution(s) in the original equation.

Example 1 (page 540, #1)

Solve $2x^2 - 10x + 12 = 0$ using the quadratic formula.

Solution

- Identify the numerical values for a, b, and c.

$$\underset{a=2}{2x^2} - \underset{b=-10}{10x} + \underset{c=12}{12} = 0$$

- Substitute the values for a, b, and c directly into the quadratic formula.

$$x = \frac{-b \pm \sqrt{b^2 - 4ac}}{2a}$$

$$x = \frac{-(-10) \pm \sqrt{(-10)^2 - 4(2)(12)}}{2(2)}$$

Continued on next page ...

- Evaluate the formula.

$$x = \frac{-(-10) \pm \sqrt{(-10)^2 - 4(2)(12)}}{2(2)}$$

$$x = \frac{10 \pm \sqrt{100 - 96}}{4}$$

$$x = \frac{10 \pm \sqrt{4}}{4}$$

$$x = \frac{10 \pm 2}{4}$$

$$x = \frac{10 + 2}{4} \quad \bigg| \quad x = \frac{10 - 2}{4}$$

$$x = \frac{12}{4} \quad \bigg| \quad x = \frac{8}{4}$$

$$x = 3 \quad \bigg| \quad x = 2$$

The solution set is {3, 2}. (The check is left to the reader.)

Example 2 (page 540, #9)

Solve $x^2 - 3x - 10 = 0$ using the quadratic formula.

Solution

- Identify the numerical values for a, b, and c.

$$\underset{a=1}{x^2} \ - \ \underset{b=-3}{3x} \ - \ \underset{c=-10}{10} \ = \ 0$$

- Substitute the values for a, b, and c directly into the quadratic formula.

$$x = \frac{-b \pm \sqrt{b^2 - 4ac}}{2a}$$

$$x = \frac{-(-3) \pm \sqrt{(-3)^2 - 4(1)(-10)}}{2(1)}$$

- Evaluate the formula.

$$x = \frac{-(-3) \pm \sqrt{(-3)^2 - 4(1)(-10)}}{2(1)}$$

$$x = \frac{3 \pm \sqrt{9 + 40}}{2}$$

$$x = \frac{3 \pm \sqrt{49}}{2}$$

$$x = \frac{3 \pm 7}{2}$$

Continued on next page ...

$$x = \frac{3+7}{2} \quad \bigg| \quad x = \frac{3-7}{2}$$

$$x = \frac{10}{2} \quad \bigg| \quad x = \frac{-4}{2}$$

$$x = 5 \quad \bigg| \quad x = -2$$

The solution set is $\{5, -2\}$. (The check is left to the reader.)

Example 3 (page 540, #15)

Solve $6x^2 = 1 - x$ using the quadratic formula.

Solution

- First express the given equation in standard form.

$$6x^2 = 1 - x$$
$$6x^2 + x - 1 = 0$$

- Now identify the numerical values for a, b, and c.

$$\underset{a=6}{6x^2} + \underset{b=1}{x} - \underset{c=-1}{1} = 0$$

- Substitute the values for a, b, and c directly into the quadratic formula.

$$x = \frac{-b \pm \sqrt{b^2 - 4ac}}{2a}$$

$$x = \frac{-(1) \pm \sqrt{(1)^2 - 4(6)(-1)}}{2(6)}$$

- Evaluate the formula.

$$x = \frac{-(1) \pm \sqrt{(1)^2 - 4(6)(-1)}}{2(6)}$$

$$x = \frac{-1 \pm \sqrt{1 - (-24)}}{12}$$

$$x = \frac{-1 \pm \sqrt{25}}{12}$$

$$x = \frac{-1 \pm 5}{12}$$

$$x = \frac{-1 + 5}{12} \quad \bigg| \quad x = \frac{-1 - 5}{12}$$

$$x = \frac{4}{12} \quad \bigg| \quad x = \frac{-6}{12}$$

$$x = \frac{1}{3} \quad \bigg| \quad x = \frac{-1}{2}$$

The solution set is $\{-\frac{1}{2}, \frac{1}{3}\}$. (The check is left to the reader.)

Example 4 (page 540, #21)

Solve $x^2 - 2 = 2x$ using the quadratic formula.

Solution

$$x^2 - 2 = 2x \text{ is equal to } x^2 - 2x - 2 = 0$$

$$x = \frac{-b \pm \sqrt{b^2 - 4ac}}{2a}$$

$$x = \frac{-(-2) \pm \sqrt{(-2)^2 - 4(1)(-2)}}{2(1)}$$

$$x = \frac{-(-2) \pm \sqrt{(-2)^2 - 4(1)(-2)}}{2(1)}$$

$$x = \frac{2 \pm \sqrt{4 - (-8)}}{2}$$

$$x = \frac{2 \pm \sqrt{12}}{2}$$

$$x = \frac{2 \pm \sqrt{(4)(3)}}{2}$$

$$x = \frac{2 \pm 2\sqrt{3}}{2}$$

$$x = 1 \pm \sqrt{3}$$

The solution set is $\{1 + \sqrt{3},\ 1 - \sqrt{3}\}$. (The check is left to the reader.)

SUPPLEMENTARY EXERCISE 9.13

In problems 1 through 15 solve the given quadratic equation by using the quadratic formula.

1. $x^2 - 4x + 4 = 0$
2. $2x^2 - x - 6 = 0$
3. $x^2 - 2x - 15 = 0$
4. $x^2 - 2x - 35 = 0$
5. $3x^2 + 5x + 2 = 0$
6. $4x^2 + 7x - 2 = 0$
7. $x^2 - 4 = 0$
8. $4x^2 + 8x + 3 = 0$
9. $x^2 - 2x = 3$
10. $x^2 + 3x = 4$
11. $x^2 = 5x - 4$
12. $2x^2 + 9x = -10$
13. $x^2 = 6x$
14. $x^2 + 4x = 21$
15. $3x^2 + 9x = 0$

9.14 CHAPTER 9 TEST

In problems 1 through 12 solve the given equation for x.

1. $3x = -18$
2. $\frac{x}{3} = 9$
3. $x + 6 = -8$
4. $x - 9 = -18$
5. $3x - 6 = 15$
6. $\frac{2x}{7} + 8 = 20$
7. $6x - 4x = 18$
8. $-20 = 3x - 8x$
9. $5x + 16 = 21x$
10. $3x - 6 - 5x = 2x + 6$
11. $2(x - 4) - 2x = 3x + 8 + x$
12. $5(2x - 3) = 4(x + 3) - 3$

Continued on next page ...

In problems 13 and 14 graph the solution on a number line for the given inequality.

13. $x + 6 \leq 3$ 14. $5 < x \leq 8$

15. Julia's age is four less than five times Jack's age. If three times Jack's age is added to Julia's age, the sum is equal to twice Julia's age. Find the ages of Julia and Jack.

16. A change machine contains dimes and quarters. The number of quarters is 30 less than the number of dimes. If the total amount of money in the machine is $31.00, how many coins of each kind are there?

In problems 17 through 20 graph the given equation or inequality on the Cartesian Plane.

17. $3x + 2y = 18$ 18. $2x - 3y = 12$ 19. $y = x^2 - 7$ 20. $2x - y > 3$

21. Find the slope of the line that passes through the points A(3, −7) and B(−5, 0).

22. Write the equation of the line that passes through the point A(8, 6) and has a slope of $m = -2$.

23. Write the equation of the line that passes through the points A(−4, 2) and B(−3, −1).

24. Find the slope and y-intercept of the line whose equation is $5x + 2y = -6$.

25. Find the equation of the line perpendicular to $x + 2y = -5$ and passes through the point (0, −3).

26. Given A(−2, 4) and B(7, 4), find the equation of the line parallel to $\overleftrightarrow{AB}$ and passes through the point (−6, 1).

In problems 27 through 29 solve the given quadratic equation.

27. $x^2 - 8x + 7 = 0$ 28. $3x^2 - 10x + 3 = 0$ 29. $5x^2 = -1 - 6x$

10. AN INTRODUCTION TO GEOMETRY

10.1 INTRODUCTION

Geometry is one of the oldest branches of mathematics. In a general sense, geometry pertains to the concepts of form, shape, size, and position. In this chapter we address all of these general concepts. Our discussion includes the location of points and lines; the measurement of angles; the shapes and forms of polygons and solids; perimeter, area, and volume formulas for various polygons and solids; and the ideas of congruency and similarity.

10.2 POINTS AND LINES

10.2.1 Point

In geometry, a **point** is considered to be an undefined term. From a physical perspective, a geometric point has no length, width, or thickness. A point also has no size; it has only position. A point is represented as a dot, and is named by a capital letter. For example, *point P* is denoted • P.

10.2.2 Line

In geometry, a **line** also is considered to be an undefined term. A geometric line is considered to consist of a set of points. From a physical perspective, a line does not have any width or thickness, but it does have length. It is assumed that all lines are straight lines (unless otherwise noted), and that all lines extend indefinitely in opposite directions. A line is named by two points that are on the line. For example, the line shown below is named *line AB*, and is denoted $\overleftrightarrow{AB}$.

A line also is named by placing a lowercase letter next to it. The line shown below is named *line m*.

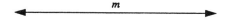

10.2.3 Line Segment

A **line segment** is a part of a line that consists of two points on the line (called endpoints) and all the points on the line between the endpoints. A line segment is named by its endpoints. For example, the line segment below is named *line segment AB*, and is denoted $\overline{AB}$. A line segment cannot be extended beyond its endpoints.

10.2.4 Half-line

Any point on a line separates the line into three distinct parts: the point itself, and two **half-lines**, one on each side of the point. An example of a half-line is shown below.

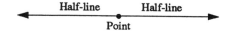

In the half-line figure, note that point P is not a part of either half-line. As a result, a half-line consists of an open endpoint followed by a set of points that extend indefinitely in one direction. A half-line is named by its open endpoint and a point on the half-line. For example, the half-line below is named *half-line PQ*, and is denoted $\overrightarrow{PQ}$.

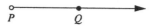

10.2.5 Ray

A **ray** is a part of a line that consists of an endpoint and the half-line that extends from the endpoint in one direction. A ray is named by its endpoint and a point on the half-line through which the ray extends. For example, the ray shown below is named *ray AB*, and is denoted $\overrightarrow{AB}$.

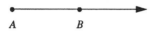

10.2.6 Angle

The union of two distinct rays that have a common endpoint form a figure called an **angle**. The rays are called the **sides** of the angle and the common endpoint is called the **vertex** of the angle. For example, in the figure below, ray AC ($\overrightarrow{AC}$) and ray AB ($\overrightarrow{AB}$) form an angle with sides AC and AB, and vertex A.

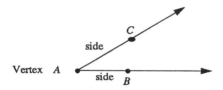

An angle is named by using three capital letters. The first and third letters name the points on either side of the angle, and the middle letter names the vertex. The symbol used for an angle is $\angle$. Thus the angle in the figure above is named *angle CAB* (denoted $\angle CAB$), or *angle BAC* (denoted $\angle BAC$).

10.2.7 Adjacent Angles

Two angles that have both a common side between them and a common vertex are called **adjacent angles**. For example, in the figure below $\angle PQR$ and $\angle RQS$ are adjacent angles since they have a common side between them ($\overrightarrow{QR}$) and a common vertex (Q).

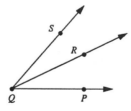

10.2.8 Vertical Angles

When two straight lines intersect they form **vertical angles**, which are two nonadjacent (or opposite) angles. For example, the figure below shows the intersection of $\overleftrightarrow{PQ}$ and $\overleftrightarrow{RT}$. As a result of the intersection of these two lines, $\angle RSP$ and $\angle QST$ are vertical angles. $\angle RSQ$ and $\angle PST$ also are vertical angles.

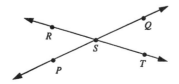

10.2.9 Straight Angle

If the sides of an angle form a straight line then the angle is referred to as a **straight angle**. For example, in the figure below, $\angle PQR$, with sides $\overrightarrow{QP}$ and $\overrightarrow{QR}$ and vertex Q, is a straight angle.

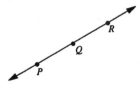

10.2.10 Perpendicular Lines

Two lines that intersect and form two *congruent* adjacent angles are called **perpendicular lines**. (Note: Two angles are congruent if they coincide throughout.) In the figure below, $\overleftrightarrow{AB}$ and $\overleftrightarrow{CD}$ are perpendicular. This is denoted $\overleftrightarrow{AB} \perp \overleftrightarrow{CD}$. (Note: The symbol $\perp$ is used to represent perpendicularity and is reads as *is perpendicular to*.)

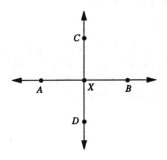

10.2.11 Right Angle

A **right angle** is an angle whose sides consist of perpendicular lines. In the figure above, $\angle CXB$, $\angle AXC$, $\angle AXD$, and $\angle BXD$ are all right angles. (Right angles are discussed in more detail in section 10.4.)

10.2.12 Acute Angle

An **acute angle** is an angle that is *narrower* than a right angle. In the figure below, $\angle BXE$ is an acute angle. (Acute angles are discussed in more detail in section 10.4.)

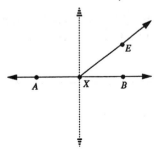

10.2.13 Obtuse Angle

An **obtuse angle** is an angle that is *wider* than a right angle, but narrower than a straight angle. In the figure below, $\angle BXF$ is an obtuse angle. (Obtuse angles are discussed in more detail in section 10.4.)

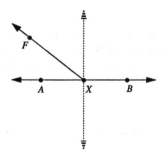

AN INTRODUCTION TO GEOMETRY   231

In Examples 1, 2, and 3 the following figure is used.

Example 1 (page 553, #1)

Evaluate $\overline{QI} \cap \overline{IK}$.

Solution

To find the intersection of the given two line segments, draw the second line segment under the first line segment. (See the figure below.) Note how the common point (point I) of the two line segments is aligned under each other. From the figure it is clear that the intersection of the two line segments is point I.

Example 2 (page 553, #5)

Evaluate $\overrightarrow{QI} \cap \overline{QU}$.

Solution

In the figure below we have drawn $\overline{QU}$ under $\overrightarrow{QI}$. Since the intersection consists of all points that are common to both the ray and the line segment, the intersection is simply $\overline{QU}$.

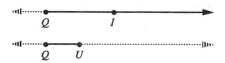

Example 3 (page 553, #7)

Evaluate $\overrightarrow{IU} \cup \overrightarrow{IK}$.

Solution

In the figure below the two rays have been drawn under each other with all common points in alignment. Since the union of these two rays consists of all the points that make up both rays, the result is the line UK. (Alternatively, the result can also be expressed as the straight angle UIK.)

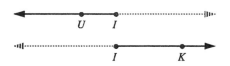

In Examples 4 and 5 the following figure is used.

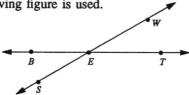

Example 4 (page 554, #21)

Evaluate $\overrightarrow{ET} \cup \overrightarrow{EW}$.

Solution

- Extract ray ET from the given figure.

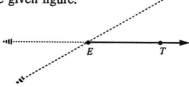

- Extract ray EW from the given figure.

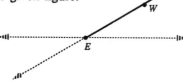

- The union of these two rays consists of all the points that make up both rays. If the two rays are *combined* at their common point E, the angle WET results.

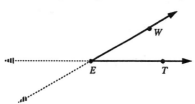

Example 5 (page 554, #27)

Evaluate $\angle BEW \cap \angle WET$.

Solution

- Extract angle BEW from the given figure.

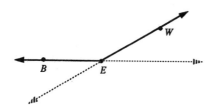

Continued on next page ...

- Extract angle WET from the given figure.

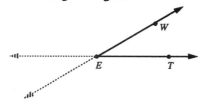

- The intersection of these two angles consists of all points that are common to both of the angles. Focusing on only these common points, the intersection is the ray EW.

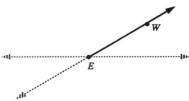

SUPPLEMENTARY EXERCISE 10.2

Refer to the figure below to answer problems 1 through 10.

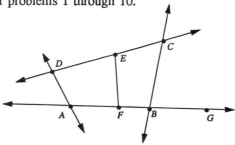

1. $\overleftrightarrow{AD} \cap \overleftrightarrow{DE}$
2. $\overrightarrow{CD} \cup \overrightarrow{CB}$
3. $\overrightarrow{DA} \cap \overline{EF}$
4. $\overrightarrow{GB} \cap \overline{EF}$
5. $\overline{DE} \cup \overline{EC}$
6. $\overrightarrow{CB} \cap \overleftrightarrow{DC}$
7. $\overrightarrow{AG} \cup \overrightarrow{AD}$
8. $\overrightarrow{AB} \cup \overrightarrow{BF}$
9. $\angle AFB \cap \overrightarrow{AF}$
10. $\overrightarrow{BA} \cup \overrightarrow{BG}$

Refer to the following figure to answer problems 11 through 15.

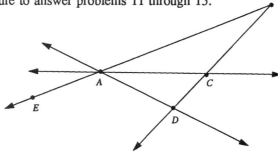

11. Name two line segments that contain point E.

12. Name two rays that contain point A as their endpoint.

13. Name the two geometric quantities that intersect at point D.

14. Name the angle formed by the union of ray CA and ray CD.

15. Name two rays that have a common endpoint.

10.3 PLANES

A plane, like a point and a line, is an undefined term. Conceptually, a plane is thought of as a set of points that forms a completely flat surface. This flat surface is endless; that is, the surface extends indefinitely in all directions. From a physical perspective, a plane is represented by a flat (i.e., no thickness) four sided figure, which has the dimensions of length and width only. Following are some important properties of planes.

Some Properties of Planes

1. A unique plane is determined by any three noncollinear points. (**Note:** Noncollinear points are points that are not on the same line.)

2. Any line on a given plane divides the plane into three distinct sets of points.
   - the set of points that make up the line
   - the set of points for each *half plane*, one on each side of the line

3. Two different planes either intersect or not intersect. If the planes intersect, then their intersection is a line. If the two planes do not intersect, then they are parallel.

4. Two lines that are on the same plane but do not intersect regardless of how far they (the lines) are extended, are called **parallel lines**. It is important to note that one criterion for two lines to be parallel is that they are in the same plane. The symbol used to denote parallelism is ∥ and is read *is parallel to*.

5. Two lines that do not lie in the same plane are called **skew** lines. Since skew lines are contained in different planes, it should be noted that skew lines do not intersect.

6. An angle on a plane divides the plane into three distinct sets of points.
   - the set of points that make up the angle
   - the set of points that are contained inside the angle (called **interior points**)
   - the set of points that are contained outside the angle (called **exterior points**)

The following figure is used for Examples 1 and 2.

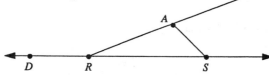

Example 1 (page 558, #3)

Evaluate $\overline{AS}$ ∩ (interior of ∠ARS).

Solution

- Extract $\overline{AS}$ from the given figure.

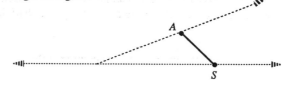

Continued on next page ...

AN INTRODUCTION TO GEOMETRY 235

- Identify the interior of $\angle ARS$ by shading it.

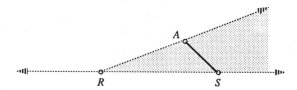

- Mentally superimpose the first figure over the second. It should be observed that the only points common to the line segment and the interior of the angle are those that make up line segment AS less its endpoints. (This is denoted $\overset{\circ\circ}{AS}$.)

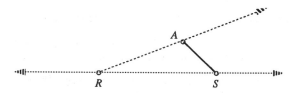

Example 2 (#7: Page 523)

Evaluate $\overrightarrow{RS} \cap$ (exterior of $\angle DRA$).

Solution

- Extract $\overrightarrow{RS}$ from the given figure.

- Identify the exterior of $\angle DRA$ by shading it.

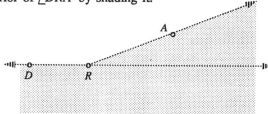

- Mentally superimpose the first figure over the second. The only points common to the ray and the exterior of the angle are those that make up the half-line RS. This is denoted $\overset{\circ}{\overrightarrow{RS}}$.

The following figure is used for Examples 3 and 4.

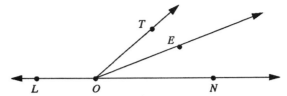

Example 3 (page 558, #13)

Evaluate (exterior ∠TOL) ∩ (interior ∠EON).

Solution

- Identify the exterior of ∠TOL by shading it.

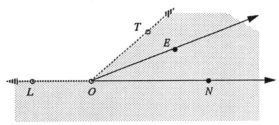

- Identify the interior of ∠EON by shading it.

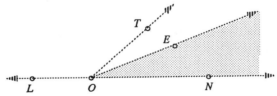

- Mentally superimpose the first figure over the second. The only points common to the two angles are all the points that make up the interior of ∠EON.

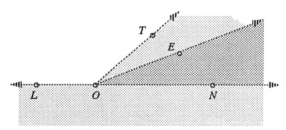

Example 4 (page 559, #17)

Evaluate (exterior ∠TOE) ∩ (interior ∠EON).

Solution

The interior of ∠EON is the region of intersection.

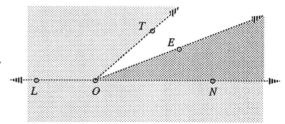

AN INTRODUCTION TO GEOMETRY   237

SUPPLEMENTARY EXERCISE 10.3

Refer to the following figure to answer problems 1 through 12.

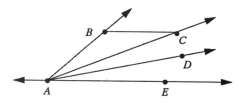

1. $\overrightarrow{AE} \cap \angle BAD$
2. $\overrightarrow{AC} \cup \overrightarrow{AE}$
3. $\overline{BC} \cap$ (interior $\angle BAC$)
4. $\angle BAD \cap \angle BAC$
5. (interior $\angle BAD$) $\cap \angle BAC$
6. $\angle DAE \cap \overleftrightarrow{AE}$
7. (exterior $\angle BAD$) $\cap \overrightarrow{AC}$
8. $\overrightarrow{AD} \cap$ (interior $\angle CAE$)
9. (interior $\angle BAC$) $\cap$ (exterior $\angle CAD$)
10. (exterior $\angle CAD$) $\cap$ (exterior $\angle BAD$)
11. The interior of $\angle BAD$ includes vertex A. (True/False).
12. Name the angle that lies on or in the interior of $\angle BAE$.

10.4 ANGLES

An angle is generated by rotating a ray about a point. The two sides of the generated angle are referred to as the **initial side** and the **terminal side**. (See below.)

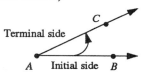

The amount of rotation from an angle's initial side to its terminal side is referred to as the **measure of the angle**. Angles are usually measured in degrees, denoted °. Degrees are divided into 60 equal parts called **minutes** (denoted '), and minutes are divided into 60 equal parts called **seconds** (denoted "). As a result, one degree is equal to 60 minutes (1° = 60 '), and one minute is equal to 60 seconds (1' = 60").

Consider for a moment the angle whose initial side is equal to its terminal side following one rotation. (See the figure below.) Note that the path of the rotation resembles a circle. Such an angle is called a **round angle** and consists of 360°, and the rotation is referred to as a complete revolution. One degree is equal to $\frac{1}{360}$ of a complete revolution. With this in mind we now classify various angles according to their measure.

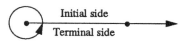

10.4.1 Right Angle

A right angle is formed by $\frac{1}{4}$ of a complete revolution. A right angle contains 90° ($\frac{1}{4}$ of 360) and is denoted by the symbol □.

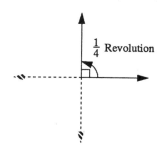

238 STUDENT STUDY GUIDE

10.4.2 Straight Angle

A straight angle is formed by $\frac{1}{2}$ of a complete revolution.
A straight angle contains 180° ($\frac{1}{2}$ of 360).

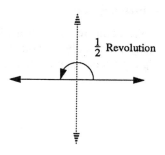

10.4.3 Acute Angle

An acute angle is an angle whose measure is greater that 0°, but less than 90°.

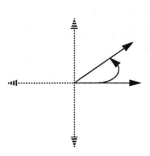

10.4.4 Obtuse Angle

An obtuse angle is an angle whose measure is greater than 90°, but less than 180°.

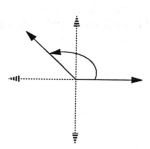

10.4.5 Reflex Angle

A reflex angle is an angle whose measure is greater than 180°, but less than 360°.

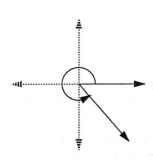

10.4.6 Conjugate Angles

If the sum of the measures of two angles is equal to 360° then the two angles are conjugate angles. In the figure at the right ∠1 and ∠2 are conjugate angles.

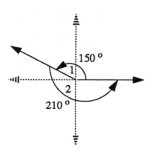

10.4.7 Supplementary Angles

If the sum of the measures of two angles is equal to 180° then the two angles are called supplementary angles. In the figure at the right ∠1 and ∠2 are supplementary angles.

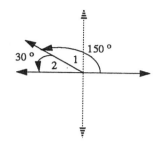

10.4.8 Complementary Angles

If the sum of the measures of two angles is equal to 90° then the two angles are called complementary angles. In the figure at the right ∠1 and ∠2 are complementary angles.

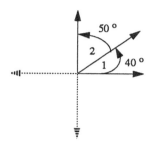

In addition to the above named angles, other angles are formed by a **transversal**. A transversal is a line that intersects two or more lines in different points. When two lines are cut by a transversal eight angles are formed. In the figure below lines *a* and *b* are cut by transversal *c*. Note that angles 3, 4, 5, and 6 are interior angles since they are inside the two lines cut by the transversal, and angles 1, 2, 7, and 8 are exterior angles for a similar reason.

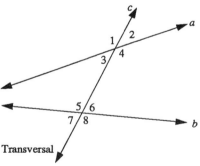

Angles that are on the opposite side of a transversal and do not have the same vertex are called **alternate angles**. In the figure above showing a transversal, ∠3 and ∠6, and ∠4 and ∠5, are **alternate interior angles**. Similarly, the two angle pairs of ∠1 and ∠8, and ∠2 and ∠7, are **alternate exterior angles**.

Another type of angle that is formed when a transversal intersects two or more lines in different points is a **corresponding angle**. Two angles are corresponding angles if they satisfy the following three criteria.

- the angles are on the same side of the transversal
- the angles do not share a common vertex
- one angle is an interior angle and the other is an exterior angle

Note that in the figure above the angle pairs of ∠1 and ∠5, ∠3 and ∠7, ∠2 and ∠6, and ∠4 and ∠8, are corresponding angles.

Finally, it can be shown that if two parallel lines are cut by a transversal, then

1. the alternate interior angles are congruent;
2. the corresponding angles are congruent; and
3. the interior angles on the same side of the transversal are supplementary.

Example 1 (page 570, #5c,d)

Find the measure of the supplement of the angle that has the given measure (parts c and d only).

Solution

Note: Two angles are supplementary angles if the sum of their measures is 180°.

c. 60°

The supplement of 60° is equal to 180°− 60°, which is equal to 120°.

d. 120° 20′

To find the supplement of 120° 20′ we subtract this measure from 180°. Note that we first reexpress 180° in terms of degrees and minutes so that we can subtract like quantities.

$$
\begin{aligned}
&\ 180° &-&\ 120°\ 20'\\
&=\ 179°\ 60' &-&\ 120°\ 20'\\
&=\ 59°\ 40'
\end{aligned}
$$

Example 2 (page 570, #9)

Two angles are complementary and one angle measures 30° less than 3 times the other. How many degrees are there in each angle?

Solution

If we let x equal one of the angles we can express the other angle in terms of x.

One angle measures 30° less than 3 times the other
$3x - 30$

As a result,

x = the first angle
$3x - 30$ = the second angle

Since the angles are complementary, the sum of their measures is 90°. This leads to the following equation, which is solved below.

$$
\begin{aligned}
\textit{first angle} + \textit{second angle} &= 90°\\
(x) + (3x - 30) &= 90\\
4x - 30 &= 90\\
4x &= 90 + 30\\
4x &= 120\\
x &= \frac{120}{4}\\
x &= 30
\end{aligned}
$$

One angle has a measure of 30° and the other has a measure of 60°.

Example 3 (page 571, #11)

In the figure below, what value of x will make AB a straight line?

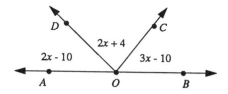

Solution

A straight line forms a straight angle, and a straight angle has a measure of 180°. This implies

$$m \angle BOC + m \angle COD + m \angle DOA = 180°$$

Substituting the algebraic expressions for the angle measures we have the following equation, which we then solve for x.

$$
\begin{aligned}
(3x - 10) + (2x + 4) + (2x - 10) &= 180 \\
3x - 10 + 2x + 4 + 2x - 10 &= 180 \\
3x + 2x + 2x + 4 - 10 - 10 &= 180 \\
7x - 16 &= 180 \\
7x &= 180 + 16 \\
7x &= 196 \\
x &= \frac{196}{7} \\
x &= 28
\end{aligned}
$$

If $x = 28$ then $m \angle BOC = 74°$, $m \angle COD = 60°$, and $m \angle DOA = 46°$

Example 4 (page 571, #23)

Find the measures of all of the angles shown in the figure below. Assume $\overleftrightarrow{CD} \parallel \overleftrightarrow{EF}$ and $m \angle 3 = 50°$.

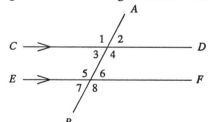

Solution

There are many ways in which the measures of these angles are determined. In order to answer this question successfully, however, the following properties must be known.

1. Supplementary angles are angles whose measures sum to 180°.
2. Vertical angles are congruent.
3. If two parallel lines are cut by a transversal then the
    a. alternate interior angles are congruent;
    b. corresponding angles are congruent; and
    c. interior angles on the same side of their transversal are supplementary.

Continued on next page ...

The approach we use to solve this problem is summarized in the chart below. (**Note:** The properties referred to in the chart are the three properties discussed above. Alternative approaches are given within braces { }.)

Solution for Example 4

Angle Measure	Reason For Measure
1. m∠3 = 50°	1. Given
2. m∠2 = m∠3 = 50°	2. Property 2
3. m∠4 = (180° − m∠3) = 130°	3. Property 1
{m∠4 = (180° − m∠2)}	{Property 1}
4. m∠1 = m∠4 = 130°	4. Property 2
{m∠1 = (180 − m∠3)}	{Property 1}
{m∠1 = (180 − m∠2)}	{Property 1}
5. m∠6 = m∠3 = 50°	5. Property 3a
{m∠6 = m∠2}	{Property 3b}
{m∠6 = m(180 − ∠4)}	{Property 3c}
6. m∠5 = m∠4 = 130°	6. Property 3a
{m∠5 = (180 − m∠6)}	{Property 1}
{m∠5 = (180 − m∠3)}	{Property 3c}
{m∠5 = m∠1}	{Property 3b}
7. m∠7 = m∠6 = 50°	7. Property 2
{m∠7 = m∠3}	{Property 3b}
{m∠7 = (180 − m∠5)}	{Property 1}
8. m∠8 = m∠5 = 130°	8. Property 2
{m∠8 = (180 − m∠6)}	{Property 1}
{m∠8 = (180 − m∠7)}	{Property 1}
{m∠8 = m∠4}	{Property 3b}

Summarizing,

$$m\angle 1 = m\angle 4 = m\angle 5 = m\angle 8 = 130°$$
$$m\angle 2 = m\angle 3 = m\angle 6 = m\angle 7 = 50°$$

Example 5 (page 572, #27)

Given the figure below, find m $\angle 3$ and m $\angle 5$. Assume m $\angle 3 = (x + 20)°$ and m $\angle 5 = x°$.

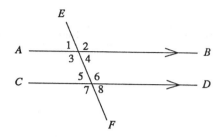

Solution

From the figure note that $\overleftrightarrow{AB} \parallel \overleftrightarrow{CD}$ and that the two lines are cut by the transversal $\overleftrightarrow{EF}$. As a result, $\angle 3$ and $\angle 5$ are supplementary. This leads to the following equation, which we then solve.

$$
\begin{aligned}
m \angle 5 &= (180 - m \angle 3) \\
x &= (180 - (x + 20)) \\
x &= 180 - x - 20 \\
x &= 180 - 20 - x \\
x &= 160 - x \\
x + x &= 160 \\
2x &= 160 \\
x &= 80
\end{aligned}
$$

Therefore, m $\angle 5 = 80°$ and m $\angle 3 = 100°$.

SUPPLEMENTARY EXERCISE 10.4

Use the following figure to answer problems 1 through 5.

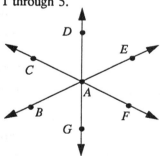

1. Name all angles with a measure of $x°$ where $0° < x < 180°$.
2. Name three straight angles.
3. Name two pairs of vertical angles formed by lines CF and BE.
4. Name an angle supplement to $\angle CAG$.
5. Name two angles adjacent to $\angle GAF$.

Continued on next page ...

6. An angle measures 23° 15′ 23″. Find its complement.

7. An angle measures 68° 29′. Find its supplement.

8. One angle is 30° less than twice a second angle. If the two angles are complementary how many degrees does the angle with the larger measure contain?

9. An angle is equal to its complement. What is the measure of the angle?

10. One angle is one-third the measure of a second angle. If the two angles are supplementary find the measure of the acute angle.

11. Using the figure at the right,
    a. find x so that $\overleftrightarrow{AB} \perp \overleftrightarrow{CD}$
    b. determine the measures of $\angle COE$, $\angle EOF$, and $\angle FOB$.

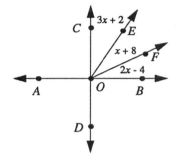

12. In the figure at the right $\overleftrightarrow{AB} \parallel \overleftrightarrow{CD}$, m $\angle 9 = 23°$, and m $\angle 11 = 48°$. Find the measure of the remaining angles.

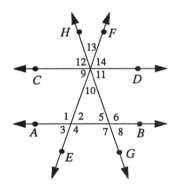

10.5 POLYGONS

A **polygon** is a simple closed broken line in a plane. The connected line segments that make up a polygon are called the **sides** of the polygon. The endpoints of these line segments (i.e., the points at which the line segments are connected) are called the **vertices** of the polygon.

Polygons are classified by their number of sides. The names of some common polygons are listed in Table 10.5.1. We now focus our attention on two specific types of polygons, namely triangles and quadrilaterals.

10.5.1 Triangles

A triangle is a polygon that has three sides. One physical attribute of a triangle is that the sum of the lengths of any two of its sides is greater than the length of the third side. Triangles are classified by the characteristics of their sides or the characteristics of their angles. This classification is summarized in Table 10.5.2.

Table 10.5.1 Names of Common Polygons

Number of Sides	Name of Polygon
3	Triangle
4	Quadrilateral
5	Pentagon
6	Hexagon
7	Heptagon
8	Octagon
9	Nonagon
10	Decagon
12	Dodecagon
20	Icosagon

One triangle that receives special attention is the right triangle. A right triangle consists of two sides that form a right angle. These sides are called the **legs** of the triangle and the third side is called the **hypotenuse**. In triangle ABC below, $\overline{AC}$ and $\overline{CB}$ are the legs and $\overline{AB}$ is the hypotenuse. The **Pythagorean Theorem** expresses a relationship that exists among the three sides of a right triangle. Stated in words, this relationship is *the square of the hypotenuse is equal to the sum of the square of the legs*. Using the figure below, this relationship is expressed mathematically as

$$(\overline{AB})^2 = (\overline{BC})^2 + (\overline{AC})^2$$
or
$$c^2 = a^2 + b^2$$

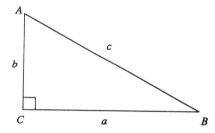

10.5.2 Quadrilaterals

A quadrilateral is a polygon that has four sides. All quadrilaterals are classified according to the characteristics of their sides. This classification is summarized in Table 10.5.3. In addition to the descriptions provided in Table 10.5.3, a parallelogram, rhombus, rectangle, and square have certain properties. Some of these properties are summarized here. In addition to these properties, a **regular polygon** is a polygon that is both equilateral and equiangular. Thus a square is considered to be a regular polygon.

Properties of a Parallelogram	Properties of a Rhombus
• The opposite sides are parallel	• All parallelogram properties hold for a rhombus
• The opposite sides are equal	• All sides are equal (this is called **equilateral**)
• The opposite angles are equal	• The diagonals are perpendicular to each other
• Two consecutive angles are supplementary	
• The diagonals bisect each other	

Properties of a Rectangle	Properties of a Square
• Parallelogram properties hold for a rectangle	• Rectangle and rhombus properties hold for a square
• All angles are equal (this is called **equiangular**)	
• The diagonals are equal	

Table 10.5.2 Triangle Classification Chart

Type of Classification	Name of Triangle	Description	Sample Figure
Triangles classified according to the characteristics of their sides. (**Note:** Hash marks are used in the sample figures to represent sides that are of equal length.)	Scalene	No sides are equal	
	Isosceles	Two sides are equal	
	Equilateral	All three sides are equal	
Triangles classified according to the characteristics of their angles. (**Note:** The sum of the measures of the interior angles of a triangle is equal to 180°.)	Acute	All three angles are acute	
	Obtuse	One angle is obtuse	
	Right	One angle is a right angle	
	Equiangular	All three angles are of equal measure	

Table 10.5.3 Quadrilateral Classification Chart

Name of Quadrilateral	Description	Sample Figure
Trapezoid	• Two sides are parallel In the sample figure $\overline{AB} \parallel \overline{CD}$.	
Isosceles Trapezoid	• Two sides are parallel • Two nonparallel sides are equal in length In the sample figure $\overline{AB} \parallel \overline{CD}$. Also, $m(\overline{AC}) = m(\overline{BD})$.	
Parallelogram	• Both pairs of opposite sides are parallel In the sample figure $\overline{AB} \parallel \overline{CD}$, and $\overline{AC} \parallel \overline{BD}$.	
Rhombus	• Both pairs of opposite sides are parallel • The adjacent sides are of equal length, which implies that all four sides are of equal length. In the sample figure $\overline{AB} \parallel \overline{CD}$, and $\overline{AC} \parallel \overline{BD}$. Also, $m(\overline{AB}) = m(\overline{BD}) = m(\overline{CD}) = m(\overline{AC})$.	
Rectangle	• Both pairs of opposite sides are parallel • One right angle, which implies that all the angles are right angles In the sample figure $\overline{AB} \parallel \overline{CD}$, and $\overline{AC} \parallel \overline{BD}$. Also, the measure of all angles is 90°.	
Square	• All four sides are of equal length • All four interior angles are of equal measure, which implies that all angles are right angles and that the opposite sides are parallel.	

Example 1 (page 581, #7)

Find the length of the hypotenuse $\overline{AB}$ in right triangle ABC. Assume that $m(\overline{AC}) = 9$ units and $m(\overline{BC}) = 12$ units.

Solution

- First sketch a picture using the given information.

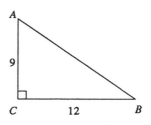

- Next use the Pythagorean Theorem.

$$(\overline{AB})^2 = (\overline{AC})^2 + (\overline{BC})^2$$
$$(\overline{AB})^2 = 9^2 + 12^2$$
$$(\overline{AB})^2 = 81 + 144$$
$$(\overline{AB})^2 = 225$$
$$\overline{AB} = \sqrt{225}$$
$$m(\overline{AB}) = 15 \text{ units}$$

Example 2 (page 581, #11)

Mary rode her moped 3 miles south and then 4 miles east. How far was she from her starting point?

Solution

- First construct a diagram of Mary's journey.

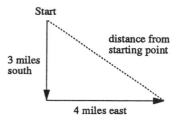

- Now use the Pythagorean Theorem.

$$(distance\ start)^2 = 3^2 + 4^2$$
$$= 9 + 16$$
$$= 25$$
$$(distance\ from\ start) = \sqrt{25}$$
$$= 5$$

Mary was 5 miles from her starting point.

Example 3 (page 582, #25)

In parallelogram ABCD, which is shown below, m $\angle A = 3x$ and m $\angle B = x + 40$. What is the value of x?

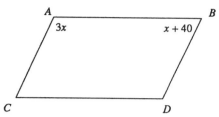

Solution

A parallelogram is a quadrilateral that has both pairs of opposite sides parallel. One of the properties of a parallelogram is that two consecutive angles are supplementary. Thus,

$$
\begin{aligned}
3x + (x + 40) &= 180 \\
3x + x + 40 &= 180 \\
4x + 40 &= 180 \\
4x &= 180 - 40 \\
4x &= 140 \\
x &= \frac{140}{4} \\
x &= 35
\end{aligned}
$$

SUPPLEMENTARY EXERCISE 10.5

In problems 1 through 10 answer true or false.

1. A parallelogram is a quadrilateral that has both pairs of opposite sides parallel.
2. A quadrilateral that is equilateral is called a square.
3. A square is equiangular.
4. The diagonals of a rhombus are equal.
5. Two consecutive angles of a parallelogram are supplementary.
6. A quadrilateral whose angles are equal, but whose adjacent sides are unequal, is a rectangle.
7. A quadrilateral whose diagonals are equal is either a rectangle or a square.
8. The nonparallel sides of an isosceles trapezoid are not equal in length.
9. A right triangle is also an acute triangle.
10. The Pythagorean Theorem describes a relationship that exists for the length of the sides of an isosceles triangle.

Continued on next page ...

250   STUDENT STUDY GUIDE

11. Using the figure at the right, find the measure of each base angle. (Note: m $\angle ABC = 12°$.)

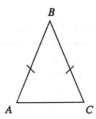

12. The measure of one angle of an equilateral triangle is represented by $(3x - 9)°$. What is the value of x?

13. Find the length of the diagonal of a rectangle whose sides are 8 and 15 units.

14. A ladder is leaning against a wall. The bottom of the ladder is 15 feet from the wall. If the ladder is 25 feet in length, at what height does it touch the wall?

10.6 PERIMETER AND AREA

The **perimeter** of a polygon is the sum of the lengths of its sides, and the perimeter of a circle is the distance around the circle. Perimeter and circumference is always measured in linear units (e.g., inches, feet, yards, centimeters, meters, etc.).

The **area** of a polygon represents the number of square units that are contained in the region enclosed within the sides of the polygon, and the area of a circle is the area of the plane surface enclosed within the circle. Area is always measured in square units (e.g., square inches ($in.^2$), square feet ($ft.^2$), square yards ($yd.^2$), square centimeters (cm^2), square meters (m^2), etc.).

Table 10.6.1 contains a summary of some common perimeter and area formulas.

Example 1 (page 593, #9)

Find the perimeter of a rectangle given that its length is 12 and the length of its diagonal is 13.

Solution

First construct and label a picture of the figure described.

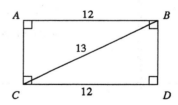

Since the width is unknown, we use the Pythagorean Theorem to find it.

$$
\begin{aligned}
W^2 + 12^2 &= 13^2 \\
W^2 + 144 &= 169 \\
W^2 &= 169 - 144 \\
W^2 &= 25 \\
W &= 5
\end{aligned}
$$

Using 5 units as the measure of the width, we now find the perimeter.

$$
\begin{aligned}
P &= 2l + 2w \\
P &= 2(12) + 2(5) \\
P &= 24 + 10 \\
P &= 34
\end{aligned}
$$

Table 10.6.1 Common Perimeter and Area Formulas

Name of Figure	Formulas (P = Perimeter) (A = Area)	Sample Figure
Triangle	$P = a + b + c$ $A = \frac{1}{2}bh$	
Square	$P = 4s$ $A = s^2$	
Trapezoid	$A = \frac{1}{2}h(b_1 + b_2)$	
Rectangle	$P = 2l + 2w$ $A = lw$ or bh	
Circle	Circumference $= 2\pi r = \pi d$ $A = \pi r^2$	
Parallelogram	$P = 2a + 2b$ $A = bh$	

Example 2 (page 594, #35a)

Find the area and perimeter of the following figure.

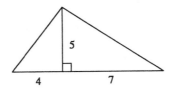

Solution

To find the area we multiply the product of the base and height by one-half. Note that the base of this triangle is 4 + 7, which is 11 units.

$$A = \frac{1}{2}bh$$
$$A = \frac{1}{2}(11)(5)$$
$$A = \frac{1}{2}(55)$$
$$A = 27.5 \text{ square units}$$

To find the perimeter of this triangle we need to first determine the length of each of its sides. We use the Pythagorean Theorem. (**Note:** The length of one side is given in the problem. Thus we will be finding the lengths of the remaining two sides.)

$$5^2 + 4^2 = x^2$$
$$25 + 16 = x^2$$
$$41 = x^2$$
$$6.4 \approx x$$

$$7^2 + 5^2 = y^2$$
$$49 + 25 = y^2$$
$$74 = y^2$$
$$8.6 \approx y$$

The perimeter is approximately equal to 11 + 6.4 + 8.6, which is equal to 26 units.

Example 3 (page 594, #35d)

Find the area and perimeter of the following figure.

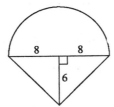

Continued on next page ...

Solution

The area of this figure consists of the sum of the area of the semicircle and the area of the triangle.

- The area of the semicircle is one half the area of a circle.

$$A_{semicircle} = (\tfrac{1}{2})\pi r^2$$
$$A_{semicircle} \approx (\tfrac{1}{2})(3.14)(8^2)$$
$$A_{semicircle} \approx (\tfrac{1}{2})(3.14)(64)$$
$$A_{semicircle} \approx (\tfrac{1}{2})(200.96)$$
$$A_{semicircle} \approx 100.48$$

- The area of the triangle is

$$A_{triangle} = \tfrac{1}{2}bh$$
$$A_{triangle} = \tfrac{1}{2}(16)(6)$$
$$A_{triangle} = \tfrac{1}{2}(96)$$
$$A_{triangle} = 48$$

Area of the figure is approximately equal to 100.48 + 48, or 148.48 square units.

The perimeter of the figure is the sum of the distance around the semicircle and the hypotenuse of each right triangle.

- The circumference of the semicircle is one half the circumference of the circle.

$$C_{semicircle} = \tfrac{1}{2}(2\pi r)$$
$$C_{semicircle} = \pi r$$
$$C_{semicircle} \approx (3.14)(8)$$
$$C_{semicircle} \approx 25.12$$

- The length of each hypotenuse is determined using the Pythagorean Theorem where the length of the sides are 8 and 6 units respectively. As a result, the length of each hypotenuse is equal to 10 units.

The perimeter is approximately equal to 25.12 + 10 + 10, which is equal to 45.12 units.

SUPPLEMENTARY EXERCISE 10.6

1. Find the area of a parallelogram with a base of eight inches and a height of four inches.

2. Find the area of a parallelogram with a base of $2\tfrac{1}{2}$ feet and a height of 18 inches.

Continued on next page ...

3. Find the area of a rectangle with a length of 40 cm and a width of 10 cm.

4. Find the area of a rectangle with a length of three feet and a width of 18 inches.

5. Find the area of a square with a side of 30 cm.

In problems 6 through 8, find the area of the triangle with the given dimensions. (Round answers to the nearest tenth.)

6. $b = 6$ in., $h = 4$ in.   7. $b = 3.9$ in., $h = 4.2$ in.   8. $b = 1\frac{1}{2}$ ft., $h = 16$ in.

9. Find the area of a right triangle with a leg of five feet and a hypotenuse of 13 feet.

10. Find the area of a rhombus with diagonals of seven and ten feet in length.

11. Find the area of a trapezoid with bases of seven and 13 inches, and a height of four inches.

For problems 12 through 14 refer to the following figure.

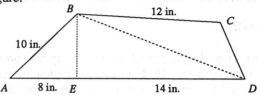

12. Find the length of $\overline{BE}$.   13. Find the area of triangle ABD.   14. Find the area of trapezoid $ABCD$.

15. Find the circumference and area of a circle with a radius of six cm. Use $\pi = 3.14$ and round your answers to the nearest tenth.

16. Find the circumference and area of a circle with a diameter of three feet. Use $\pi = 3.14$ and round your answers to the nearest tenth.

In problems 17 through 20 find the area of each figure.

17.
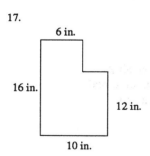

18.
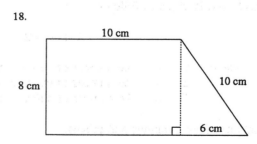

19.
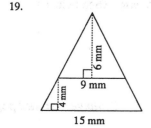

20.
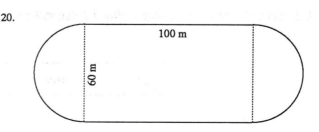

Continued on next page ...

21. Find the area of the shaded region in the figure below. Use $\pi = 3.14$ and round your answers to the nearest tenth.

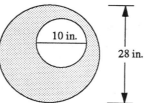

22. A wheel with a radius of three inches is rolled. If the wheel makes five complete revolutions before it is stopped how far will it have traveled?

23. A walkway two feet wide surrounds a rectangular grass strip that measures 20 feet long and 14 feet wide. Find the area of the walkway.

10.7 SOLIDS

A **polyhedron** is a simple closed surface in space comprised of polygonal regions. The figure below is an example of a polyhedron. Each polygonal region is called a **face**. When two faces intersect they form a boundary (i.e., a line segment) called an **edge**. The point at which two or more edges intersect is called a **vertex**.

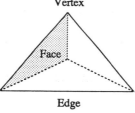

A relationship exists among the number of vertices, edges, and faces of a polyhedron. This relationship is known as **Euler's formula** and is stated as follows.

$$V - E + F = 2$$

where V = the number of vertices of the polyhedron
E = the number of edges of the polyhedron, and
F = the number of sides of the polyhedron

Two special kinds of polyhedra are pyramids and prisms.

10.7.1 Pyramids

A pyramid is a polyhedron that has the following attributes.

- A pyramid is formed by a simple closed polygonal region called the base. (See below.)

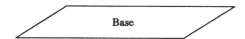

Continued on next page ...

- There exists a point that is not in the same plane in which the base is contained. (See below.)

Point

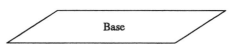

- The sides (i.e., the faces) of a pyramid are formed by joining the point and the vertices of the base with line segments. These sides will always be triangular regions. (See below.)

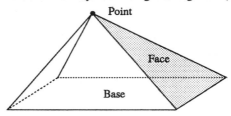

A pyramid is classified according to the polygonal region that forms its base. For example, if the base is a triangle then the solid is called a triangular pyramid; if the base is square then the solid is called a square pyramid; and so forth.

10.7.2 Prism

A prism is a polyhedron that has the following attributes.

- A prism is formed by two congruent polygonal regions (called bases), which are located in parallel planes. (See below.)

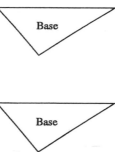

- The sides (i.e., the faces) of a prism are formed by joining the vertices of these bases with line segments. The regions formed will always be parallelograms. (See below.)

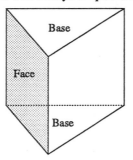

A prism is classified according to the polygonal regions forming its bases. If the bases are triangles then the solid is called a triangular prism; if the bases are rectangles then the solid is called a rectangular prism; if the bases are squares then the solid is called a square prism, or cube; and so forth.

10.7.3 Solids Involving Circular Regions

Three types of solids that involve circular regions are cylinders, cones, and spheres.

A **circular cylinder** is a simple closed surface that is bounded on two ends by circular regions (called bases). A **circular cone** (or simply, cone) is a simple closed surface similar to a pyramid, except its base is a circular region. A **sphere** consists of a set of points in space that are of equal distance from some fixed point. An example of each of these solids is shown below.

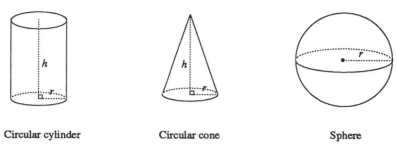

Circular cylinder   Circular cone   Sphere

10.7.4 Surface Area and Volume Formulas for Solids

The formulas for surface area and volume that are discussed in your textbook in this section are summarized in Table 10.7.1.

Table 10.7.1 Surface Area/Volume Formulas for Solids

Type of Solid	Surface Area Formula	Volume Formula
Pyramid	Find the area of all the faces and bases and add them together	$V = \frac{1}{3}$(area of base)(height)
Prism	Find the area of all the faces and bases and add them together	$V =$ (area of base)(height)
Cylinder	$S.A. = 2\pi r^2 + 2\pi rh$	$V = \pi r^2 h$
Cone	$S.A. = \pi r^2 + \pi rs$	$V = \frac{1}{3}\pi r^2 h$
Sphere	$S.A. = 4\pi r^2$	$V = \frac{4}{3}\pi r^3$

Example 1 (page 605, #9)

Find the total surface area and volume of the figure at the right.

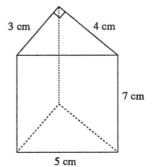

Continued on next page ...

Solution

<u>Finding Surface Area</u>

- Find the area of the triangular bases.

$$A_{top\ base} = A_{bottom\ base} = \frac{1}{2}bh$$
$$= \frac{1}{2}(3)(4)$$
$$= 6 \text{ sq. cm}$$

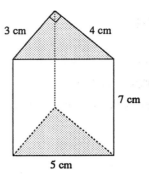

As a result, the area of the two bases is 6 + 6, which is equal to 12 sq. cm.

- Find the area of the left and right faces.

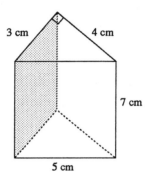

$$A_{left\ face} = bh$$
$$= (3)(7)$$
$$= 21 \text{ sq. cm}$$

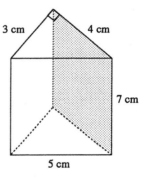

$$A_{right\ face} = bh$$
$$= (4)(7)$$
$$= 28 \text{ sq. cm}$$

- Find the area of the front face.

To find the area of the front face we need to first determine the width of this side. Note that the width is equal to the hypotenuse of the right triangle (from one of the bases). Using the Pythagorean Theorem with sides of 3 and 4, the width is equal to 5 cm. The area of the front face can now be determined.

$$A_{front\ face} = bh$$
$$= (5)(7)$$
$$= 35 \text{ sq. cm}$$

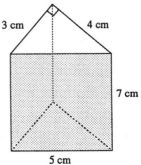

Continued on next page ...

The total surface area is the sum of the individual areas of the faces and bases. This is equal to 12 + 21 + 28 + 35, which is 96 square centimeters.

Finding Volume

$$V = (\text{Area of base})(\text{height}) = (6)(7) = 42 \text{ cubic cm}$$

Example 2 (page 605, #13)

Find the surface area and volume of the figure below.

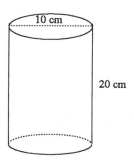

Solution

$$\begin{aligned}
Surface\ Area_{cylinder} &= 2\pi r^2 + 2\pi rh \\
&\approx (2)(3.14)(5^2) + (2)(3.14)(5)(20) \\
&\approx (2)(3.14)(25) + (2)(3.14)(5)(20) \\
&\approx 157 + 628 \\
&\approx 785 \text{ sq. cm}
\end{aligned}$$

$$\begin{aligned}
Volume_{cylinder} &= \pi r^2 h \\
&\approx (3.14)(5^2)(20) \\
&\approx (3.14)(25)(20) \\
&\approx 1{,}570 \text{ cubic cm}
\end{aligned}$$

SUPPLEMENTARY EXERCISE 10.7

In problems 1 and 2 find (a) the total surface area and (b) the volume for each solid. Let $\pi = 3.14$ and express decimal answers to the nearest tenth.

1.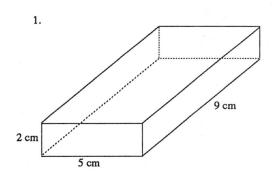

2.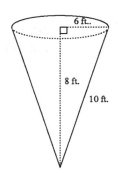

Continued on next page ...

In problems 3 and 4 find the volume of each solid.

3.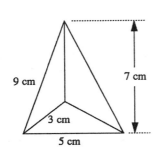

4.

5. A tuna fish can has a diameter of 3.25 inches and a height of 2.5 inches. Find the surface area and volume of this can. Use $\pi = 3.14$ and round your answers to the nearest tenth.

6. What is the surface area of a sphere that has a radius of 5 cm? Use $\pi = 3.14$ and round your answers to the nearest tenth.

7. What is the volume of a sphere that has a diameter of seven inches? Use $\pi = 3.14$ and round your answers to the nearest tenth.

8. In the figure below a right cylinder is inscribed in a cube. If the edge of the cube is 8 cm in length, what is the volume and total surface area of the cylinder? Use $\pi = 3.14$ and round your answers to the nearest tenth.

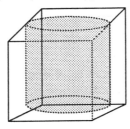

9. A cardboard box that houses a stereo speaker measures 30 inches wide, six inches deep, and four feet high. Determine how many of these boxes can fit in a space that measures 10 feet high, 12 feet wide, and six feet deep.

10. Determine the number of faces, bases, vertices, and edges of a hexagonal pyramid. Also verify Euler's formula for this pyramid.

10.8 CONGRUENT AND SIMILAR TRIANGLES

Congruent figures are figures that coincide; that is, they have exactly the same shape and size. The symbol used in this study guide to represent congruence is ≅, and is read *is congruent with*. Whenever two polygons are congruent their corresponding parts (i.e., their sides and angles) are also congruent.

It can be proven that two triangles are congruent in any one of three ways. The conditions that must be met to show such a congruence for each method are summarized here.

The Included Angle Method (Side-Angle-Side)

If the measures of two sides and the included angle of one triangle are equal to the respective measures of two sides and the included angle of the second triangle, then the two triangles are congruent by Side-Angle-Side (*SAS*).

The Included Side Method (Angle-Side-Angle)

If the measures of two angles and the included side of one triangle are equal to the respective measures of two angles and the included side of the second triangle, then the two triangles are congruent by Angle-Side-Angle (*ASA*).

The Three Sides Method (Side-Side-Side)

If the measures of three sides of one triangle are equal to the respective measures of three sides of the second triangle, then the two triangles are congruent by Side-Side-Side (*SSS*).

Two polygons that have the same shape but not necessarily the same size are called **similar polygons**. Similar polygons are polygons whose corresponding angles are equal, and whose corresponding sides are in proportion. The symbol for similarity is ~ and is read as *is similar to*. Triangles can be proven similar by using applicable theorems from geometry. These theorems are stated here without proof.

Theorem 1

If the measures of the three angles of one triangle are equal to the respective measures of the three angles of the second triangle, then the two triangles are similar.

Theorem 2

If the sides of two triangles are respectively proportional, then the triangles are similar.

Example 1 (page 610, #9)

In triangles *MNO* and *RST*, angles *M* and *N* are equal respectively to angles *R* and *S*. If side *MN* is represented by $6x - 12$, and side *RS* is represented by $2x + 16$, what value of x will make the triangles congruent?

Solution

- First construct a sketch of the triangles based on the given information.

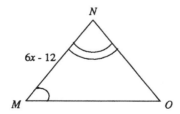 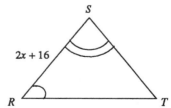

- If these triangles are to be congruent then we must use the angle-side-angle method. (Why?) The measures of the sides *MN* and *RS* are set equal to each other.

$$
\begin{aligned}
6x - 12 &= 2x + 16 \\
6x - 2x &= 16 + 12 \\
4x &= 28 \\
x &= 7
\end{aligned}
$$

Example 2 (page 610, #13, 15, 17)

In the figure below, triangles *ABC* and *DEC* are congruent. Find the measures of $\overline{DE}$, $\overline{AC}$, and $\angle D$.

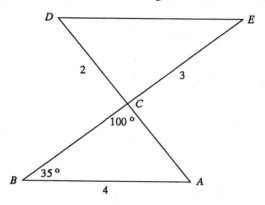

Solution

Since the triangles are congruent their corresponding parts also are congruent.

- $\overline{DE}$ and $\overline{BA}$ are corresponding parts. Thus m($\overline{DE}$) = m($\overline{BA}$) = 4.
- $\overline{AC}$ and $\overline{DC}$ are corresponding parts. Thus m($\overline{AC}$) = m($\overline{DC}$) = 2.
- $\angle D$ and $\angle A$ are corresponding parts. Thus m $\angle D$ = m $\angle A$ = (180 − (100 + 35)) = 45°.

Example 3 (page 611, #27)

Find the missing length of the two similar triangles shown below.

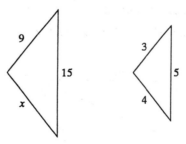

Solution

Since the two triangles are similar their respective sides are proportional (Theorem 2).

$$\frac{9}{3} = \frac{x}{4}$$
$$3x = (4)(9)$$
$$3x = 36$$
$$x = 12$$

The length of the missing leg is 12 units.

SUPPLEMENTARY EXERCISE 10.8

In the figure below triangle ABC is similar to triangle BCD. Use this figure to find the measures given in problems 1 through 6.

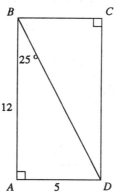

1. m($\overline{BD}$) 2. m($\overline{CD}$) 3. m($\overline{BC}$) 4. m∠CBD 5. m∠ACB 6. m∠BCD

7. In triangles ABC and DEF the measure of angles A and B are equal respectively to the measures of angles D and E. If the length of side AB is represented by $3x - 4$, and the length of side DE is represented by $2x + 1$, find x so that the triangles are congruent.

8. In triangles ABC and DEF, angles B and E have a measure of 90°. Also, the length of side AB is equal to the length of side DE. If the measure of angle A is represented by $(5x - 12)°$ and the measure of angle D is represented by $(3x + 8)°$, what must the measure of these angles be so that the triangles are congruent?

9. In the figure below m($\overline{AB}$) = m($\overline{CD}$), and m∠1 = m∠2.
    a. Is triangle ABC congruent to triangle CDA? Why or why not?
    b. Find the length of $\overline{BD}$ and $\overline{AC}$.

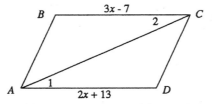

In problems 10 through 13 each pair of triangles is similar. Find the missing length.

10.
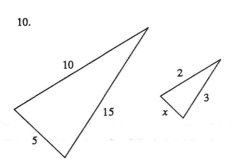

11.
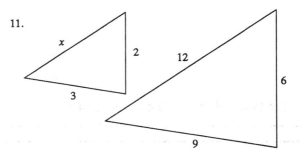

Continued on next page ...

264 STUDENT STUDY GUIDE

12.

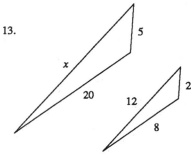

13.

10.9 NETWORKS

A network consists of a set of points, called vertices, and a set of line segments or curves, called edges, that connect the points. A network is traversable if each edge can be traced exactly one time. Conceptually, this implies that the network can be drawn by tracing each edge exactly once without lifting the tracing instrument from the paper.

If a vertex has an odd number of edges passing through it or connected to it, then the vertex is called an **odd vertex**. Similarly, an **even vertex** has an even number of edges passing through it or connected to it.

According to the work of Leonard Euler:

- If a network contains only even vertices then the network is traversable by a route in which the initial vertex (starting point) and terminal vertex (ending point) are the same.

- If a network contains exactly two odd vertices, with the remaining vertices being even, the network can be traversed, but the trace will not begin and end at the same vertex. Rather, the trace will start at one odd vertex and it will end at the other odd vertex.

- If a network has more than two odd vertices then it is not traversable.

Example 1 (page 615, #7)

Using the network below, determine (a) the number of even vertices, (b) the number of odd vertices, (c) if the network is traversable, and (d) the possible starting points if the network is traversable.

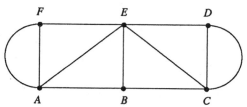

Solution

An analysis of the vertices yields the following information.

- Vertex A is even since it has four edges passing through it.

- Vertex B is odd since it has three edges passing through it.

- Vertex C is even since it has four edges passing through it.

Continued on next page ...

- Vertex D is odd since it has three edges passing through it.
- Vertex E is odd since it has five edges passing through it.
- Vertex F is odd since it has three edges passing through it.

As a result, there are two even vertices and four odd vertices. Since there are more than two odd vertices the network is not traversable.

SUPPLEMENTARY EXERCISE 10.9

In problems 1 through 6 determine (a) the number of even vertices, (b) the number of odd vertices, (c) whether the network is traversable, and (d) the possible starting points if the network is traversable.

1.
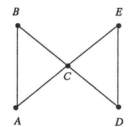

2.
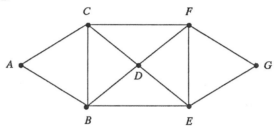

3.
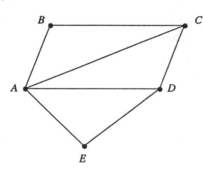

4.
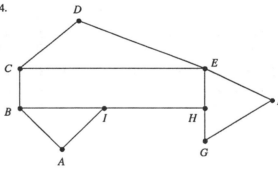

5.
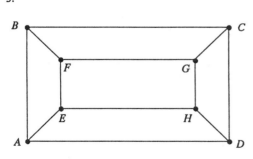

6.
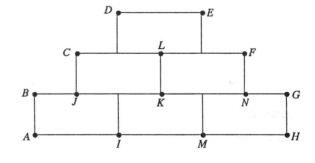

10.10 CHAPTER 10 TEST

For problems 1 through 8 match Column A with the best answer from Column B.

A

1. collinear
2. line segment AB
3. coplanar
4. the measure of an angle
5. complementary angles
6. is parallel to
7. is congruent to
8. two lines that do not lie in the same plane

B

a. m $\angle a$ + m $\angle b$ = 90°
b. set of points that lie in the same plane
c. AB
d. ∥
e. skew lines
f. set of points that lie on the same plane
g. $\overline{AB}$
h. ≅
i. ~
j. m $\angle ABC$

Refer to the figure below for problems 9 through 13. Assume that $\overline{ER} \perp \overline{AC}$.

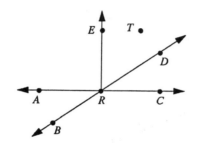

9. Name the two rays that form $\angle DRC$.
10. Name a pair of vertical angles.
11. Name an angle that is adjacent to $\angle DRC$.
12. Find (interior $\angle ERC$) ∩ (exterior $\angle DRC$)
13. If m $\angle ARD$ = 10x + 30 and m $\angle DRC$ = 3x + 7, find x.

In problems 14 through 17 refer to the following figure. Assume that lines *a* and *b* are parallel.

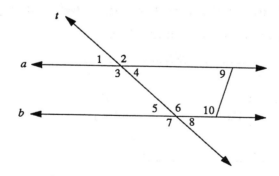

14. Name three angles congruent to $\angle 1$.
15. If m $\angle 6$ = 120°, find the m $\angle 2$.
16. If m $\angle 8$ = 58°, find the m $\angle 3$.
17. If m $\angle 5$ = (7x + 4)° and m $\angle 4$ = (10x − 20)°, find x.

Continued on next page ...

In problems 18 through 20 classify the given triangle according to its angles.

18.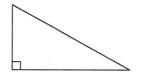

19.

20.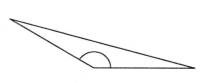

In problems 21 through 26 refer to the following plane figure. Assume $\overleftrightarrow{AB} \perp \overleftrightarrow{BD}$, m ∠1 = 57°, m ∠2 = 64°, and m ∠3 = 41°.

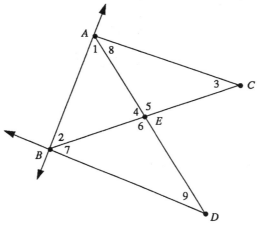

21. m ∠8 = _____
22. m ∠4 = _____
23. m ∠7 = _____
24. m ∠9 = _____
25. m ∠CED = _____
26. m ∠5 = _____

27. Find the area of a triangle with a base of 12 inches and a height of 10 inches.

28. Find the area of a rectangle whose width is 10 cm and length is 14 cm.

29. Find the area of a trapezoid with bases of 12 cm and 18 cm, respectively, and height 10 cm.

30. Find the perimeter and area of the parallelogram shown below.

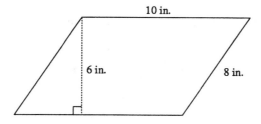

31. A triangle with sides of 2 inches, 4 inches, and 6 inches is given. Determine if this triangle is a right triangle.

Continued on next page ...

32. Triangles *ACG* and *HYM* are similar triangles. If $\overline{AC}$ measures 10 units, $\overline{AG}$ measures 25 units, $\overline{GC}$ measures 20 units, and $\overline{HY}$ measures 28 units, find the measures of the remaining two sides of triangle *HYM*.

33. Find the volume of a backyard wading pool that has a diameter of 12 feet and is 24 inches high. Use $\pi = 3.14$ and round your answer to the nearest tenth.

34. Find the area of the shaded region of the figure below. Use $\pi = 3.14$ and round your answer to the nearest tenth.

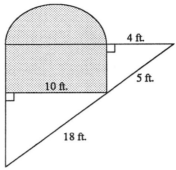

35. Find the perimeter and area of a circle with diameter 40 cm. Use $\pi = 3.14$ and round your answer to the nearest tenth.

36. A rectangular pyramid is 4 m high and has a base with length 1.5 m and width 1 m. Find the volume of this pyramid.

37. Determine if the network below is traversable. If it is then indicate the possible starting points.

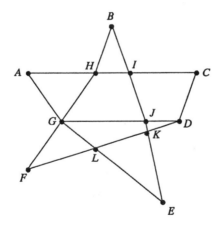

11. CONSUMER MATHEMATICS

11.1 INTRODUCTION

In this chapter we discuss various topics relating to mathematics applied to the world of business. Two topics discussed in detail are percents and interest calculations. The intent of this chapter is to provide you with the necessary tools to become an informed and intelligent consumer.

11.2 RATIO AND PROPORTION

A **ratio** is a comparison of two numbers by division. In general, there are three symbols that are used to express a ratio: (1) a division sign ($\div$); (2) a fraction bar (—); and (3) a colon (:). For example, the following all express the same relationship. In the context of ratios, all are read as *x is to y*.

$$x \div y \;=\; \frac{x}{y} \;=\; x : y$$

If two equivalent ratios are set equal to one another, the resulting equation is called a **proportion**. For example, $\frac{4}{6} = \frac{2}{3}$ is a proportion and is read as *4 is to 6 as 2 is to 3*. (Note that in the context of a proportion, the equal sign is read *as*.) In a proportion, the denominator of the left ratio and the numerator of the right ratio are called the **mean** terms, and the numerator of the left ratio and the denominator of the right ratio are called the **extreme** terms.

$$\frac{a}{b} = \frac{c}{d} \qquad\qquad \frac{a}{b} = \frac{c}{d}$$

mean terms (*b* and *c*)   extreme terms (*a* and *d*)

The product of the means is equal to the product of the extremes.

$$\frac{a}{b} = \frac{c}{d}$$
$$(c)(b) = (a)(d)$$

Example 1 (page 628, #3c,e)

Express the following as ratios in simplest form (parts c and e only).

Solution

To express a ratio in simplest form we express it first as a fraction and then reduce the fraction to lowest terms.

c. $\dfrac{18}{12}$

$$\frac{18}{12} = \frac{3 \times 6}{2 \times 6} = \frac{3 \times \cancel{6}^{1}}{2 \times \cancel{6}_{1}} = \frac{3}{2}$$

Continued on next page ...

e. $3\frac{1}{2}$ to $4\frac{2}{3}$

$$3\frac{1}{2} \text{ to } 4\frac{2}{3} = \frac{3\frac{1}{2}}{4\frac{2}{3}} = \frac{\frac{7}{2}}{\frac{14}{3}}$$

$$= \frac{7}{2} \div \frac{14}{3}$$

$$= \frac{7}{2} \times \frac{3}{14}$$

$$= \frac{21}{28}$$

$$= \frac{3}{4}$$

Example 2 (page 629, #5)

A mathematics class has 30 students in it. There are 18 women and 12 men.

    a. What is the ratio of men to women?

    b. What is the ratio of women to men?

    c. What is the ratio of the number of men to the number of students in the class?

Solution

    a. The ratio of men to women is

$$\frac{number\ of\ men}{number\ of\ women} = \frac{12}{18} = \frac{2}{3}$$

    b. The ratio of women to men is

$$\frac{number\ of\ women}{number\ of\ men} = \frac{18}{12} = \frac{3}{2}$$

    c. The ratio of men to students in the class is

$$\frac{number\ of\ men}{number\ of\ students} = \frac{12}{30} = \frac{2}{5}$$

Example 3 (page 629, #9a,b,c)

Find the value of x in each proportion (parts a, b, and c).

Solution

The following general procedure will be observed in solving these problems.

1. Cross multiply; that is, set the product of the mean terms equal to the product of the extreme terms.

2. Working with the side of the equation that consists of numerical factors only, find the product of these factors.

3. Divide the product from step 2 by the numerical factor that is on the opposite side of the equal sign. (The resulting quotient is the missing term.)

Continued on next page ...

a. $x : 7 = 6 : 21$

$$\frac{x}{7} = \frac{6}{21} \quad \text{(Cross multiply.)}$$
$$(6)(7) = (x)(21) \quad \text{(Solve for } x.\text{)}$$
$$42 = 21x$$
$$\frac{42}{21} = x$$
$$2 = x$$

b. $3 : x = 14 : 28$

$$\frac{3}{x} = \frac{14}{28} \quad \text{(Cross multiply.)}$$
$$(14)(x) = (3)(28) \quad \text{(Solve for } x.\text{)}$$
$$14x = 84$$
$$x = \frac{84}{14}$$
$$x = 6$$

c. $5 : 1 = x : 6$

$$\frac{5}{1} = \frac{x}{6} \quad \text{(Cross multiply.)}$$
$$(x)(1) = (5)(6) \quad \text{(Solve for } x.\text{)}$$
$$x = 30$$

Example 4 (page 629, #13)

A tree casts a shadow 6 feet long. Gerry, whose height is 5 feet 6 inches, is standing next to the tree. If she casts a shadow 16 inches long, what is the height of the tree?

Solution

- Set up a proportion so that each ratio of the proportion is comparing similar quantities.

$$\frac{\text{shadow of tree}}{\text{shadow of Gerry}} = \frac{\text{height of tree}}{\text{height of Gerry}}$$

- Replace the quantities with their respective measurements. All measurements are in inches.

$$\frac{72 \text{ inches}}{16 \text{ inches}} = \frac{x \text{ inches}}{66 \text{ inches}}$$

- Solve the proportion.

$$\frac{72}{16} = \frac{x}{66}$$
$$(16)(x) = (72)(66)$$
$$x = \frac{4{,}752}{16}$$
$$x = 297 \text{ inches}$$

The height of the tree is 297 inches, which is equal to 24 feet 9 inches.

SUPPLEMENTARY EXERCISE 11.2

In problems 1 through 10 express each ratio in simplest form.

1. 4 to 12
2. 6 : 30
3. 21 is to 5
4. $\frac{39}{66}$
5. 16 to 4
6. 72 : 110

7. 200 miles to 3 hours
8. 6 children to 3 families
9. 10 months to 3 years
10. 1 dollar to 1 dime

In problems 11 through 19 find the missing term in each proportion.

11. $\frac{7}{8} = \frac{x}{64}$
12. $\frac{x}{5} = \frac{21}{35}$
13. $\frac{108}{x} = \frac{18}{6}$
14. $\frac{100}{25} = \frac{22}{x}$
15. $\frac{30}{x} = \frac{15}{8}$
16. $\frac{x}{90} = \frac{12}{20}$
17. 10 : 12 = 30 : x
18. 104 : 16 = x : 16
19. 78 : x = 15 : 20

20. If one pound of trigger fish costs $3.07, how much will five ponds cost?
21. Jane can travel 68 miles in her car using 2 gallons of gasoline. How far can she travel in her car using 10 gallons of gasoline?
22. Grandma's tomato sauce sells for three cans for $1.98. At this price how much will one can cost?
23. Bill's electric bill for the month of July is $110.00 based on 120 kilowatts of electricity used. If his kilowatt usage for the month of August is 90 kilowatts, how much will his August electric bill be based on the July bill?
24. If you travel at an average rate of 58 miles per hour, how long will it take you to travel 812 miles?

11.3 PERCENTS, DECIMALS, AND FRACTIONS

11.3.1 Expressing a Percent As a Fraction

Recall that a ratio is the quotient of two expressions. A percent is a special way of writing the ratio of two numbers when the second number (i.e., the denominator) is 100. For example, the ratio $\frac{35}{100}$ can be expressed in percent form as 35%. (The symbol % is read as *percent* and means per hundred.) As a result, any number expressed as a ratio with a denominator of 100 can be written in the form of a percent. The numerator, N, is the number of the percent, and the denominator is represented by the percent symbol.

$$\frac{N}{100} = N\%$$

11.3.2 Expressing a Decimal As a Percent

Any decimal number can be expressed as a fraction in which its denominator is equal to 100. For example

- $0.24 = \frac{24}{100}$

- $0.6 = 0.60 = \frac{60}{100}$

- $0.125 = \frac{125}{1000} = \frac{12.5}{100}$

Every ratio with a denominator of 100 also can be expressed as a percent. For example

- $\dfrac{24}{100} = 24\%$

- $\dfrac{60}{100} = 60\%$

- $\dfrac{12.5}{100} = 12.5\,\%$

Combining these two concepts we conclude that every decimal can be expressed as a percent by first expressing the given decimal as a ratio with a denominator of 100, and then converting this ratio to a percent.

- $0.24 = \dfrac{24}{100} = 24\%$

- $0.6 = .60 = \dfrac{60}{100} = 60\%$

- $0.125 = \dfrac{125}{100} = 12.5\,\%$

The procedure usually used to convert decimals to percents is to multiply the percent by 100 and append the percent symbol. Appending a percent symbol to a number is equivalent to dividing the number by 100. As a result, the numerical value of the number is not altered; rather, it is expressed in a different but equivalent form. This procedure eliminates the intermediate step of first expressing the decimal as a ratio. (**Note:** Multiplying a number by 100 is tantamount to moving the decimal point of the number being multiplied two places to the right.)

11.3.3 Expressing a Percent As a Decimal

A percent can be written in decimal form by first changing the percent to a ratio with a denominator of 100, and then changing the fraction to a decimal. For example

- $24\% = \dfrac{24}{100} = 0.24$

- $6\% = \dfrac{6}{100} = 0.06$

- $125\% = \dfrac{125}{100} = 1.25$

- $0.5\% = \dfrac{0.5}{100} = 0.005$

The procedure usually used to convert percents to decimals is to divide the percent by 100 and remove the percent symbol. Removing the percent symbol from a number is equivalent to multiplying the number by 100. As a result, the numerical value of the number is not altered; rather, it is expressed in a different but equivalent form. (**Note:** Dividing a number by 100 is tantamount to moving the decimal point of the number being divided two places to the left.)

11.3.4 Expressing a Fraction As a Percent

To express a fraction as a percent, first convert the fraction to a decimal (see section 8.6), and then use the procedure given in section 11.3.2. For example

- $\frac{3}{5} = 3 \div 5 = 0.6 = 60\%$
- $\frac{5}{16} = 5 \div 16 = 0.3125 = 31.25\%$

Example 1 (page 634, #1a)

Express 15% as a fraction in simplest terms.

Solution

Express the percent as a ratio with a denominator of 100 and reduce the fraction.

$$15\% = \frac{15}{100} = \frac{3}{20}$$

Example 2 (page 634, #3b)

Express $2\frac{1}{3}\%$ as a fraction in simplest terms.

Solution

Express the percent as a ratio with a denominator of 100 and reduce the fraction.

$$2\frac{1}{3}\% = \frac{2\frac{1}{3}}{100} = \frac{\frac{7}{3}}{100} = \frac{7}{3} \div \frac{100}{1}$$
$$= \frac{7}{3} \times \frac{1}{100}$$
$$= \frac{7}{300}$$

Example 3 (page 634, #5a)

Express 6.5% as a fraction in simplest terms.

Solution

$$6.5\% = \frac{6.5}{100} = \frac{65}{1000} = \frac{13}{200}$$

Example 4 (page 635, #7c)

Express $4\frac{1}{2}\%$ as a decimal.

Solution

Move the decimal point of the number two places to the left and remove the percent sign.

$$4\frac{1}{2}\% = 4.5\% = 0.045$$

Example 5 (page 635, #9c)

Express 0.25% as a decimal.

Solution

$$0.25\% = 0.0025$$

Example 6 (page 635, #11a)

Express 0.09 as a percent

Solution

Move the decimal point two places to the right and append a percent sign.

$$0.09 = 9\%$$

Example 7 (page 635, #15c)

Express $\frac{5}{8}$ as a percent.

Solution

Convert $\frac{5}{8}$ to a decimal and then convert the decimal to a percent.

$$\frac{5}{8} = 0.625 = 62.5\%$$

SUPPLEMENTARY EXERCISE 11.3

In problems 1 through 20 complete the charts by expressing each of the given values in equivalent form.

Number	Fraction (reduced form)	Decimal	Percent
1	$\frac{1}{2}$		
2		0.3	
3			80%
4	$\frac{1}{8}$		
5		0.625	
6			100%
7	$\frac{1}{5}$		
8		0.002	
9			1%
10		3.004	

Number	Fraction (reduced form)	Decimal	Percent
11		0.08	
12			3.2%
13	$\frac{11}{16}$		
14	$\frac{11}{25}$		
15		0.10	
16			0.8%
17		5.2	
18			340%
19			$37\frac{1}{2}\%$
20	$2\frac{7}{8}$		

11.4 MARKUPS AND MARKDOWNS

The difference between the selling price of an item (i.e., how much an item costs a consumer), and the dealer's cost (i.e., how much an item costs a dealer) is called **markup**. The follwoing three equations show the relationship among markup, dealer's cost, and selling price.

- Markup = Selling Price − Dealer's Cost
- Selling Price = Markup + Dealer's Cost
- Dealer's Cost = Selling Price − Markup

The difference between the original price and the sale price is called **markdown**. Markdown is more popularly known as *discount*. The following three equations show the relationship among markdown, sale price, and original price.

- Markdown = Original Price − Sale Price
- Sale Price = Original Price − Markdown
- Original Price = Sale price + Markdown

Markups are usually given as a percentage of the cost, or as a percentage of the selling price. A *percent markup on cost* is that amount of money that is added to the dealer's cost. This sum yields the selling price. A *percent markup on selling price* is the amount of money a dealer earns (i.e., gets to keep) from the selling price. Markdowns also can be expressed in terms of percent reduction. As a result, solving business-related problems involving markups and markdowns require finding the percent of a number or finding what percent one number is of another. Such problems can be solved by the general proportion

$$\frac{is}{of} = \frac{\%}{100}$$

The first ratio in the proportion, $\frac{is}{of}$, represents the comparison of the two numbers. When comparing two numbers we always compare percentage to base, which is expressed as $\frac{percentage}{base}$. We choose to use the descriptive words *is* and *of* in lieu of the formal terms *percentage* and *base*. In a percent problem *is* usually appears adjacent to the percentage, and *of* usually appears prior to the base. Thus when setting up a proportion to solve a percent problem, the numerator of the first ratio is the number that represents *is* and the denominator of the first ratio is the number that represents *of*. This is illustrated in Example 1. The second ratio of the proportion, $\frac{\%}{100}$, is the actual percentage changed to fractional form. Thus the numerator of this second ratio represents the actual percentage. The denominator of the ratio (100) remains constant.

Example 1 (page 643, #1a,c,e)

Solve the given percent problems (parts a,c, and e only).

Solution

a. What percent of 48 is 24?

$$is = 24 \text{ and } of = 48$$

$$\frac{is}{of} = \frac{\%}{100}$$

$$\frac{24}{48} = \frac{x}{100}$$

$$(48)(x) = (24)(100)$$

Continued on next page ...

$$48x = 2400$$
$$x = \frac{2400}{48}$$
$$x = 50$$

24 is 50% of 48

c. What percent of 30 is 40?

$is = 40$ and $of = 30$

$$\frac{is}{of} = \frac{\%}{100}$$

$$\frac{40}{30} = \frac{x}{100}$$

$$(30)(x) = (40)(100)$$
$$30x = 4000$$
$$x = \frac{4000}{30}$$
$$x = 133.\overline{3}$$

40 is $133.\overline{3}$% of 30

e. What percent of 216 is 54?

$is = 54$ and $of = 216$

$$\frac{is}{of} = \frac{\%}{100}$$

$$\frac{54}{216} = \frac{x}{100}$$

$$(216)(x) = (54)(100)$$
$$216x = 5400$$
$$x = \frac{5400}{216}$$
$$x = 25$$

54 is 25% of 216

Example 2 (page 643, #3c)

Given cost = $25.00 and percent markup on cost is 30%, find the dollar amount of markup to the nearest cent.

Solution

- To find the dollar amount of markup solve the percent problem *30% of $25.00 is = ?*

Continued on next page ...

- Use the $\frac{is}{of}$ proportion where $is = x$, $of = 25$, and $\% = 30$.

$$\frac{is}{of} = \frac{\%}{100}$$

$$\frac{x}{25} = \frac{30}{100}$$

$$(30)(25) = (x)(100)$$
$$750 = 100x$$
$$\frac{750}{100} = x$$
$$7.5 = x$$

The markup is $7.50.

Example 3 (page 644, #5a)

Assume the selling price of an item is $50.00 and the dealer's cost is $20.00. Find the markup and the percent markup on the cost.

Solution

- First find the markup.

$$\begin{aligned} \text{Markup} &= \text{Selling Price} - \text{Dealer's Cost} \\ &= \$50.00 - \$20.00 \\ &= \$30.00 \end{aligned}$$

- Determine the percent markup on the cost. That is,

What percent of the cost is the markup?

Translating this statement, we get

What percent of $20.00 is $30.00?

To answer this question use the $\frac{is}{of}$ proportion where $is = 30$, $of = 20$, and $\% = x$.

$$\frac{is}{of} = \frac{\%}{100}$$

$$\frac{30}{20} = \frac{x}{100}$$

$$(20)(x) = (30)(100)$$
$$20x = 3000$$
$$x = \frac{3000}{20}$$
$$x = 150$$

The percent markup is 150%.

Example 4 (page 644, #7)

A coat sells for $125. The markup is 40% of the cost. What is the cost of the coat to the nearest cent?

Solution

To find the dealer's cost we determine the amount of money the coat was marked up. Once we know the dollar amount of the markup we subtract this amount from the selling price ($125) to find the dealer's cost. Unfortunately, we cannot determine the dollar amount of the markup from the problem as it is stated. The markup can be represented in terms of *percent of cost*, though.

Selling Price	=	Markup	+	Dealer's Cost
$125	=	40% of cost	+	(100% of the Dealer's Cost)
$125	=	140% of cost		

We now know that the selling price (which is the dealer's cost for the coat and the markup) represents 140% of the dealer's cost. This translates to *$125 is 140% of the dealer's cost.* Using the $\frac{is}{of}$ proportion with $is = 125$, $of = x$, and $\% = 140$ we get

$$\frac{is}{of} = \frac{\%}{100}$$

$$\frac{125}{x} = \frac{140}{100}$$

$$(140)(x) = (125)(100)$$

$$140x = 12500$$

$$x = \frac{12500}{140}$$

$$x = 89.285$$

Dealer's cost is approximately $89.29.

(**Note:** Use this problem as a reference for problems 8 through 11 in your textbook.)

Example 5 (page 644, #13a)

Assume the selling price of an item is $400.00 and the percent markup on the selling price is 40%. Find the markup and the cost.

Solution

- The percent markup on the selling price is 40%. This implies *40% of the selling price is markup*.
- Find the markup.

40% of 400 is markup

$$\frac{is}{of} = \frac{\%}{100}$$

$$\frac{x}{400} = \frac{40}{100}$$

$$(40)(400) = (x)(100)$$

Continued on next page ...

280 STUDENT STUDY GUIDE

$$(40)(400) = (x)(100)$$
$$16000 = 100x$$
$$\frac{16000}{100} = x$$
$$160 = x$$

The markup is $160.

- Find the dealer's cost.

$$\text{Dealer's Cost} = \text{Selling Price} - \text{Markup}$$
$$= \$400 - \$160$$
$$= \$240$$

The dealer's cost is $240.

Example 6 (page 644, #15)

Al's Appliance Outlet had a clearance sale on refrigerators. A certain model, originally priced at $395, was advertised at 20% off. What was the dollar markdown? What was the sale price?

Solution

- Express the mark down in words.

markdown is 20% of 395

- Calculate the markdown.

$$\frac{is}{of} = \frac{\%}{100}$$
$$\frac{x}{395} = \frac{20}{100}$$
$$(20)(395) = (x)(100)$$
$$7900 = 100x$$
$$\frac{7900}{100} = x$$
$$79 = x$$

The markdown is $79.

- Find the sale price.

$$\text{Sale Price} = \text{Original Price} - \text{Markdown}$$
$$= \$395 - \$79$$
$$= \$316$$

The sale price is $316.

SUPPLEMENTARY EXERCISE 11.4

1. A&M Collision Shop bought a truck for $150.00. The truck was then sold at a markup of 35%. What was the selling price of the truck?

2. ForeArms Gym buys a complete set of barbells for $88.00, which are then sold for $100.00. What is the percent of markup?

3. Andy's Appliance buys electric stoves for $190.00 each and sells them for $250.00 each. What is the percent of markup on cost?

4. An antique dealer bought a rolltop desk for $500.00. If she sold the desk for $850.00 what was the amount of markup and the percent of markup on the cost?

5. A dining room table sold for $480.00. If there were a 45% markup on the cost how much did the table cost the person who sold it?

In problems 6 through 10 find (a) the markup and (b) the selling price to the nearest cent.

6. A camera costing $119.20 with an 18% markup on cost

7. A stereo receiver costing $469 with a 20% markup on cost

8. A television set costing $425 with a 32% markup on cost

9. A washing machine costing $290 with a 9% markup on cost

10. An automobile costing $4,250 with a 24% markup on cost

11. A dress that normally sells for $88.00 has been marked down 30%. What is the new selling price? How much was the markdown?

12. Kim receives a discount of 15% as an employee of a department store. If she buys a pair of shoes that regularly sell for $28.00, how much of a discount will she receive? What will the shoes cost her?

13. A television set regularly selling for $399 is now on sale for $330. What is the percent of markdown?

14. Chuck bought a camera on sale for $195.00. If he saved 25% what was the percent of markdown?

15. Milt gave a businessman an estimate of $1300 to have a parking lot paved. Milt offered the man a 2% markdown if the man paid cash for the job. How much would the businessman save if he paid cash?

16. Ron was offered a $750 markdown on a $6200 automobile. What was the percent of the markdown?

17. Which is the better buy: A stereo tape deck that sells for $269 and discounted 20%, or the same tape deck on sale for $215.99?

18. A used car advertised for $1500 was bought for $1200. If the buyer received an additional 10% markdown, what was the total amount of money saved from the advertised price?

19. Driveway sealer sells for $8.50 per gallon. If 500 gallons are bought, a discount of $85 from the total price is given. What is the percent of markdown on a purchase of 500 gallons?

20. A salesman indicates that all refrigerators in his store have been marked down 20% and the sale price on the price tag reflects this discount. If the price tag shows the regular price to be $889.99 and the sale price to be $719.89, is the tag marked correctly? Why or why not?

11.5 SIMPLE INTEREST

The amount of money paid for the use of another person's money is called **interest**. The amount of money we use, on which interest is being paid, is called **principal**. The amount of interest is a certain percentage, called the **rate of interest**, of the principal over a fixed period of time. Simple interest (I), is the product of principal (P), rate of interest (r), and time (t) expressed in years. This formula is expressed mathematically as $I = Prt$.

Example 1 (page 649, #5)

Find the simple interest on a $5,000 loan for 6 months at $11\frac{1}{2}\%$.

Solution

The principal (P) is equal to $5,000; the rate of interest (r) is equal to $11\frac{1}{2}\%$, which is equal to 0.115 expressed as a decimal; and the time period (t) is equal to 6 months, or 0.5 year (since time must be expressed in terms of years). As a result,

$$I = P \times r \times t$$
$$I = (5000) \times (0.115) \times (0.5)$$
$$I = \$287.50$$

Note that the total amount that is paid at the end of the six-month period is equal to $5,287.50, which is the principal plus the amount of interest paid. This total amount is known as the **maturity value**.

Example 2 (page 649, #11)

Connie has a balance due of $240 on her charge account. The rate is $1\frac{1}{2}\%$ simple interest per month on the unpaid balance. If Connie decided not to pay anything toward her balance this month, and she does not charge anything to her account next month, what will be her new balance due next month?

Solution

- Principal = $240
- Rate of interest = $1\frac{1}{2}\%$ per month, which is 18% per year ($1\frac{1}{2} \times 18$), or 0.18
- Time = 1 month, which is $\frac{1}{12}$ of one year

$$I = P \times r \times t$$
$$I = (240) \times (0.18) \times \left(\frac{1}{12}\right)$$
$$I = \$3.60$$

Example 3 (page 650, #23)

If a certificate of deposit pays 12.8% simple interest, find the maturity value of a $5,000 certificate at the end of 26 weeks. How much interest will it earn?

Solution

The formula $A = P(1 + rt)$ can be used for this problem. Note that 12.8% expressed as a decimal is 0.128, and that 26 weeks is equal to one-half year.

Continued on next page ...

$$
\begin{aligned}
A &= P(1+rt) \\
&= 5000(1 + (0.128)(0.5)) \\
&= 5000(1 + 0.064) \\
&= 5000(1.064) \\
&= 5320
\end{aligned}
$$

The maturity value is $5,320, and the interest earned is $5,320 − $5,000, which is $320.

SUPPLEMENTARY EXERCISE 11.5

1. Ralph bought a house for $45,000. How much interest must be paid if the rate of interest is 11% for 30 years?

2. Elaine deposited $10,000 in a certificate of deposit (CD) for six months at 14.8% per year. How much interest will she receive on her original investment?

3. Shannon bought a used car for $5,300. If she put $1,800 down and financed the balance for three years at 9.75%, how much interest must she pay?

4. What is the amount due on a loan of $925.00 for two years at 18%?

5. Cousin Mike owed $8,500 in gambling debts to a loan shark. He paid 25% of his debt but had to finance the balance for nine months at 30% interest. How much interest did he have to pay?

In problems 6 through 10 complete the chart below.

	Principal	Rate	Time	Interest
6.	$700.00	8%	4 months	?
7.	$1,600.00	12%	6 months	?
8.	$8,000.00	9.75%	2 years	?
9.	$10,000.00	12.5%	3 years, 4 months	?
10.	$50,000.00	$14\frac{1}{4}$%	25 years	?

11.6 COMPOUND INTEREST

When the interest due at the end of a certain period is added to the principal, and the sum earns interest for the next period, the interest paid is called **compound interest**. The interest for each succeeding period is greater than the previous one because the interest keeps increasing.

To compute compound interest we use Table 11.6.1, which is based on a $1.00 investment. It should be noted that the table entries are not the amount of interest $1.00 earns based on a particular rate for a specified period of time. Rather, the entries represent the sum of the $1 investment and the interest earned on this investment. This total accumulation is called the **compound amount**. To find the compound amount on a principal other than $1.00, first find the compound amount for $1.00, based on the specified period of time and interest rate, and then multiply this amount by the given principal.

Table 11.6.1 Compounded Amount of $1.00

Periods	2%	3%	4%	6%	8%	10%	12%	14%	16%
1	1.020	1.030	1.040	1.060	1.080	1.100	1.120	1.140	1.160
2	1.040	1.061	1.082	1.124	1.166	1.210	1.254	1.300	1.346
4	1.082	1.126	1.170	1.262	1.360	1.464	1.574	1.689	1.811
6	1.126	1.194	1.265	1.419	1.587	1.772	1.974	2.195	2.436
8	1.172	1.267	1.369	1.594	1.851	2.144	2.476	2.853	3.278
10	1.219	1.344	1.480	1.791	2.159	2.594	3.106	3.707	4.411
12	1.268	1.426	1.601	2.012	2.518	3.138	3.896	4.818	5.936
14	1.319	1.513	1.732	2.261	2.937	3.797	4.887	6.261	7.988
16	1.373	1.605	1.873	2.540	3.426	4.595	6.130	8.137	10.748
20	1.486	1.806	2.191	3.207	4.661	6.728	9.646	13.743	19.461
24	1.608	2.033	2.563	4.049	6.341	9.850	15.179	23.212	35.236
28	1.741	2.288	2.999	5.112	8.627	14.421	23.884	39.204	63.800

Example 1 (page 655)

Find the compound amount for $700 at 6% for 4 years if interest is compounded **annually**.

Solution

Since interest is compounded annually there is one period per year, or four periods for four years. Using Table 11.6.1, the compound amount for $1 based on a 6% rate of interest for four periods is 1.262. To find the compound amount for $700 multiply 1.262 by 700. Thus the compound amount is $700 \times 1.262 = \$883.40$.

Example 2 (page 655, #7)

Find the compound amount for $500 at 8% for 10 years if interest is compounded **semiannually**.

Solution

If interest is compounded semiannually, then there are two periods per year and 20 periods for 10 years. An 8% annual rate of interest is equivalent to a 4% semiannual rate of interest. Using Table 11.6.1, the compound amount for $1 based on a 4% semiannual rate of interest for 20 periods is 2.191. The compound amount for $500 is 500×2.191, or $1,095.50.

Example 3 (page 655, #13)

Find the compound amount for $500 at 8% for two years if interest is compounded **quarterly**.

Solution

- Quarterly implies four periods per year. Thus two years has 8 periods.
- An 8% annual rate of interest implies $\frac{8}{4}$%, which is a 2% quarterly rate.
- The compound amount for $1.00 based on 8 periods at 2% is 1.172.
- $500 \times 1.172 = 586.00$

The compound amount is $586.00.

Example 4 (page 655, #19)

Find the compound amount and the compound interest on $3,000 invested for five years at 12% compounded quarterly.

Solution

- There are 20 periods.
- The (quarterly) rate of interest is 3%.
- The compound amount for $1.00 based on 20 periods at 3% is 1.806.

The compound amount for $3,000 is $5,418.00 (3,000 × 1.806).

The compound interest is $2,418.00 ($5,418 − $3,000).

Example 5 (page 655, #23)

In 1988, the price of a medium-sized automobile was $9,000. What can we expect the price of this same type of car to be in the year 2000, if we assume that the annual rate of inflation is 10%?

Solution

This problem implies that interest (in this case rate of inflation) is compounded annually for 12 years (from 1988 to 2000). The number of periods involved is 12, and the rate of interest (inflation) is 10%. Using Table 11.6.1, the compounded amount for $1 based on a rate of 10% for a period of 12 years is 3.138. Thus the compounded amount for $9,000 is 9,000 × 3.138, which is equal to 28,242. The price of a medium-sized automobile that costs $9,000 in 1988 will cost $28,242 in the year 2000, based on annual rate of inflation of 10%.

SUPPLEMENTARY EXERCISE 11.6

In problems 1 through 8 find the compound amount for each investment if interest is (a) compounded annually, (b) compounded semiannually, and (c) compounded quarterly.

1. $3,000 for four years at 12%
2. $5,500 for two years at 8%
3. $250 for one year at 16%
4. $10,500 for six years at 10% (Answer part (a) only.)
5. $25,000 for 20 years at 14% (Answer part (a) only.)
6. $50,000 for 10 years at 12% (Answer parts (a) and (b) only.)
7. $75,000 for 12 years at 6% (Answer parts (a) and (b) only.)
8. $92,155 for 14 years at 8% (Answer parts (a) and (b) only.)
9. Howard Prince invested $20,000 for five years at a simple interest rate of 8%. Mike Pascale invested the same amount for the same period of time, but his rate of interest was 8% compounded semiannually. What is the total difference in interest earned for the two investments?
10. Dr. Alberg invested $500 seven years ago at 8% compounded quarterly. How much is his investment worth today?
11. Kit Taylor put $800 in a savings account today earning 6% compounded annually. How long will it take for the $800 to double itself?
12. Heidi Horowitz borrowed $2,000 from her mother to buy a car. She agreed to repay the loan, plus interest, in two years. The interest rate she and her mother agreed upon was 4% compounded semiannually. How much interest did Heidi pay?

11.7 EFFECTIVE RATE OF INTEREST

Effective rate is the annual interest rate that gives the same yield as the nominal interest rate compounded more than once a year. Effective rate is often used to compare interest rates that are compounded at different intervals. If interest is paid (or charged) more than once a year then the effective rate is greater than the stated annual rate; otherwise the two are equal. The following formula is used to determine effective rate:

$$E = (1 + r)^n - 1$$

where E = effective rate
n = the number of payment periods per year
r = the interest rate per period (expressed as a decimal)

To illustrate the use of this formula, let us compare the following two offers.

[1] Bank A offers 5% interest compounded quarterly on all of their savings accounts.

[2] Bank B offers 5½% interest compounded semiannually on all of their savings accounts.

Shown in the table below is a comparison of these two offers.

Bank A	Bank B
n = four periods	n = two periods
$r = \dfrac{5\%}{4} = 1.25\% = 0.0125$	$r = \dfrac{5\frac{1}{2}\%}{2} = 2.75\% = 0.0275$
$E = (1 + r)^n - 1$	$E = (1 + r)^n - 1$
$E = (1 + 0.0125)^4 - 1$	$E = (1 + 0.0275)^2 - 1$
$E = (1.0125)^4 - 1$	$E = (1.0275)^2 - 1$
$E = 1.050945337 - 1$	$E = 1.05575625 - 1$
$E = 0.050945337$	$E = 0.05575625$
$E \approx 5.09\%$	$E \approx 5.58\%$

Bank B offers the higher effective rate.

Example 1 (page 660, #3)

What is the effective rate if money is invested at 12% compounded semiannually?

Solution

- The term "compounded semiannually" implies that $n = 2$ and $r = \dfrac{12\%}{2}$, which is 6%.

- Solve for E.

$$\begin{aligned}E &= (1 + r)^n - 1 \\&= (1 + 6\%)^2 - 1 \\&= (1 + 0.06)^2 - 1 \\&= 1.06^2 - 1 \\&= 1.1236 - 1 \\&= 0.1236\end{aligned}$$

The effective rate of interest is 12.36%

Example 2 (page 660, #13)

Which is the higher interest rate: 5% compounded quarterly or 5.5% compounded semiannually?

Solution

5% compounded quarterly	5.5% compounded semiannually
n = four periods	n = two periods
$r = \dfrac{5\%}{4} = 1.25\% = 0.0125$	$r = \dfrac{5.5\%}{2} = 2.75\% = 0.0275$
$E = (1 + r)^n - 1$	$E = (1 + r)^n - 1$
$= (1 + 0.0125)^4 - 1$	$= (1 + 0.0275)^2 - 1$
$= (1.0125)^4 - 1$	$= (1.0275)^2 - 1$
$= 1.050945337 - 1$	$= 1.05575625 - 1$
$= 0.050945337$	$= 0.05575625$
$E \approx 5.10\%$	$E \approx 5.58\%$

The second investment has the higher rate of interest.

SUPPLEMENTARY EXERCISE 11.7

In problems 1 through 5 find the effective rate of interest for each nominal rate. Round fractional percentages to the nearest hundredth of a percent.

1. 7% compounded semiannually
2. 11% compounded quarterly
3. 8.8% compounded annually
4. 9½% compounded quarterly
5. 16% compounded monthly

6. Which bank offers the higher interest rate on savings?

   Bank A, whose rate is 5½% compounded semiannually
   Bank B, whose rate is 5¼% compounded monthly

7. A bank advertises that money invested at 9.75% compounded semiannually has an effective annual rate of 10.13%. Is this correct? Why or why not?

11.8 LIFE INSURANCE

Life insurance can be purchased in many different forms. Some of the types of policies are reviewed here.

11.8.1 Term Life

A term life insurance policy provides protection for a specified number of years. The insurance company will pay the face value of the policy to a beneficiary provided that the policy holder dies during the period of time for which the policy is in effect. At the end of the term the policy is considered null and void, which means the insurance company is no longer liable for payment. Term insurance usually can be renewed for another term, but often at a higher rate. This type of policy is the least expensive of the four types of life insurance discussed in this section.

11.8.2 Straight Life

A straight life insurance policy is one in which the policyholder pays premiums until death. The premiums for this type of policy are higher than term life because part of the premiums are saved by the company for the policyholder. In the event the policyholder cancels the policy these savings, called **cash surrender value,** are paid to the policyholder.

11.8.3 Limited Payment Life

A limited payment life insurance policy is similar to straight life in that there is a savings factor incorporated into the premium. The difference, though, is rather than making payments until death (as in straight life), the policyholder is required to make only a limited number of premium payments. Nevertheless, the policy is in effect until it is either canceled or the policyholder dies, at which time the beneficiary receives the face value of the policy. The premiums for this type of policy are higher than those of term or straight life.

11.8.4 Endowment Life

An endowment life insurance policy has both a limited number of premium payments and a savings factor. However, the savings factor is usually larger than that of a limited payment policy. If the policyholder survives the term of the policy, then the face value of the policy is paid to him/her. The premium payments for this type of policy are the most expensive of the various types of policies discussed here.

Premiums for life insurance can be paid monthly, quarterly, semiannually, or annually. Premiums paid annually will always result in an economic savings to the policyholder. Annual premium costs are usually stated per $1,000 of life insurance. An example of these rates is provided in Table 11.8.1. To determine semiannual, quarterly, or monthly premiums, multiply the annual premium by a certain percentage. These percentages are provided in Table 11.8.2.

Table 11.8.1 Annual Premium Rate per $1,000 of Life Insurance

Age	5-Year Term	Straight Life	Limited Payment (20-year)	Endowment (20-year)
20	$6.99	$15.48	$25.96	49.32
25	7.57	17.50	28.83	49.73
30	8.60	20.15	32.10	50.17
35	9.98	23.84	35.92	51.13
40	12.44	28.37	40.74	52.69
45	15.59	33.90	46.57	55.33
50	19.89	41.23	53.01	59.00
55	27.09	50.88	59.73	64.55
60	—	63.95	70.00	—

Table 11.8.2 Premium Rates

Payment Period	Percentage of Annual Premium
Semiannually	51%
Quarterly	26%
Monthly	9%

Example 1 (page 665, #1a)

Find the annual premium for a $25,000 straight life insurance policy at age 25.

Solution

Using Table 11.8.2, it is found that the annual premium per $1,000 of coverage for a straight life policy at age 25 is $17.50. The annual premium for a $25,000 policy is 25 × $17.50, which is $437.50.

Example 2 (page 665, #3)

Laurie Adams is 25 years old and wishes to purchase a $20,000, five-year term life insurance policy. Determine the annual cost if she pays (a) annually, (b) semiannually, (c) quarterly, or (d) monthly.

Solution

a. Using Table 11.8.1, it is found that the annual premium per $1000 of coverage for a five-year term policy at age 25 is $7.57. Thus the annual cost for a $20,000 policy is 20 × $7.57, which is $151.40.

b. If premiums are paid semiannually, then the annual cost is calculated by first finding the semiannual premium and then doubling this amount. (Use Table 11.8.2.)

- Multiply the annual premium by 51%. This product is the semiannual cost.

$$\$151.40 \times 0.51 = \$77.21$$

- Multiply the semiannual premium by 2 to determine the annual cost.

$$\$77.21 \times 2 = \$154.42$$

The annual cost, based on semiannual payments, is $154.42.

c. The annual cost based on quarterly payments is determined in a similar manner as part (b).

- Multiply the annual premium by 26% (see Table 11.8.2) to get the quarterly cost.

$$\$151.40 \times 0.26 = \$39.36$$

- Multiply the quarterly premium by 4 to get the annual cost.

$$\$39.36 \times 4 = \$157.44$$

The annual coat based on quarterly payments is $157.44.

d. The annual cost based on monthly payments is determined in a similar manner as parts (b) and (c).

- Multiply the annual premium by 9% (see Table 11.8.2) to get the monthly cost.

$$\$151.40 \times 0.09 = \$13.63$$

- Multiply the monthly premium by 12 to get the annual cost.

$$\$13.63 \times 12 = \$163.56$$

The annual cost based on monthly payments is $163.56.

SUPPLEMENTARY EXERCISE 11.8

In problems 1 through 8 calculate the premium for each of the given insurance policies.

	Type of Policy	Age at Issue	Face Value	Type of Payments
1.	5-year term	35	$22,000	semiannual
2.	straight life	30	$42,000	monthly
3.	limited payment (20-year)	45	$10,000	annual
4.	endowment (20-year)	20	$35,000	quarterly
5.	straight life	25	$25,000	semiannual
6.	endowment (20-year)	40	$50,000	quarterly
7.	5-year term	50	$100,000	monthly
8.	limited payment (20-year)	60	$500,000	annual

9. Jerry took out a $20,000 straight life insurance policy at age 30. Ten years later he canceled the policy. If his premiums were paid on a monthly basis, how much would have he saved each year if he had paid his premiums annually?

10. Sylvia took out a $33,000 five-year term life insurance policy at age 50 and renewed the policy at age 55. She paid her premiums quarterly for both terms. How much more per quarter were her premiums for the second term than the first term?

11.9 INSTALLMENT PLANS AND MORTGAGES

11.9.1 Installment Plans

Financial institutions and credit merchants occasionally disguise the true costs of credit or the true rate of interest (or both). Such practices prompted the U.S. Government to enact the Truth in Lending Act. This act requires anyone who extends credit to a consumer to reveal the *true annual interest rate*. All charges that make up the difference between the amount borrowed and the amount repaid on installment loans, and the difference between the cash purchase price of an item and the installment price of the same item also must be made known to the purchaser.

The true annual interest rate (also know as the **annual percentage rate**, or **APR**), is approximated by the following formula.

$$i = \frac{2nr}{n + 1}$$

where i = the true interest
n = the number of payments
r = the stated rate of interest

Charge account and credit card purchases usually have associated with them an interest rate of 1.5% or 1.75% per month. Typically, no interest is charged if the balance is paid by the due date, which is anywhere from 25 to 30 days from the time the bill is received. In the event a nonzero balance exists after the due date then the unpaid balance is subjected to an interest charge, which is added to the unpaid balance.

11.9.2 Mortgages

Most people who purchase a house do so with the expectation that they will pay for it on a periodic basis for a fixed period of time. To determine this periodic payment **amortization schedules** have been compiled. An amortized payment includes partial payments on the principal and the interest payments on the declining balance of the principal. An example of an amortized schedule is given in Table 11.9.1. The entries in this table are typical monthly payments per $1,000 of a mortgage for the given interest rates and time periods. (**Note:** This table lists payments per $1,000 for traditional fixed-rate mortgages.)

Table 11.9.1 Monthly Mortgage Payments per $1,000

Rate (%)	Length of Mortgage		
	10 Years	20 Years	30 Years
8	$12.13	$8.36	$7.34
9	12.67	9.00	8.05
10	13.22	9.65	8.78
10½	13.49	9.98	9.15
12	14.35	11.01	10.29
12½	14.64	11.36	10.67
14	15.53	12.44	11.85
15	16.13	13.17	12.64
16	16.75	13.91	13.45
18	18.02	15.43	15.07
20	19.33	16.99	16.71

Example 1 (page 672, #1)

Carl Thomas purchased a television set advertised for $500.00. He bought the television on an installment plan, paying $100.00 down and agreeing to pay the balance in 12 monthly payments. The store charged a finance charge of 12% simple interest on the balance. (a) What was the amount of each payment? (b) What was the true annual interest rate (correct to the nearest tenth of a percent)?

Solution

a. To find the amount of each payment we first determine the total amount of money Carl pays and then divide this amount by 12.

- The total amount financed is the difference between the purchase price and the down payment. For this problem, the amount financed is $500 − $100 = $400.

- The total amount of interest Carl pays is determined by the simple interest formula (since the store's management charged 12% simple interest).

$$I = Prt$$
$$I = (400)(0.12)(1)$$
$$I = 48$$

- The total amount of money Carl must pay is the sum of the amount financed and the interest charge. This total is $400 + $48, which is equal to $448.

- Now divide $448 by 12 to determine his monthly payments for the one year period. Thus Carl's monthly payments are $448 ÷ 12 = $37.33

b. The true annual interest rate is found by using the APR formula with $n = 12$ and $r = 12\% = 0.12$.

$$i = \frac{2nr}{n+1}$$

$$i = \frac{(2)(12)(0.12)}{12+1}$$

$$i = \frac{2.88}{13} = 0.2215$$

The true annual interest rate (to the nearest tenth) is 22.2%.

Example 2 (page 673, #5)

Sandra West purchased a coat for $100.00 and charged it. The store charges an interest rate of $1\frac{1}{2}$% per month on the unpaid balance. If Sandra decides to make payments of $30 per month, how much interest will she pay by the time the coat is paid for? (Assume no additional purchases and no interest charged on the first payment.)

Solution

The following table illustrates Sandra's monthly payments.

Month	Amount Owed	Finance Charge	Unpaid Balance	Payment	New Balance
1	100	0	100	30	100 − 30 = 70
2	70	70 × 0.015 = 1.05	70 + 1.05 = 71.05	30	71.05 − 30 = 41.05
3	41.05	41.05 × 0.015 = 0.62	41.05 + 0.62 = 41.67	30	41.67 − 30 = 11.67
4	11.67	11.67 × 0.015 = 0.18	11.67 + 0.18 = 11.85	11.85	11.85 − 11.85 = 0

The total amount of interest is the sum of the finance charges and is $1.05 + $0.62 + $0.18 = $1.85.

Example 3 (page 673, #9)

What is the monthly payment for a $20,000 mortgage at 9% amortized for 10 years?

Solution

Using Table 11.9.1 the monthly mortgage payment per $1,000 at 9% for 10 years is $12.67. The monthly mortgage payment for $20,000 at 9% for 10 years is the product of $12.67 and $20,000, which is $253.40.

Example 4 (page 673, #15)

The Smiths assumed a $40,000 mortgage for 20 years at $10\frac{1}{2}$%. (a) What is their monthly payment? (b) How much total interest will the Smiths pay on their mortgage?

Solution

a. Using Table 11.9.1 the monthly payment per $1,000 for $10\frac{1}{2}$% for 20 years is $9.98. As a result, the monthly payment for $40,000 is $9.98 × 40, which is equal to $399.20.

b. The total interest that will be paid is the difference between the total amount of money paid and the amount mortgaged.
- Since the mortgage is for 20 years there are 20 × 12, or 240 payments.
- If one payment is $399.20, then 240 payments total 240 × $399.20, or $95,808.
- The total amount of money paid less the amount of money mortgaged yields the total amount of interest. Thus the interest paid is $95,808 − $40,000, or $55,808.

SUPPLEMENTARY EXERCISE 11.9

1. Crazy Lar charged $500.00 worth of novelty items from Beansy's Supply House. Lar was charged 8% simple interest and agreed to pay for the merchandise in 10 equal monthly payments. Find (a) the amount of each payment and (b) the true annual interest rate.

2. Tony bought an assortment of household cleaning appliances for $1,200.00. He put $100.00 down and agreed to pay for the balance in 24 equal monthly payments at 12% simple interest. Find (a) the amount of each payment and (b) the true annual interest rate.

3. Jane charged $850 worth of clothes from a mail order house. If Jane makes payments of $100.00 per month, how much interest will she pay by the time her balance is zero? (Assume (1) no additional purchases are made, (2) the interest rate is $1\frac{1}{4}$% per month on the unpaid balance, and (3) no interest is charged on the first payment.)

4. Joe financed a one year membership of $500.00 at a health club and agreed to pay $22.00 per month with a 2% per month interest charge on the unpaid balance. (a) How many months will it take to pay for the membership? (b) How much interest will Joe pay when the membership is paid in full? (Assume no interest charged on the first payment.)

In problems 5 through 8 find the monthly payment for principal and interest for each mortgage.

	Amount of Mortgage	Interest Rate (%)	Terms of Mortgage (years)
5.	$42,000	10½	30
6.	$75,000	14	20
7.	$18,500	8	10
8.	$300,000	12½	10

11.10 CHAPTER 11 TEST

In problems 1 through 6 solve for the missing term in each proportion.

1. $\dfrac{x}{45} = \dfrac{7}{9}$ 2. $\dfrac{15}{60} = \dfrac{x}{20}$ 3. $\dfrac{8}{x} = \dfrac{120}{75}$

4. $3 : 20 = x : 15$ 5. $2 : 8.4 = x : 67.2$ 6. $21 : x = 20 : 18$

In problems 7 through 12 complete the chart by expressing each of the given entries in equivalent forms.

	Fraction	Decimal	Percent
7.	$\frac{1}{2}$		
8.	$\frac{4}{5}$		
9.		0.002	
10.		2.25	
11.			2%
12.			25%

Continued on next page ...

13. A used car dealer buys a used car for $1,500. What is the percent of markup on the selling price if the dealer sells the car at a $900.00 profit?

14. The selling price of an item is $5.00. If the percent of markup based on the cost is 30%, how much did the item cost the seller?

15. How much simple interest will $5,000.00 earn in six months if the rate of interest is 8.75%?

16. A deposit of $500.00 is made into an account that pays 8% compounded quarterly. How much money will there be in this account at the end of one year?

17. What is the effective rate, to the nearest hundredth of a percent, if money is invested at 9% compounded semiannually?

18. How much would the premiums be on a five-year term $28,000.00 life insurance policy, taken out at age 30, if payments are to be made quarterly?

19. Mary bought a $350.00 VCR at Byrons Department store. She paid $25.00 down and agreed to pay the balance in 12 equal monthly payments. The finance charge was 12% simple interest. Determine (a) the amount of her monthly payments and (b) the true annual interest rate to the nearest tenth.

20. A $39,000.00 mortgage was assumed for 20 years at $10½%. Determine (a) the monthly mortgage payment and (b) the total interest that will be paid on the mortgage.

12. INTRODUCTION TO COMPUTERS

12.1 INTRODUCTION

A computer is an electronic machine capable of receiving, storing, and manipulating data. The manipulation of such data, called **processing**, involves computations as well as sorting, storing, and retrieving data.

Computers can perform only three basic functions: they can perform arithmetic calculations; they can compare or test the relationship between two quantities; and they can store and retrieve data. What makes computers so useful is that they can perform these functions very quickly and accurately. Computers also can be programmed to execute many different sets of instructions and hence are capable of performing a variety of tasks.

Computers, however, do not always function correctly. This can be the result of any one (or more) of three types of problems: an error in the computer program, incorrect input, or an actual machine malfunction.

In this chapter we present a brief overview of computers and how they function.

12.2 HISTORY OF COMPUTERS

{Table 1, which is located on pages 691-692 of your textbook, provides a summary of the development of computing devices and electronic computers. Table 2, which is located on page 693 of your textbook, summarizes the various types of computers. We will neither review this information nor discuss any of the exercises found on pages 693-694 of your textbook.}

12.3 HOW A COMPUTER SYSTEM WORKS

A computer system consists of a central processing unit (CPU), two forms of memory (main and auxiliary), an input unit, and an output unit. A simple configuration of these pieces of equipment is shown in the figure below.

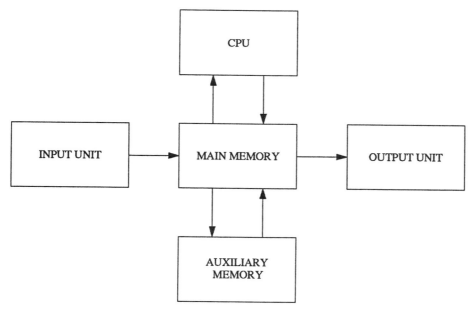

12.3.1 How a Computer System Functions

Basically, a computer system works in the following manner.

- Data are transferred to main memory for processing via a designated input device.
- The CPU calls the data (i.e., it *fetches* the data) from main memory for processing.
- Once processed, the data are returned to main memory. From here a designated output device makes the results available to the user of the system.

12.3.2 What The Individual Components Do

The function of the individual components is as follows.

The CPU

The CPU performs all the arithmetic and logic functions, and oversees the operation of the entire system. The CPU is the brain of the system and consists of three separate units: the **control** unit, the **arithmetic/logic** unit, and **registers**.

Main Memory

Main memory is an extension of the CPU and is the primary storage unit of the computer.

Auxiliary Memory

Auxiliary memory is external from the CPU and supplements main memory. Although auxiliary memory is much slower in operation than main memory, auxiliary memory is less expensive, is larger in capacity, and is permanent. Examples of auxiliary storage units include magnetic tape and disks.

Input/Output Devices

Input/output devices (commonly called I/O devices), enable communication to occur between a user and a computer. I/O units are external to the CPU and must be connected to a computer to be functional. An example of an input device is a keyboard, and an example of an output device is a video monitor.

12.3.3 Computer Languages

Machine instructions are written using special languages. These languages can be classified into one of three groups: machine language, assembly language, and high-level language.

A **machine language** is the basic language of a computer. Programs written in machine language require no further interpretation by the computer.

An **assembly language** is a symbolic form of machine language. Programs written in assembly language must first be converted to machine language prior to being executed by a computer. This translation process is done by a program called an **assembler**.

A **high-level language** is a programming language that is English-like in nature, and designed for convenience of use. Programs written in a high-level language must be converted to machine language prior to being executed by a computer. This translation process can be accomplished by a program called a **compiler** or a program called an **interpreter**.

The design of a computer program (i.e., the logical sequence of actions that need to be performed) can be developed via the use of a **flowchart**. This design stage should be independent of the language in which a program will be written. Flowchart symbols and two examples of their use are presented on pages 697 and 698 of your textbook.

12.4 USING BASIC

A computer program written in Basic consists of a group of statements containing instructions to the computer. Each statement consists of a **line number** followed by a specific instruction. Statements are executed by the computer in line number order. This means that the computer executes the first line of the program (i.e., the one with the lowest line number), and continues executing each succeeding line in order unless otherwise directed. Line numbers are usually written in multiples of ten. A summary of the statements discussed in this section of your textbook is provided here.

REM	a statement used primarily for program documentation;
PRINT	a statement that can be used to (a) print out a string of characters, (b) evaluate expressions, and (c) provide blank lines in an output;
LET	a statement that is used to assign a specific numerical value to a variable;
READ/DATA	A READ statement instructs a computer to assign a datum (a single piece of data) from a corresponding DATA statement to a variable in the READ statement.
INPUT	a statement that assigns data to variables from outside a program;
END	a statement that indicates the physical end of a program

The symbols used in Basic for arithmetic operations are summarized in Table 12.4.1, and the relational operators are summarized in Table 12.4.2. When evaluating mathematical expressions it is important that a prescribed order of operations is followed. This order of operations is summarized below.

<u>Order of Operations</u>

Working in sequence from left to right

1. remove all parenthetical expressions;
2. evaluate all exponential expressions;
3. perform all multiplications and divisions; and
4. perform all additions and subtractions.

Table 12.4.1 Arithmetic Operators Used in Basic

Symbol	Meaning	Example	Answer
+	addition	5 + 7	12
−	subtraction	7 − 5	2
*	multiplication	7 * 2	35
/	division	8/2	4
=	equals	12/2 = 3 * 2	6 = 6
↑	exponentiation	2 ↑ 3	8
( )	parentheses	(1 + 2) ↑ 2	9

Table 12.4.2 Relational Operators Used in Basic

Symbol	Meaning	Common Algebraic Expression
>	greater than	$x > 2$
<	less than	$x < 3$
≥	greater than or equal to	$x \geq 2$
≤	less than or equal to	$x \leq 3$
<>	not equal to	$x \neq 5$

298   STUDENT STUDY GUIDE

If two or more operations of equal priority (i.e., multiplication and division, or addition and subtraction) occur in any part of an expression, evaluate them as they occur working from left to right.

Example 1 (page 708, #3)

Evaluate $5 + 3 * 2$

Solution

- Multiply first.

$$\begin{aligned} & 5 + 3 * 2 \\ = & 5 + 6 \end{aligned}$$

- Now add.

$$5 + 6 = 11$$

Example 2 (page 708, #13)

Evaluate $9 * 4 / 3 + 8 - 3$

Solution

Working from left to right, follow the order of operations.

$$\begin{aligned} & 9 * 4 / 3 + 8 - 3 && \textit{(Multiply first.)} \\ = & 36 / 3 + 8 - 3 && \textit{(Now divide.)} \\ = & 12 + 8 - 3 && \textit{(Next add.)} \\ = & 20 - 3 && \textit{(Finally, subtract.)} \\ = & 17 \end{aligned}$$

Example 3 (page 708, #25)

Express $7(3 + 4)^2$ in Basic notation.

Solution

Use the symbols for addition, multiplication, and exponentiation from Table 12.4.1.

$$7(3 + 4)^2$$
$$= 7 * (3 + 4) \uparrow 2$$

Example 4 (page 709, #35)

Express $(2 * x - 1) * (3 * x + 1) / (2 * x + 1)$ in algebraic notation.

Solution

Use the fact that $*$ means multiplication, and $/$ means division.

$$(2 * x - 1) * (3 * x + 1) / (2 * x + 1)$$
$$= \frac{(2x - 1)(3x + 1)}{(2x + 1)}$$

Example 5 (page 709, #39)

Express $\dfrac{x^2 - 8x + 16}{x + 4}$ in Basic notation.

Solution

Use the fact that exponentiation is represented by ↑ and division is represented by /.

$$\dfrac{x^2 - 8x + 16}{x + 4}$$
$$= (x \uparrow 2 - 8 * x + 16) / (x + 4)$$

Note that parentheses are needed for both the numerator and denominator. If they were not used, then the expression would not be equivalent.

Example 6 (page 709, #45)

Determine the output for the program below.

```
 5 READ X, Y
10 LET Z = X ↑ 3 + Y
15 PRINT Z
20 DATA 2,4
25 END
```

Solution

- In line 5, 2 and 4 are assigned from the DATA statement in line 20 to the variables X and Y, respectively.

$$X \leftarrow 2 \quad Y \leftarrow 4$$

- In line 10, the expression X ↑ 3 + Y is evaluated and the result is assigned to the variable Z.

$$\begin{aligned} & X \uparrow 3 + Y \\ =\ & 2 \uparrow 3 + 4 \\ =\ & 8 + 4 \\ =\ & 12 \end{aligned}$$

- In line 15, the currently assigned value of Z (which is 12) is printed.
- Line 20 was used with line 5.
- Line 25 signifies the end of the program.

The final output is 12.

SUPPLEMENTARY EXERCISE 12.4

In problems 1 through 10 evaluate each expression.

1. 20 + 4 * 2
2. 3 + 2 * 6 − 5
3. 14 − 12 / 3
4. 6 − 10 / 5 + 1 / 1
5. 3 * 8 + 4 * 6
6. 8 / 4 + 2 * 3 + 1
7. 2 ↑ 3 + 3 ↑ 2
8. 9 * 4 + 6 / 3 * 2 − 28
9. 8 + (3 + 6) ↑ 2
10. 6 * 2 ↑ 4 − (4 − 1) ↑ 3 + 9

Continued on next page ...

In problems 11 through 20 write each expression in Basic notation. (Do not evaluate any expressions.)

11. $2(3 + 5) - 8$
12. $\dfrac{x + y}{5}$
13. $6 + 2 + 4 \times 8$
14. $2x^2 + 3$

15. $5^3(3^2 + 4)$
16. $\dfrac{A - B}{C + D}$
17. $100 - 4(6 - 2)^2$
18. $B^2 - 4AC$

19. $5^3 + (6 + 3)^2 + 10^2$
20. $AX^2 + BX + C$

In problems 21 through 24 determine the printed output for each program.

21.
```
10 PRINT
20 PRINT "2 + 2 = ";
30 LET A = 2
40 LET B = 3
50 LET X = A + B
60 PRINT X
70 END
```

22.
```
10 REM PRINT "HELLO"
20 PRINT "GOODBYE"
30 END
```

23.
```
10 PRINT "BEGIN"
20 LET A = 2
30 LET B = 3
40 LET C = 14
50 LET D = 16
60 LET X = A + B - C * D / A ↑ 3
70 PRINT X
80 END
```

24.
```
10 READ A, B, C
20 READ D, E
30 LET X = (A - B) ↑ C + D * E
40 PRINT X
50 DATA 7, 5, 4, 3, 1.5
60 END
```

12.5 MORE BASIC STATEMENTS

A summary of the statements presented in this section of your textbook is provided here.

GOTO — a statement that allows *unconditional branching*, that is, control of the program is transferred from one part to another unconditionally;

IF-THEN — a decision making statement that allows for *conditional branching*; if a certain condition is true then a prescribed action will occur;

FOR-NEXT — two statements that form a *controlled loop*, which will enable a set of instructions to be repeatedly executed a specific number of times

Example 1 (page 715, #7)

Determine the printed output for the following program.

```
 5 REM - MIDWAY
10 READ X, Y
15 IF X < 0 THEN 40
20 LET Z = (X + Y) / 2
25 PRINT Z
30 GOTO 10
35 DATA 22, 35, 41, 37, 80, 92, -1, -1
40 END
```

Continued on next page ...

Solution

- Line 5 is for documentation.
- In line 10, the first two data items (22 and 35) are assigned from the DATA statement in line 35 to the variables X and Y, respectively.

$$X \leftarrow 22 \quad Y \leftarrow 35$$

- In line 15, the current value of X is compared to zero. Since the expression 22 < 0 is false, control of the program continues to line 20.
- In line 20, the expression (X + Y) / 2 is evaluated and the result is assigned to the variable Z.

$$\begin{aligned} & (X + Y) / 2 \\ = \ & (22 + 35) / 2 \\ = \ & 57 / 2 \\ = \ & 28.5 \\ & (Z \leftarrow 28.5) \end{aligned}$$

- In line 25, print the current value of Z is printed.
- In line 30, control of the program is transferred to line 10.

....

In line 10, the data items 41 and 37 are assigned to X and Y, respectively.

- In line 15, 41 is compared to zero. Since 41 < 0 is false the action described in the conditional part of the IF-THEN statement is not performed.
- In line 20. the expression (X + Y) / 2 is evaluated and the result is assigned to the variable Z.

$$\begin{aligned} & (X + Y) / 2 \\ = \ & (41 + 37) / 2 \\ = \ & 78 / 2 \\ = \ & 39 \\ & (Z \leftarrow 39) \end{aligned}$$

- In line 25, the current value of Z is printed.
- In line 30, control of the program is transferred to line 10.

....

- Lines 10 through 25 are executed as before, except now X is assigned 80, Y is assigned 92, and the result of the evaluation of the expression is 86, which is assigned to Z.
- In line 30, control of the program is transferred to line 10.

....

- In line 10, −1 is assigned to both X and Y, and in line 15, −1 is compared to zero. Since −1 < 0 is true, control of the program is transferred to line 40, which in turn terminates the execution of the program.

The final output is:

28.5
39
86

Example 2 (page 715, #11)

Determine the printed output for the following program.

```
10  REM - PAIRS
20  FOR X = 1 TO 10
30  LET Y = X ↑ 2 - 4
40  PRINT X, Y
50  NEXT X
60  END
```

Solution

This program demonstrates the use of the FOR and NEXT statements. According to the program, the *body of the loop* (lines 30 and 40) is executed exactly 10 times. The flow of control of this program is illustrated in Table 12.5.1, and the final form of the output is given to the right of the table.

Table 12.5.1 Summary of Program Execution (Example 2)

Line Number	Value of X	Value of Y (X ↑ 2 − 4)	Output	
20	1			
30		$1^2 - 4 = -3$		
40			1	−3
20	2			
30		$2^2 - 4 = 0$		
40			2	0
20	3			
30		$3^2 - 4 = 5$		
40			3	5
20	4			
30		$4^2 - 4 = 12$		
40			4	12
20	5			
30		$5^2 - 4 = 21$		
40			5	21
20	6			
30		$6^2 - 4 = 32$		
40			6	32
20	7			
30		$7^2 - 4 = 45$		
40			7	45
20	8			
30		$8^2 - 4 = 60$		
40			8	60
20	9			
30		$9^2 - 4 = 77$		
40			9	77
20	10			
30		$10^2 - 4 = 96$		
40			10	96
20	11*	*program ends		

Output of Program

1	−3
2	0
3	5
4	12
5	21
6	32
7	45
8	60
9	77
10	96

INTRODUCTION TO COMPUTERS 303

SUPPLEMENTARY EXERCISE 12.5

In problems 1 through 4 determine the printed output for the given program.

1.
   ```
   10 READ X
   20 IF X > 0 THEN 40
   30 GOTO 20
   40 PRINT X
   50 DATA 5,-3,0,16,4,-9,3,-1,6,7
   60 END
   ```

2.
   ```
   10 LET S = 0
   20 READ X
   30 IF X < 0 THEN 80
   40 LET S = S + X
   50 PRINT X, S
   60 GOTO 20
   70 DATA 1,2,3,4,5,-1
   80 END
   ```

3.
   ```
   10 FOR X = 1 TO 3
   20 PRINT X, X ↑ 2
   30 NEXT X
   40 END
   ```

4.
   ```
   10 FOR X = 1 TO 4
   20 IF X = 2 THEN 50
   30 PRINT X
   40 GOTO 60
   50 PRINT "ELEPHANT"
   60 NEXT X
   70 END
   ```

5. Rewrite the program in problem 2 above using FOR-NEXT statements.

12.6 CHAPTER 12 TEST

In problems 1 through 10 match Column A with the best description from Column B.

A	B
1. CPU	a. produces output
2. output unit	b. assignment statement
3. input unit	c. makes a program readable
4. LET	d. data is entered from outside the program
5. GOTO	e. the brain of a computer system
6. FOR-NEXT	f. a controlled looping mechanism
7. REM	g. data is provided from within the program
8. READ-DATA	h. keyboard
9. PRINT	i. unconditional branching
10. INPUT	j. printer

In problems 11 through 13 evaluate the given expression.

11. $2 \uparrow 2 + 2 * 5 \uparrow 2 - 3$ 12. $5 + 5 * 5 - 5 / 5 + 5 * 5 / 5$ 13. $2 * (8 - (3 * 2) * 5 / 5 / 2) + 9$

14. Rewrite the following expression in Basic notation.

$$\frac{5-3}{2} + \frac{18}{3 \times 2} + 6$$

15. Use the flowchart on the next page to determine which choice (a, b, c, d, or e) best describes the contents of the printed output.

   a. names of 19 year olds   b. names of 19 year old females   c. names of 19 year old males
   d. names of males   e. none of the above

Continued on next page ...

Flowchart for Problem 15

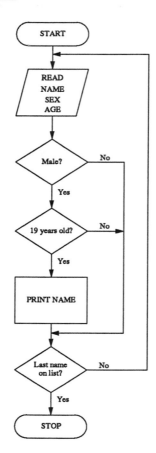

In problems 16 and 17 determine the printed output for the given program.

16. 10 LET A = 5
     20 LET B = 8
     30 PRINT A
     40 PRINT B
     50 LET A = B
     60 LET B = A
     70 PRINT A
     80 PRINT B
     90 LET C = (B + 8) / 2 + 10 + 1 * 4
     100 PRINT C
     110 END

17. 10 FOR I = 1 TO 4
     20 READ L, W
     30 LET A = L * W
     40 IF A > 25 THEN 60
     50 GOTO 70
     60 PRINT A
     70 NEXT I
     80 DATA 1,2,8,10,6,4,5,5
     90 END

18. Write a Basic program that will find the average of the numbers 85, 90, 84, 76, 70, and print this average.

FLORIDA CLAST REVIEW

FLORIDA COLLEGE LEVEL ACADEMIC SKILLS REVIEW
(Mathematics)

1. INTRODUCTION

This publication provides a review of the mathematics subtest of the College Level Academic Skills Test (CLAST), which is administered by the state of Florida. The material contained in this supplement was written by Dr. Eunice Everett of Seminole Community College, Sanford, Florida.

The CLAST mathematics subtest consists of 56 skills, which are grouped into the five subject areas shown below.

- A. Arithmetic
- B. Geometry and Measurement
- C. Algebra
- D. Statistics and Probability
- E. Logic and Sets

The skills within each subject area are subdivided further by the following levels of abstraction.

I. The skills at the **algorithmic process** level include memorized facts and rote processes.

II. The skills emphasizing **concepts** require students to both understand a concept and be able to apply this understanding to solve problems.

III. **Generalization** skills require students to recognize patterns, to determine appropriate rules used in a certain process, and to derive formulas.

IV. **Problem solving** skills require students to solve a variety of problems, including real world problems.

The 56 skills are listed in Table 1.1. Note that the skills are grouped by subject area and by the level of abstraction within that subject area. The number of each skill reflects both subject area and level. For example, skill IIC2 is the second skill listed in *algebra* at the *concept* level.

All 56 mathematics skills will be reviewed in this publication. Specifically, explanatory information and solved examples will be included for each of the skills. In some instances the explanation given for similar skills will be grouped together. Additionally, the examples provided will, in general, follow the form of questions published by the State of Florida as examples of questions included on the CLAST.

Table 1.1. CLAST Mathematics Skills (Revision, Fall, 1990)

	Arithmetic Skills
IA1	Adds, subtracts, multiplies, and divides rational numbers
IA2	Adds, subtracts, multiplies, and divides rational numbers in decimal form
IA3	Calculates percent increase and percent decrease
IA4	Solves the sentence "a% of b is c," where values for two of the variables are given
IIA1	Recognizes the meaning of exponents
IIA2	Recognizes the role of the base number in determining place value in the base-ten numeration system
IIA3	Identifies equivalent forms of positive rational numbers involving decimals, percents, and fractions
IIA4	Determines the order relation between real numbers
IIA5	Identifies a reasonable estimate of a sum, average, or product of numbers
IIIA1	Infers relations between numbers in general by examining particular number pairs
IVA1	Solves real-world problems which do not require the use of variables and which do not involve percent
IVA2	Solves real-world problems which do not require the use of variables and which do require the use of percent
IVA3	Solves problems that involve the structure and logic of arithmetic
	Geometry and Measurement Skills
IB1	Rounds measurements to the nearest given unit of the measuring device
IB2	Calculates distances, areas, and volumes
IIB1	Identifies relationships between angle measures
IIB2	Classifies simple plane figures by recognizing their properties
IIB3	Recognizes similar triangles and their properties
IIB4	Identifies appropriate units of measurement for geometric objects
IIIB1	Infers formulas for measuring geometric figures
IIIB2	Selects applicable formulas for computing measures of geometric figures
IVB1	Solves real-world problems involving perimeters, areas, and volumes of geometric figures
IVB2	Solves real-world problems involving the Pythagorean property
	Algebra Skills
IC1	Adds, subtracts, multiplies, and divides real numbers
IC2	Applies the order-of-operations agreement to computations involving numbers and variables
IC3	Uses scientific notation in calculations involving very large or very small measurements
IC4	Solves linear equations and inequalities
IC5	Uses given formulas to compute results when geometric measurements are not involved
IC6	Finds particular values of a function
IC7	Factors a quadratic expression
IC8	Finds the roots of a quadratic equation
IC9	Solves a system of two linear equations in two unknowns
IIC1	Uses properties of operations correctly
IIC2	Determines whether a particular number is among the solutions of a given equation or inequality
IIC3	Recognizes statements and conditions of proportionality and variation
IIC4	Identifies regions of the coordinate plane which correspond to specified conditions and vice versa
IIIC1	This skill has been deleted
IIIC2	Uses applicable properties to select equivalent equations or inequalities
IVC1	Solves real-world problems involving the use of variables, aside from commonly used geometric formulas
IVC2	Solves problems that involve the structure and logic of algebra

Table 1.1. CLAST Computation Skills (Continued)

	Statistics Skills, Including Probability
ID1	Identifies information contained in bar, line, and circle graphs
ID2	Determines the mean, median, and mode of a set of numbers
ID3	Uses the fundamental counting principle
IID1	Recognizes properties and interrelationships among the mean, median, and mode in a variety of distributions
IID2	Chooses the most appropriate procedure for selecting an unbiased sample from a target population
IID3	Identifies the probability of a specified outcome in an experiment
IIID1	Infers relations and makes accurate predictions from studying statistical data
IVD1	Interprets real-world data involving frequency and cumulative frequency tables
IVD2	Solves real-world problems involving probabilities

	Logical Reasoning
IE1	Deduces facts of set inclusion or set non-inclusion from a diagram
IIE1	Identifies statements equivalent to the negations of simple and compound statements
IIE2	Determines equivalence or nonequivalence of statements
IIE3	Draws logical conclusions from data
IIE4	Recognizes that an argument may not be valid even though its conclusion is true
IIIE1	Recognizes valid reasoning patterns as illustrated by valid arguments in everyday language
IIIE2	Selects applicable rules for transforming statements without affecting their meaning
IVE1	Draws logical conclusions when facts warrants them

2. SUGGESTIONS FOR PREPARING AND TAKING THE CLAST

Prior to taking the CLAST it will be to your advantage to actively and consciously prepare for it. By knowing the format of the exam, the types of questions that will be asked, and various test taking strategies your chances of meeting with success on the exam will be increased.

2.1 Preparing for the CLAST

To prepare for the CLAST we suggest that you do the following:

1. Familiarize yourself with the content of the CLAST by reading the list of skills presented in Table 1.1.

2. Prepare for each of the skills by studying the appropriate sections of a textbook that covers the skills and use this study guide to assist you. Moreover, study this chapter thoroughly and review all of the sample problems. These sample problems will give you an idea of what is expected for each skill and the types of questions that will be asked.

3. Become aware of the form of the test itself. Presently, there are 55 multiple choice questions, of which five are for experimental purposes. These five questions will not affect your score. The 50 questions for which you will be accountable are distributed over the five subject areas in the following manner:

    - Arithmetic - 13 questions
    - Algebra - 16 questions
    - Geometry and Measurement - 7 questions
    - Statistics and Probability - 7 questions
    - Logic and Sets - 7 questions

    Additionally, there are both advantages and disadvantages to a multiple choice format. These are listed in the box at the top of the next page.

4. Plan your study time and avoid *last minute cramming*. Waiting until the last minute to study can lead to the confusion of concepts and formulas while you are taking the test.

5. Maintain a positive attitude. Avoid negative thoughts about the test and your abilities. After you have prepared well for the test, approach the test with confidence. In short, believe in yourself and the studying you have done.

6. Get proper rest the night before the test.

7. On the day of the exam you should try to do the following.

    - Wake up early enough to avoid rushing, which can lead to a higher degree of anxiety.

    - Arrive early enough to have some time to become oriented to the testing area.

    - Avoid contact with other people who might be anxious about the exam. Such contact could increase your anxiety level.

    - Dress with flexibility. Bring a sweater with you so that you can wear it if the room is too cold, or remove it if the room is too warm.

    - Bring a blue or black pen, several #2 pencils with erasers, two forms of identification (one with a photo) and a watch. Place the watch on your desk within your view so that you can be aware of time limits. Also, do *not* bring a calculator. (Calculators are not allowed.)

	Advantages		Disadvantages
a.	When you do not know how to solve a problem, you can sometimes check each of the choices in the problem itself and thereby discover the correct choice.	a.	Checking each answer choice in a problem can be very time consuming.
b.	If your answer to a question is not among the choices, this will alert you that you either have an error or your answer is not yet in the appropriate form.	b.	When you have an error in your work, your incorrect answer might be one of the choices. Common errors are usually listed among the incorrect choices. Agreement of your answer with one of the choices does not necessarily make your answer correct.
c.	Some choices might be recognized as absurd and you could readily eliminate them. This leaves fewer reasonable choices from which to select an answer.	c.	The correct multiple choice answer might be in a different form from your answer. As a result, you might have to alter the form of your answer to get it to match one of the choices.
d.	If you have no idea how to answer a question you can still guess.	d.	No partial credit is awarded. Answers that are incorrect, regardless of why, are marked wrong.

2.2 Taking the CLAST

While taking the test you should be aware of the following points.

1. There is no penalty for guessing and you may write on the test copy.

2. Once you begin the test write down any formulas or rules you have memorized that could be forgotten or confused with others.

3. Read and follow the directions carefully. The directions will vary among the questions.

4. Read each question very carefully before you work on a problem. Also, look at the forms of the answer choices. This will give you some guidelines as to the direction to take in solving a problem. This strategy will be most helpful for those questions that do not have explicit instructions.

5. The order in which you answer the questions can sometimes improve your score. First, survey the test and make a mental time schedule. Once this is done, answer those questions that you decide are the easiest. As you read a problem, it might also be helpful if you *cross out* those answer choices that you readily recognize as being incorrect.

6. When you have completed a problem and are satisfied with its answer, mark through it on the test copy so that you know it has been completed. If you have to bypass a problem label it 1, 2, or 3 to indicate the level of difficulty (where 3 is the most difficult). Then, as you go back through the test again, work on level 1 questions first, then level 2, and finally level 3. Remember, all of the questions are of equal value, regardless of their level of difficulty. Budget your time to solve the maximum number of questions within the allotted time frame.

7. Be careful to record your answers correctly on the answer sheet. Do not miss questions because you failed to record the answer in the proper place or because you chose an answer with correct numbers, but incorrect units.

8. When time is running short, guess at those questions you have not answered. Do not leave any blanks. If time permits, recheck those questions where you might have made careless errors. For a student's first take of the test, the time limit is 90 minutes.

9. If during the test you are distracted by someone near you, ask for a change of seat.

10. Once you have completed the test do not lament over questions you might have missed. Reward yourself for doing your very best.

3. ARITHMETIC

In this section we discuss the CLAST subject area Arithmetic. Each skill within this subject will be addressed separately, and examples of the type of problems that could be asked on the CLAST for each skill will be presented.

Skill IA 1. Add, subtract, multiply, and divide rational numbers

You must be able to add, subtract, multiply, and divide fractions, some of which will be proper, improper, or mixed. The fractions could be positive or negative. The addition and subtraction problems could have unlike denominators and might require regrouping. You will also need to recall the rules for adding and subtracting signed numbers. We present five examples that demonstrate this skill.

Example 1: $-6 + 3\frac{1}{3} =$

A. $2\frac{2}{3}$ B. $-3\frac{2}{3}$ C. $-2\frac{2}{3}$ D. $-9\frac{1}{3}$

Solution

$$-6 + 3\frac{1}{3}$$
$$= \frac{-18}{3} + \frac{10}{3}$$
$$= \frac{-8}{3}$$
$$= -2\frac{2}{3}$$

As a result, choice C is correct.

Example 2: $\frac{-5}{8} - (-2) =$

A. $\frac{-7}{8}$ B. $-2\frac{5}{8}$ C. $1\frac{3}{8}$ D. $-1\frac{3}{8}$

Solution

$$\frac{-5}{8} - (-2)$$
$$= \frac{-5}{8} + 2$$
$$= \frac{-5}{8} + \frac{16}{8}$$
$$= \frac{11}{8} = 1\frac{3}{8}$$

Thus choice C is correct.

Example 3: $\frac{-1}{7} \times 2\frac{4}{5} =$

    A. $\frac{18}{35}$ B. $\frac{-2}{5}$ C. $\frac{-24}{35}$ D. $\frac{2}{5}$

Solution

$$\frac{-1}{7} \times 2\frac{4}{5} = \frac{-1}{7} \times \frac{14}{5} = \frac{-14}{35} = \frac{-2}{5}$$

Choice B is correct.

Example 4: $9 \div 3\frac{1}{3} =$

    A. $2\frac{7}{10}$ B. 30   C. $\frac{10}{27}$ D. $3\frac{1}{3}$

Solution

$$9 \div 3\frac{1}{3} = 9 \div \frac{10}{3} = 9 \times \frac{3}{10} = \frac{27}{10} = 2\frac{7}{10}$$

Choice A is correct.

Example 5: $\left[\frac{-1}{4}\right] \div \left[\frac{-3}{8}\right] =$

    A. $\frac{3}{2}$ B. $\frac{2}{3}$ C. $\frac{3}{32}$ D. $\frac{-2}{3}$

Solution

$$\left[\frac{-1}{4}\right] \div \left[\frac{-3}{8}\right]$$
$$= \left[\frac{-1}{4}\right] \times \left[\frac{-8}{3}\right]$$
$$= \frac{(-1)(-8)}{(4)(3)}$$
$$= \frac{8}{12}$$
$$= \frac{2}{3}$$

Choice B is correct.

ARITHMETIC

Skill IA 2. Add, subtract, multiply, and divide rational numbers in decimal form

The decimal numbers can be positive or negative so you will need to recall the rules of operation for signed numbers as well as the rules for adding, subtracting, multiplying, and dividing numbers in decimal form. When adding and subtracting numbers in decimal form remember to *line up* the decimal points. We present five examples.

Example 6: $1.59 + 0.235 =$

    A. 1.725   B. 3.94   C. 0.01825   D. 1.825

Solution

$$\begin{array}{r} 1.590 \\ +0.235 \\ \hline 1.825 \end{array}$$

Choice D is correct.

Example 7: $-14.34 - 2.076 =$

    A. 16.416   B. −16.316   C. −12.264   D. −16.416

Solution

To combine two negative numbers, we add their absolute values and keep the common sign for the sign of the answer. Note also in this example that we append a zero to −14.34 so that the decimals are similar.

$$\begin{array}{r} -14.340 \\ +-2.076 \\ \hline -16.416 \end{array}$$

Choice D is correct.

Example 8: $13.28 \div (-.04) =$

    A. −33.2   B. −332   C. −3.32   D. −0.0332

Solution

Note that we move the decimal point for both the divisor and dividend two places to the right.

$$13.28 \div (-.04) = 1328 \div (-4) = -332$$

The correct choice is B.

Example 9: $(-.12) \div (-2.4) =$

  A. 0.5   B. 5.0   C. 0.05   D. −0.5

Solution

Since a negative number divided by a negative number is positive, we can omit the sign when we do our computation. Note that the decimal point for both the divisor and dividend is moved one place to the right.

$$0.12 \div 2.4 = 1.2 \div 24 = 0.05$$

The correct choice is C.

Example 10: $(-.04) \times (1.36)$

  A. −5.44   B. −0.0544   C. −0.544   D. 5.44

Solution

$$\begin{array}{r} 1.36 \\ \times \quad -0.04 \\ \hline 544 \\ 000 \\ \hline -0.0544 \end{array}$$

The correct choice is B.

Skill IA 3. Calculate percent increase and percent decrease

This skill can involve two types of problems.

1. Given an original amount and a percent increase or a percent decrease you could be asked to select the new amount.
2. Given two different amounts you could be asked for the percent increase or percent decrease.

Recall that percent increase is used to show how much a quantity has increased over its original value while percent decrease shows how much a quantity has decreased from its original value. You will need to use the basic percent equation: *percent × base = amount*. Two examples are given, one for each type. The first example is of type 2.

Example 11: If 25 is decreased to 15, what is the percent decrease?

  A. $66\frac{2}{3}\%$ B. 10%   C. 40%   D. 60%

Solution

- First find the amount of decrease.

$$25 - 15 = 10$$

Continued on next page ...

- Next, write the basic percent equation. The *base* is the original amount (25), the *amount* is the amount of decrease (10), and the *percent* of decrease is P.

$$\begin{array}{ccccc} percent & \times & base & = & amount \\ P & \times & 25 & = & 10 \end{array}$$

- Now solve the equation.

$$\begin{aligned} 25P &= 10 \\ P &= \frac{10}{25} \\ P &= \frac{40}{100} \\ P &= 40\% \end{aligned}$$

The answer is choice C.

Example 12: If 36 is increased by 25% of itself, what is the result?

    A. 9   B. 27   C. 45   D. 76

Solution

$$\begin{array}{ccccc} percent & \times & base & = & amount \\ (.25) & \times & (36) & = & 9.00 \end{array}$$

Since 25% of 36 is 9, 36 is increased by 9 to obtain 45. The correct choice is C.

Skill IA 4.   Solve the sentence "a% of b is c," where values for two of the variables are given

You will be asked to find the value of one of a, b or c in a problem of the form *a% of b is c*. The given numbers will be positive and may be expressed in fractional or decimal form. Three examples are given.

Example 13: What is $13\frac{1}{2}\%$ of 96?

    A. 12.528   B. 12.96   C. 711.11   D. 1296

Solution

$13\frac{1}{2}\%$ written as a decimal is 0.135. (See Skill IIA3.) Use the basic percent equation.

$$\begin{array}{ccccc} percent & \times & base & = & amount \\ (0.135) & \times & (96) & = & 12.96 \end{array}$$

The correct choice is B.

318 CLAST REVIEW

Example 14: 133 is what percent of 76?

    A. 1.75%   B. 101.08%   C. 57.14%   D. 175%

Solution

You are looking for the percent.

$$\begin{array}{rcl} percent \times base &=& amount \\ (a\%) \times (76) &=& 133 \\ (\frac{a}{100}) \times (76) &=& 133 \\ \frac{76a}{100} &=& 133 \\ 76a &=& 13300 \\ a &=& \frac{13300}{76} = 175 \end{array}$$

133 is 175% of 76. The correct choice is D.

Example 15: 54 is 20% of what number?

    A. 104   B. 10.8   C. 270   D. 2700

Solution

You are looking for the base b.

$$\begin{array}{rcl} percent \times base &=& amount \\ (0.20) \times b &=& 54 \\ b &=& \frac{54}{0.20} = \frac{5400}{20} \\ b &=& 270 \end{array}$$

54 is 20% of 270. Note that $20\% = \frac{20}{100} = \frac{1}{5}$. In the solution above, the fraction $\frac{1}{5}$ could be used rather than the decimal 0.20. The correct choice is C.

Skill IIA 1. Recognize the meaning of exponents

For this skill you will not be asked to compute with exponents. Instead, you will need to rewrite expressions using the definition of natural number exponents. Recall that

$$\begin{array}{rcl} a^1 &=& a \\ a^2 &=& a \cdot a \\ a^3 &=& a \cdot a \cdot a \\ &\vdots& \\ a^n &=& a \cdot a \cdot a \cdots a \text{ (n factors of a)} \end{array}$$

ARITHMETIC 319

The following laws of exponents might also be involved.

$$a^m \cdot a^n = a^{m+n}$$
$$\frac{a^m}{a^n} = a^{m-n}, \; m > n$$
$$\left[a^m\right]^n = a^{mn}$$

We present three examples.

Example 16: $3^3 + 4^3 =$

A. $(3 + 4)^3$ B. $(3 + 4)^6$ C. $3(3) + (4)(3)$ D. $(3)(3)(3) + (4)(4)(4)$

Solution

- Choice A is incorrect because of the order of operations agreement that powers must be computed before addition, unless grouping symbols require a different order.

- Choice B is incorrect for the same reason as A. Furthermore, exponents are added only when you are *multiplying* powers of the same base as in $2^3 \cdot 2^4 = 2^7$.

- Choice C is incorrect because a base is not multiplied by its exponent.

- Choice D is correct because 3^3 means $(3)(3)(3)$ and 4^3 means $(4)(4)(4)$.

Example 17: $(5^2)^3 =$

A. 5^5 B. 5^8 C. $(5 \times 2)^3$ D. $5^2 \times 5^2 \times 5^2$

Solution

- Choice A is incorrect because the law $a^m \cdot a^n = a^{m+n}$ is being incorrectly applied to a power of a power.

- Choice B is incorrect since the exponent 2 is being raised to the third power.

- Choice C is incorrect since the exponent 2 is being incorrectly treated as a factor of the base.

- Choice D is correct since the base 5^2 is being used as a factor three times.

Example 18: $(6^4)(4^3) =$

A. $(6 + 6 + 6 + 6)(4 + 4 + 4)$ B. $(6 \times 6 \times 6 \times 6)(4 \times 4 \times 4)$ C. $(6 \times 4)(4 \times 3)$ D. $(6 \times 4)^7$

Solution

Choice B is the correct choice. 6^4 means that 6 is to be used as a factor four times. As a result, 6^4 means $6 \times 6 \times 6 \times 6$. Similarly, 4^3 means $4 \times 4 \times 4$.

Skill IIA 2. Recognize the role of the base number in determining place value in the base-ten numeration system

Our number system is a place value system based on the number 10. Each position has a place value as illustrated with the number 721.4603 shown below.

$$7 \quad 2 \quad 1 \quad 4 \quad 6 \quad 0 \quad 3$$
$$\uparrow \quad \uparrow \quad \uparrow \quad \uparrow \quad \uparrow \quad \uparrow \quad \uparrow$$
$$10^2 \quad 10^1 \quad 10^0 \quad \frac{1}{10^1} \quad \frac{1}{10^2} \quad \frac{1}{10^3} \quad \frac{1}{10^4}$$

(Note that 10^0 is equal to 1.)

We can also express numbers (numerals) in expanded form. For example, the standard base-ten numeral 721.4603 expressed in expanded form is

$$(7 \times 10^2) + (2 \times 10^1) + (1 \times 10^0) + (4 \times \frac{1}{10^1}) + (6 \times \frac{1}{10^2}) + (0 \times \frac{1}{10^3}) + (3 \times \frac{1}{10^4})$$

Given a standard base-ten numeral, you may be asked to determine the place value of a specified digit or to select the expanded form of the numeral. Given the expanded form, you may be asked to select the standard form. We present three examples.

Example 19: Select the place value associated with the underlined digit.

$$7.035\underline{1}$$

A. $\frac{1}{10^4}$ B. $\frac{1}{10^3}$ C. $\frac{1}{10^2}$ D. 10^3

Solution

$$7.035\underline{1} = 7 \quad 0 \quad 3 \quad 5 \quad 1$$
$$\uparrow \quad\quad \uparrow \quad \uparrow \quad \uparrow \quad \uparrow$$
$$10^0 \quad \frac{1}{10^1} \quad \frac{1}{10^2} \quad \frac{1}{10^3} \quad \frac{1}{10^4}$$

Note that the digit 5 is in the *thousandths* place, which is $\frac{1}{10^3}$. The correct choice is B.

Example 20: Select the expanded notation for 3016.024.

A. $(3 \times 10^3) + (1 \times 10) + (6 \times 10^0) + (2 \times \frac{1}{10}) + (4 \times \frac{1}{10^3})$

B. $(3 \times 10^3) + (1 \times 10^2) + (6 \times 10) + (2 \times \frac{1}{10^2}) + (4 \times \frac{1}{10^3})$

C. $(3 \times 10^3) + (1 \times 10) + (6 \times 10^0) + (2 \times \frac{1}{10^2}) + (4 \times \frac{1}{10^3})$

D. $(3 \times 10^2) + (1 \times 10) + (6 \times 10^0) + (2 \times \frac{1}{10}) + (4 \times \frac{1}{10^3})$

Continued on next page ...

Solution

$$\begin{array}{ccccccc} 3 & 0 & 1 & 6\;. & 0 & 2 & 4 \\ \uparrow & \uparrow & \uparrow & \uparrow & \uparrow & \uparrow & \uparrow \\ 10^3 & 10^2 & 10^1 & 10^0 & \frac{1}{10^1} & \frac{1}{10^2} & \frac{1}{10^3} \end{array}$$

3016.024 expressed in standard notation, showing all digits is

$$(3 \times 10^3) + (0 \times 10^2) + (1 \times 10^1) + (6 \times 10^0) + (0 \times \tfrac{1}{10^1}) + (2 \times \tfrac{1}{10^2}) + (4 \times \tfrac{1}{10^3})$$

This expression is equivalent to

$$(3 \times 10^3) + (1 \times 10) + (6 \times 10^0) + (2 \times \tfrac{1}{10^2}) + (4 \times \tfrac{1}{10^3})$$

The correct choice is C.

Example 21: Select the numeral for

$$(5 \times 10^4) + (7 \times 10^2) + (3 \times 10) + (2 \times \tfrac{1}{10}) + (2 \times \tfrac{1}{10^3})$$

A. 573.22   B. 50073.022   C. 5073.202   D. 50730.202

Solution

$$\begin{aligned} & (5 \times 10^4) & + & (7 \times 10^2) & + & (3 \times 10) & + & (2 \times \tfrac{1}{10}) & + & (2 \times \tfrac{1}{10^3}) \\ = & (5 \times 10000) & + & (7 \times 100) & + & (3 \times 10) & + & (2 \times .1) & + & (2 \times .001) \\ = & 50000 & + & 700 & + & 30 & + & .2 & + & .002 \\ = & 50730.202 \end{aligned}$$

The correct choice is D.

Skill IIA 3. Identify equivalent forms of positive rational numbers involving decimals, percents, and fractions

This skill requires you to:

1. convert decimal numbers to percents and percents to decimal numbers;
2. convert decimal numbers to fractions and fractions to decimal numbers; and
3. convert percents to fractions and fractions to percents.

When you make these conversions keep in mind that $1\% = \dfrac{1}{100} = 0.01$.

The following three examples (examples 22, 23, and 24) illustrate some of the various conversions. You will need to look at the choices given in the questions to determine which conversion is required. Several possible conversions can be involved in the answer choices.

Example 22: $0.27 =$

A. $\dfrac{27}{100}\%$ B. $2\dfrac{7}{10}\%$ C. $\dfrac{27}{10}$ D. $\dfrac{27}{100}$

Solution

As a fraction $0.27 = \dfrac{27}{100}$, which is equal to 27%. As a result, the only possible choice is D.

Example 23: $340\% =$

A. 0.340   B. 3.40   C. 340.0   D. 3400.0

Solution

Since the answer choices are all expressed as decimals, we must convert 340% to a decimal.

$$340\% = 340 \times \dfrac{1}{100}$$
$$= \dfrac{340}{100} = \dfrac{34}{10} = 3.4 = 3.40$$

As a result, the correct choice is B.

Example 24: $\dfrac{17}{20} =$

A. 0.85   B. 0.085   C. 8.5%   D. .85%

Solution

$$\dfrac{17}{20} = \dfrac{85}{100} = 0.85 = 85\%$$

Of these equivalent expressions, 0.85 is given in choice A.

Skill IIA 4. Determine the order relation between real numbers

Given a pair of numbers you will be asked to identify the correct order relation (<, >, =) between the numbers. The numbers can be positive or negative integers, fractions, decimals, including repeating decimals, or square roots of whole numbers. (Recall that a *bar* placed over a digit or group of digits implies the digit or group of digits will repeat. For example, $3.\overline{24}$ represents the repeating decimal 3.24242424...)

In examples 25 through 29 you are to identify the symbol that should be placed in the box to form a true statement.

Example 25: $\frac{14}{18} \square \frac{7}{9}$

    A. =   B. <   C. >

Solution

Note that $\frac{14}{18}$ can be reduced to $\frac{7}{9}$ by dividing both the numerator and denominator by 2. As a result choice A is correct.

Example 26: $.25 \square \frac{1}{5}$

    A. =   B. <   C. >

Solution

Convert both terms to *like fractions*.

$$0.25 = \frac{25}{100} \quad \text{and} \quad \frac{1}{5} = \frac{20}{100}$$

Clearly, $\frac{25}{100}$ is greater than $\frac{20}{100}$. As a result, $.25 > \frac{1}{5}$. Choice C is correct.

Example 27: $3.1\overline{6} \square 3.\overline{16}$

    A. =   B. <   C. >

Solution

$$3.1\overline{6} = 3.166666...$$
$$3.\overline{16} = 3.161616...$$

The digits disagree in the thousandths position. $3.1\overline{6}$ has the larger digit in that position. Therefore, $3.1\overline{6} > 3.\overline{16}$, which is choice C.

Example 28: $\frac{3}{5} \square \frac{-3}{4}$

    A. =   B. <   C. >

Solution

Since $\frac{3}{5}$ is positive and $\frac{-3}{4}$ is negative, $\frac{3}{5}$ is the larger number. Choice C is correct.

Example 29: $\sqrt{60}\ \square\ 8.1$

    A. =   B. <   C. >

Solution

$7^2 = 49$ and $8^2 = 64$. Therefore, $\sqrt{49} < \sqrt{60} < \sqrt{64}$ and $7 < \sqrt{60} < 8$. However, $8 < 8.1$. As a result, $\sqrt{60} < 8.1$. Choice B is correct.

Skill IIA 5. Identify a reasonable estimate of a sum, average, or product of numbers

You will be given a word problem that involves ranges, approximate values, and/or distributions. You will be asked to select the most appropriate sum, average, or product. Three examples follow; the second is more involved than the first; the third involves a chart.

Example 30

Two hundred students took an achievement test. All of the students scored less than 85, but more than 52. Which of the following values is a reasonable estimate of the average score of the students?

    A. 85   B. 71   C. 52   D. 48

Solution

Since we are trying to estimate an average of numbers *between* 52 and 85, the average must lie between these two numbers. Obviously, 71 (choice B) is the correct choice.

Example 31

Twenty-five people work for an advertising firm. The lowest paid person earns $200 per week and the highest paid person earns $450 per week. Which of the following values could be a reasonable estimate of the total weekly payroll for the company?

    A. $11,250   B. $7,200   C. $4,500   D. $325

Solution

The average of $200 and $450 is $325. For 25 workers an estimate of the payroll could be (25)($325), which is $7,125. Thus choice B is reasonable.

Note that the remaining choices are not reasonable.

- Choice A ($11,250) is too large an estimate. Although (25)($450) = $11,250, at least one worker earns only $200 per week.

- Choice C ($4,500) is too small an estimate since 25($200) = $5000.

- Choice D ($325) represents a possible average *salary*, not an average *payroll*.

Example 32

The chart below contains a list of the number of shares of Fleming Timber Company stock owned by each of the Fleming family members. Each share of stock is valued at $8,033. What could be a reasonable estimate of the total value of all of the stock owned by Fletcher, Paul, Tim, and Eunice.

Family Member	Number of Shares
Fletcher	22½
Mavis	18
Paul	40
Tim	33½
Gene	20
Eunice	17
Harvey	12

A. $320,000

B. $880,000

C. $9,050,000

D. $88,000

Solution

Round the numbers of stock for each of the four members to the nearest multiple of ten and add these numbers. Multiply the result by $8,000.

$$20 + 40 + 30 + 20 = 110 \quad \text{and} \quad 110 \times (\$8,000) = \$880,000$$

Choice B is a reasonable estimate.

Skill IIIA 1.   Infer relations between numbers in general by examining particular number pairs

You will be given several pairs of numbers for which a linear or quadratic relationship exists between the numbers of each pair. You need to study the pairs to determine the relationship. You will then be asked to identify a missing number of a number pair.

Example 33

Look for a common linear relationship between the numbers in each pair and then identify the missing number.

$$(3,1) \quad (.9,.3) \quad (-6,-2) \quad (\tfrac{1}{5}, \tfrac{1}{15}) \quad (33,11) \quad (\tfrac{1}{9}, \text{—})$$

A. $\dfrac{1}{3}$ B. 3   C. $\dfrac{1}{27}$ D. 27

Solution

Notice that for each of the given pairs of numbers that are complete, the first number is three times the second number. For example, in the pair (3,1) the first number (3) is three times the second number (1); in the pair (.9,.3) the first number (.9) is three times the second number (.3); and so forth. It is important to observe that this relationship holds true for *all* the completed pairs. To find the missing term, then, we must solve the equation $\tfrac{1}{9} = 3(?)$. The correct answer is $\tfrac{1}{27}$, which is choice C.

Also, for this skill, you might be asked to provide the next term in an arithmetic, geometric, or harmonic progression of numbers. The following illustrates these various progressions (or sequences).

Arithmetic Progression

In an arithmetic progression a number called the *common difference* is used to generate the terms of the sequence. For example, the sequence $-15, -12, -9, -6, -3, 0, 3, 6, \ldots$ has a common difference of 3 since each succeeding term can be found by adding 3 to the previous term. (What would be the tenth term of this sequence? Since 6 is the eighth term, $6 + 3 = 9$ is the ninth term, and $9 + 3 = 12$ is the tenth term.)

Geometric Progression

In a geometric progression a number called the *common ratio* is used to generate the terms of the sequence. For example, the sequence $-64, -16, -4, -1, \frac{-1}{4}, \frac{-1}{16}, \ldots$ has a common ratio of $\frac{1}{4}$ since each succeeding term can be found by multiplying the previous term by $\frac{1}{4}$. (What would be the ninth term of this progression? Since $\frac{-1}{16}$ is the sixth term, $(\frac{-1}{16})(\frac{1}{4}) = \frac{-1}{64}$ is the seventh term, and $(\frac{-1}{64})(\frac{1}{4}) = \frac{-1}{256}$ is the eighth term, and $(\frac{-1}{256})(\frac{1}{4}) = \frac{-1}{1024}$ is the ninth term.)

Harmonic Progression

In a harmonic progression the denominators will have a common difference. For example, the terms of the sequence $\frac{1}{4}, \frac{1}{9}, \frac{1}{14}, \frac{1}{19}, \frac{1}{24}, \ldots$ take the form $\frac{1}{4}, \frac{1}{4+5}, \frac{1}{4+2\cdot5}, \frac{1}{4+3\cdot5}, \frac{1}{4+4\cdot5}$, etc. The ninth term of this sequence would be $\frac{1}{4+8\cdot5}$, which is equal to $\frac{1}{44}$.

Example 34: Identify the missing term in the following geometric progression.

$$125, -25, 5, -1, \frac{1}{5}, —$$

A. $\frac{1}{25}$ B. 1 C. $\frac{-1}{25}$ D. -1

Solution

The common ratio of a geometric progression can be determined by dividing a term of the sequence by the term to its immediate left. In the given progression, $5 \div (-25) = \frac{-1}{5}$. (Note that this quotient will result for any two consecutive terms of the sequence.) As a result, the common ratio is $\frac{-1}{5}$ and the missing term of the sequence is $(\frac{-1}{5})(\frac{1}{5}) = \frac{-1}{25}$, which is choice C.

Example 35: Identify the missing term in the following arithmetic progression.

$$25, 17, 9, —, -7, -15$$

A. 1 B. -1 C. 8 D. -8

Continued on next page ...

Solution

The common difference of an arithmetic progression can be determined by subtracting any term of the sequence from the term to its immediate left. For example, in the given sequence $17 - 25 = -8$. (Note that this difference will be the same for any two consecutive terms of the sequence.) As a result, the common difference is -8 and the missing term is $9 + (-8) = 1$, which is choice A.

Skill IVA 1. Solve real-world problems which do not require the use of variables and which do not involve percent

You will need to solve real world problems from real world situations that might require from two to seven arithmetic operations and/or conversions. The conversions can be within the English or metric systems, or they can involve changing forms (percent, decimal, fraction). The solution of a given problem will have no steps that involve the use of percents. Problems might contain irrelevant information.

Example 36

A bookstore ordered 45 books for a class of 43 students. Each book, which cost the bookstore $25, was to be sold for $31. The bookstore must pay a $3 service charge for each unsold book returned. If the bookstore returned four books how much profit did the bookstore make?

      A. $234   B. $246   C. $296   D. $134

Solution

1. Read the problem carefully and restate it in your own words.
   - 45 books ordered costing $25 each
   - 41 books sold at $31 each
   - 4 books returned for a refund, but a $3 service charge assessed to each
   - (The 43 students is irrelevant information)

2. Select a strategy.

   Profit = (Total Income) – (Total Expenses)

3. Execute the strategy and solve the problem.

total income	=	(number of books sold)	×	(selling price of each book)
	=	41	×	$31
	=	$1271		

Continued on next page ...

$$\text{total expenses} = \left[\begin{pmatrix}\text{number}\\\text{of books}\\\text{sold}\end{pmatrix} \times \begin{pmatrix}\text{cost}\\\text{of each}\\\text{book}\end{pmatrix}\right] + \left[\begin{pmatrix}\text{number}\\\text{of books}\\\text{returned}\end{pmatrix} \times \begin{pmatrix}\text{service charge}\\\text{for}\\\text{each book}\end{pmatrix}\right]$$

$$= [(45-4) \times \$25] + [(4) \times (\$3)]$$
$$= (41 \times \$25) + (4 \times \$3)$$
$$= \$1025 + \$12$$
$$= \$1037$$

profit = (total income) − (total expenses)
= $1271 − $1037
= $234 (Choice A is correct.)

4. Check and reflect; look at the problem from a different perspective.

A total of 41 books were bought and sold at a profit of $6 each, which yields a total profit of $246. From this profit a total service charge of $12 must be subtracted, which leaves a total profit of $234.

(**Note:** This four step method for solving problems can be very useful for solving complicated problems.)

Example 37

An automobile rents for $38.00 per day plus $0.20 per mile. Insurance for each rental costs $26.00. Find the rental cost for a 1,400 mile trip that took two people through six states in seven days.

    A. $534   B. $292   C. $572   D. $546

Solution

1. Read the problem carefully and restate it in your own words.
   - $38.00 per day for 7 days
   - $0.20 per mile for 1400 miles
   - $26.00 total for insurance
   - (The number of states and the number of people are both irrelevant.)

2. Select a strategy.

$$\text{total cost} = \begin{pmatrix}\text{cost based on}\\\text{daily rental rate}\end{pmatrix} + \begin{pmatrix}\text{cost based on}\\\text{miles driven}\end{pmatrix} + \begin{pmatrix}\text{cost of}\\\text{insurance}\end{pmatrix}$$

3. Execute the strategy and solve the problem.

Cost based on daily rental rate = (number of days rental) × (rental per day)
= 7 × $38.00
= $266.00

Continued on next page ...

Cost based on miles driven	=	(number of miles)	×	(fee per mile)
	=	1400	×	$0.20
	=	$280.00		

Cost of insurance = $26.00

total cost = (cost based on daily rental rate) + (cost based on miles driven) + (cost of insurance)

= $266 + $280 + $26

= $572 (Choice C is correct.)

4. Check and reflect. Did I include all of the possible costs in my solution? Yes. Compute the numbers again.

Example 38

Two biology classes of 30 students each planned a field trip. On the day of the field trip, many students enrolled in the school had a flu virus. Only $\frac{5}{6}$ of the students in one class and $\frac{4}{5}$ of the students in the second class went on the field trip. How many students missed the field trip?

    A. 11   B. 49   C. 10   D. 19

Solution
- For the first class, $\frac{5}{6}(30) = 25$. Thus 25 students attended.
- For the second class, $\frac{4}{5}(30) = 24$. Thus 24 students attended.
- Total number attending = 25 + 24 = 49
- Total number missing the field trip = 60 − 49 = 11

Eleven students missed the field trip. The correct choice is A.

Check and reflect. Look at the problem another way.
- For the first class, $\frac{1}{6}(30) = 5$. Thus 5 students did not attend.
- For the second class, $\frac{1}{5}(30) = 6$. Thus 6 students did not attend.

A total of 11 students did not attend.

330 CLAST REVIEW

Skill IVA 2. Solve real-world problems which do not require the use of variables and which do require the use of percent

The problems for this skill will be similar in nature to those of skill IVA 1, except that their solutions will require the use of percent in at least one step.

Example 39

By putting a cap over the bed of his truck, Roger was assured he would use only 85% as much gasoline. The actual amount of gasoline used normally without a cap is 400 gallons every six months. How many gallons of gasoline would Roger save per year by putting a cap on his truck?

   A. 600 gallons   B. 120 gallons   C. 200 gallons   D. 60 gallons

Solution

With a cap, Roger will use (.85)(400) = 340 gallons every six months, or 680 gallons per year. Without a cap Roger used 800 gallons per year. Therefore he will save 120 gallons per year (choice B).

Skill IVA 3. Solve problems that involve the structure and logic of arithmetic

The problems that address this skill can involve the concepts of factor, multiple, divisibility, remainder, prime, etc. The problems will contain more than one condition or relation, but will not contain irrelevant information.

Example 40

How many whole numbers leave a remainder of three when divided into 68 and a remainder of one when divided into 27?

   A. none   B. 1   C. 2   D. 3

Solution

1. Read the problem carefully and restate it in your own words.

   We want to find all whole numbers that (1) leave a remainder of 3 when divided into 68, and (2) leave a remainder of 1 when divided into 27.

2. Select a strategy.

   Consider the conditions. List all numbers that satisfy condition 1. List all numbers that satisfy condition 2. Find all numbers greater than 1 that are common to these two lists of numbers. (These numbers will then satisfy both conditions.)

Continued on next page ...

3. Execute the strategy and solve the problem.

   - Numbers that meet condition 1 must evenly divide 65. The divisors of 65 are 1, 5, 13, and 65.

   - Numbers that meet condition 2 must evenly divide 26. The divisors of 26 are 1, 2, 13, and 26.

   - The common divisors are 1 and 13. We must eliminate 1, however, since 1 evenly divides every whole number. As a result there is only one number (13) that satisfies both conditions.

   Choice B is correct.

4. Check and reflect.

   $27 \div 13 = 2$ with a remainder of 1 and $68 \div 13 = 5$ with a remainder of 3.

4. GEOMETRY

In this section we discuss the CLAST subject area Geometry. Each skill within this subject will be addressed separately, and examples of the type of problems that could be asked on the CLAST for each skill will be presented.

Skill IB 1. Round measurements to the nearest given unit of the measuring device

The measurements in these problems can include both English and metric units of length, area, volume, mass, weight, and time. The measurements will include both decimals and fractions. Conversions of common English units of measurements such as inches to feet, ounces to pounds, and seconds to minutes might be required.

Measurements are always approximations and the degree of accuracy depends upon the unit of measure. It is often necessary to round a measurement to the nearest given unit of measure. If the measurement is between two of the given units of measure, and the fractional part (or decimal part) is less than half of the given unit, *round down* by eliminating the fractional part. If the fractional part is one half or more of the given unit, *round up*. For example, to round $21\frac{2}{3}$ pounds to the nearest pound, round to 22 pounds since two thirds of a pound is more than one half pound.

Example 1: Round the measurement 3917 kilograms to the nearest ten kilograms.

    A. 3900 kilograms   B. 3910 kilograms   C. 3920 kilograms   D. 4000 kilograms

Solution

There are 3 thousand, 9 hundred, 17 kilograms. Since 17 kilograms is closer to 20 kilograms than to 10 kilograms, we *round up*. That is, 3917 kilograms rounded to the nearest 10 kilograms is 3920 kilograms, which is choice C.

Alternatively, this problem can be thought of as rounding to the *tens* position. Since the number seven, which is in the *ones* position, is greater than five, we round to the next higher *tens* position. This leads to 3920.

Example 2

Round the measurement of the length of the object pictured below to the nearest ¼ inch.

    A. 3 in.   B. $3\frac{1}{2}$ in.

    C. $3\frac{3}{4}$ in.   D. $3\frac{1}{4}$ in.

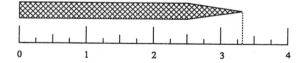

Solution

The length of the object is between 3¼ in. and 3½ in. Note that it is closer to 3¼ in. As a result, the length rounds to 3¼ in., which is choice D.

Skill IB 2. Calculate distances, areas, and volumes

For these problems, English-English and metric-metric conversions might be required without conversion factors or formulas being provided.

In general, the perimeter of a plane figure (i.e., a two-dimensional figure), is the *distance around* the figure. Distance (perimeter and circumference included) is always measured in linear units (e.g., inches, feet, yards, centimeters, meters, etc.). Area is always measured in square units (e.g., square inches ($in.^2$), square feet ($ft.^2$), square yards ($yd.^2$), square centimeters (cm^2), square meters (m^2), etc.). An acre is a square unit itself so we do not say *square acres*. You will need the formulas for perimeter and area given in Table 4.1. (Other formulas might be needed as well.)

If you are asked to find the perimeter of a figure that does not readily fit one of the figures shown in Table 4.1, you should find the length of each of its sides (boundaries) and add these lengths together. To illustrate, consider the problem of finding the perimeter of the figure shown below.

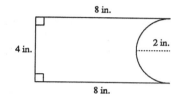

- Find the total length of the *straight* sides.

$$4 \text{ in.} + 8 \text{ in.} + 8 \text{ in.} = 20 \text{ in.}$$

- Find the length of the *curved* side.
  (**Note:** Circumference of a semicircle is one-half that of a circle.)

$$\begin{aligned} C_{semicircle} &= \frac{1}{2}(2\pi r) \\ &= \frac{1}{2} \cdot 2 \cdot \pi \cdot 2 \\ &= 2\pi \text{ in.} \end{aligned}$$

Adding the two lengths we find the perimeter is $(20 + 2\pi)$ in.

Example 3: What is the distance, in meters, around the polygon shown below?

A. 6.35 m   B. 63.5 m
C. 635 m   D. 63500 m

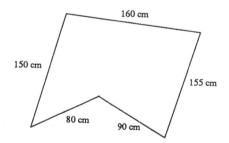

Solution

The perimeter is equal to $(150 + 160 + 155 + 90 + 80)$ cm, which is equal to 635 cm. We must now convert this answer to meters. Recall that 100 cm is equal to 1 m. As a result,

$$635 \text{ cm} = \frac{635}{100} \text{ m} = 6.35 \text{ m}$$

Choice A is correct.

Table 4.1. Common Perimeter and Area Formulas

Name of Figure	Formulas (P = Perimeter) (A = Area)	Sample Figure
Triangle	$P = a + b + c$ $A = \frac{1}{2}bh$	
Square	$P = 4s$ $A = s^2$	
Trapezoid	$A = \frac{1}{2}h(b_1 + b_2)$	
Rectangle	$P = 2l + 2w$ $A = lw$ or bh	
Circle	Circumference $= 2\pi r = \pi d$ $A = \pi r^2$	
Parallelogram	$A = bh$	

Example 4

What is the distance around a circular lake if the diameter of the lake is ½ mile?

A. $\frac{1}{4}\pi$ sq. mi.   B. $\frac{1}{2}\pi$ mi.   C. π mi.   D. π sq. mi.

Solution

$$\text{diameter} = \frac{1}{2} \text{ mile} \quad \text{and} \quad \text{radius} = \left(\frac{1}{2}\right)\left(\frac{1}{2}\right) = \frac{1}{4} \text{ mile}$$

$$\begin{aligned} C &= 2\pi r \\ &= 2\pi\left(\frac{1}{4}\right) \\ &= \frac{1}{2}\pi \end{aligned}$$

Since circumference is in linear units, the distance around the lake is $\frac{1}{2}\pi$ miles. Choice B is correct.

Example 5

What is the area of a trapezoid with bases 8 inches and 12 inches, and height 6 inches?

A. 120 sq. in.   B. 30 sq. in.   C. 60 in.   D. 60 sq. in.

Solution

The area formula for a trapezoid is $A = \frac{1}{2}h(b_1 + b_2)$. For this problem $h = 6$ in., $b_1 = 8$ in., and $b_2 = 12$ in.

$$\begin{aligned} A &= \frac{1}{2}h(b_1 + b_2) \\ &= \frac{1}{2}(6)(8 + 12) \\ &= 3(20) \\ &= 60 \text{ square inches} \end{aligned}$$

Choice D is correct.

The Pythagorean Theorem can also be involved in the solution of the problems at this skill level. This theorem states a relationship among the three sides of a right triangle.

For any right triangle, $c^2 = a^2 + b^2$.

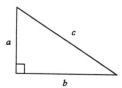

Note that c is the length of the hypotenuse, which is the side opposite the right angle.

To find the area of a plane figure made up of other figures, try to separate the figure into other figures for which you can find the area. To illustrate this, consider the problem of finding the area of the shaded region in the figure below.

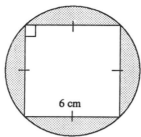

First note that the area of the shaded region is equal to the difference between the area of the circle and the area of the square.

<u>Find the area of the circle</u>

$$A = \pi r^2$$

We need to find the radius of the circle. Note that if we construct a diagonal of the square we obtain a right triangle as shown below. Moreover, this diagonal is the diameter of the circle. Using the Pythagorean Theorem we can determine the length of this diameter.

$$
\begin{aligned}
d^2 &= 6^2 + 6^2 \\
d^2 &= 36 + 36 \\
d^2 &= 72 \\
d &= \sqrt{72} \\
d &= \sqrt{36}\sqrt{2} \\
d &= 6\sqrt{2} \text{ cm}
\end{aligned}
$$

Since the radius is half of the diameter, $r = 3\sqrt{2}$ cm. We can now determine the area of the circle.

$$
\begin{aligned}
A &= \pi r^2 \\
&= \pi (3\sqrt{2})^2 \\
&= \pi(9)(2) \\
&= 18\pi \text{ cm}^2
\end{aligned}
$$

<u>Next we find the area of the square</u>

$$A = s^2 = 6^2 = 36 \text{ cm}^2$$

As a result, the area of the shaded region is equal to $(18\pi - 36)$ cm^2.

To find the surface area of a solid having polygons as faces, we find the area of each face and add these areas together. Consider the next example.

Example 6

What is the surface area of a rectangular solid that is 12 inches long, 10 inches wide, and 6 inches high?

   A. 720 cubic inches   B. 504 square inches   C. 504 inches   D. 252 square inches

Solution

1. Read carefully and restate the problem in your own words.

    A rectangular solid is a box. The lengths of the edges are 12 in., 10 in., and 6 in. We want to find the total area of its faces. There are 6 faces.

2. Explore and draw a diagram.

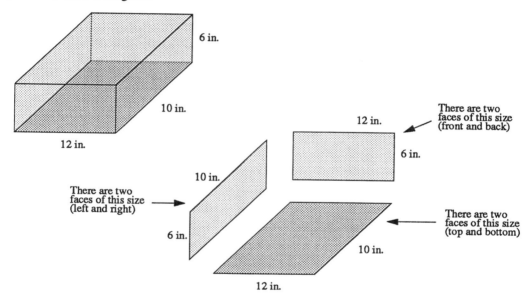

3. Select a strategy.

    Each face is a rectangle. Find the area of each of the three faces of different sizes. Multiply each area by two. Add the results.

4. Carry out the strategy and solve the problem.

    - Area of first rectangle is $12 \times 6 = 72$ sq. in.
    - Area of second rectangle is $10 \times 6 = 60$ sq. in.
    - Area of third rectangle is $12 \times 10 = 120$ sq. in.

    $$\begin{aligned}\text{Surface area of solid} &= (2 \times 72) + (2 \times 60) + (2 \times 120) \\ &= 144 + 120 + 240 \\ &= 504 \text{ sq. in.}\end{aligned}$$

The correct choice is B.

Continued on next page ...

5. Check and reflect.

   Compute the area a different way.

   $$\begin{aligned}\text{Surface area of solid} &= 2[(12 \times 6) + (10 \times 6) + (12 \times 10)] \\ &= 2(72 + 60 + 120) \\ &= 2(252) \\ &= 504 \text{ sq. in.}\end{aligned}$$

The volume of a solid (which is a three-dimensional object) is measured in cubic units (e.g., cubic feet (ft.^{3}), cubic centimeters (cm^3), cubic meters (m^3), etc.). Table 4.2 provides some of the formulas that will be needed.

Table 4.2. Common Volume Formulas

Name of Figure	Volume Formula	Sample Figure
Rectangular Solid	$V = lwh$	
Right Circular Cylinder	$V = \pi r^2 h$	
Right Circular Cone	$V = \dfrac{1}{3}\pi r^2 h$	
Sphere	$V = \dfrac{4}{3}\pi r^3$	

To illustrate the use of one of these formulas, consider the problem of finding the volume of the right circular cone shown below.

$$V = \frac{1}{3}\pi r^2 h$$
$$= \frac{1}{3}\pi(5)^2(10)$$
$$= \frac{1}{3}\pi(25)(10)$$
$$= \frac{250}{3}\pi \text{ cu. in.}$$

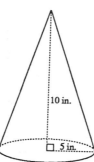

Example 7: What is the volume in centiliters of a 3.25 liter bottle?

    A. 3.25 *cl*   B. 32.5 *cl*   C. 325 *cl*   D. 3250 *cl*

Solution

$$1 \text{ liter} = 100 \text{ centiliters}$$
$$3.25 \, l = (3.25)(100) \, cl$$
$$= 325 \, cl$$

The correct choice is C.

Example 8

A rectangular solid is three meters long, two meters wide, and one meter high. Find the volume in cubic centimeters.

    A. 60 m^3 B. 6000 cm^3 C. 6,000,000 cm^3 D. 22,000,000 cm^3

Solution

- Substitute the values for length, width, and height into the correct volume formula and evaluate the formula.

$$V = lwh$$
$$= (3 \text{ meters})(2 \text{ meters})(1 \text{ meter})$$
$$= 6 \text{ cubic meters}$$

- Now convert to the desired unit of measure.

$$1 \, m = 100 \, cm$$
$$1 \, m^3 = 100^3 \, cm^3$$
$$= 1,000,000 \, cm^3$$

Therefore, V = 6 × 1,000,000 = 6,000,000 cm^3. Choice C is correct.

340 CLAST REVIEW

Example 9: A sphere has a diameter of 12 inches. Find its volume.

    A. 288 cu. in.   B. 288π cu. in.   C. 2304π cu. in.   D. 36π sq. in.

Solution

A diameter of 12 inches implies a radius of 6 inches.

$$\begin{aligned} V &= \frac{4}{3}\pi r^3 \\ &= \frac{4}{3}\pi (6)^3 \\ &= \frac{4}{3}\pi (216) \\ &= 288\pi \text{ cu. in.} \end{aligned}$$

The correct choice is B.

Skill IIB 1. Identify relationships between angle measures

Following are some relationships between angle measures that might be tested.

1. <u>Properties of Vertical, Supplementary, and Complementary Angles</u>

    a. Vertical angles are congruent (i.e., they have the same measure).

    b. Two angles are supplementary if and only if the sum of their measures is 180°.

    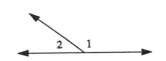
        $m\angle 1 + m\angle 2 = 180°$
    So $\angle 1$ and $\angle 2$ are supplementary.
    (Note: The notation $m\angle$ is read as *measure of angle*.)

    c. Two angles are complementary if and only if the sum of their measures is 90°.

    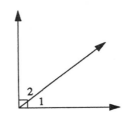
        $m\angle 1 + m\angle 2 = 90°$
    So $\angle 1$ and $\angle 2$ are complementary.

2. <u>Properties of Plane Figures, Including Isosceles Triangles, Equilateral Triangles, and Parallelograms</u>

    These properties are displayed in Table 4.3. The symbol ≅ is read *is congruent to*, which means "has the same measure as." For example, if $\angle 1 \cong \angle 2$, then $m\angle 1 = m\angle 2$ and vice versa. Also, if $\overline{AB} \cong \overline{CD}$, then length AB = length CD, that is AB = CD.

Table 4.3. Properties of Plane Figures

Name of Figure	Properties	Sample Figure
Triangle	$m\angle A + m\angle B + m\angle C = 180°$	
Isosceles Triangle	If $AB = AC$ then $\angle B \cong \angle C$	
Equilateral Triangle	If $BC = AC = AB$ then $\angle A \cong \angle B \cong \angle C$	
Parallelogram	$\angle A \cong \angle C$ and $\angle B \cong \angle D$ $m\angle A + m\angle B + m\angle C + m\angle D = 360°$ $m\angle A + m\angle D = 180°$ $\angle 1 \cong \angle 2$	

3. <u>Parallel Line Properties</u>

Consider the figure below with line $l_1 \parallel l_2$, and l_3 a transversal. (Note: The symbol $\parallel$ is read as *is parallel to*.)

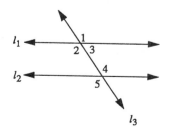

a. Corresponding angles are congruent
$\angle 1 \cong \angle 4$ (i.e., $m\angle 1 = m\angle 4$)

b. Vertical angles are congruent
$\angle 1 \cong \angle 2$ and $\angle 4 \cong \angle 5$

c. Alternate interior angles are congruent
$\angle 2 \cong \angle 4$

d. Interior angles on the same side of the transversal are supplementary
$m\angle 3 + m\angle 4 = 180°$

342 CLAST REVIEW

4. Perpendicular Line Properties

   a. If two lines are perpendicular, four right angles are formed at their point of intersection. (A right angle has measure 90°.)

   b. In a plane, if two lines are perpendicular to the same line, then the two lines are parallel.

   c. In a plane, if one of two parallel lines is perpendicular to a transversal, then the second line is also perpendicular to the transversal.

(Note: The properties discussed in 1, 2, 3, and 4 above are not necessarily exhaustive.)

Within the questions on the test itself, lower case letters will often be used to refer to the measures of angles, as in the following examples.

Example 10

Which of the statements below is true for the figure shown? (The measure of angle ABC is represented by x.)

A. $z = 150°$ B. $y = 100°$

C. $v = 30°$ D. $u \neq v$

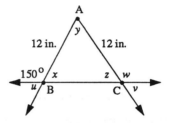

Solution

- A is false.

  $x = 180° - 150° = 30°$. Since $\overline{AB} \cong \overline{AC}$, $x = z$. Therefore $z = 30°$.

- B is false.

  $x + y + z = 180°$. Also, $x = z = 30°$. Therefore, $30° + 30° + y = 180°$, which implies $y = 120°$.

- C is true.

  $z = 30°$. Since vertical angles have the same measure, $z = v$. Therefore, $v = 30°$.

- D is false.

  Since $\overline{AB} \cong \overline{AC}$, $x = z$. Since vertical angles have the same measure, $u = x$ and $z = v$. Thus, $u = v$.

Choice C is correct.

GEOMETRY 343

Example 11: Which of the statements below is *not* true for the figure shown in [1]?

        A. $l_1 \parallel l_2$ B. $e = 50°$

        C. $b = e$ D. $b + a = c + e$

[1]   [2]

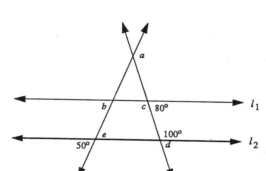 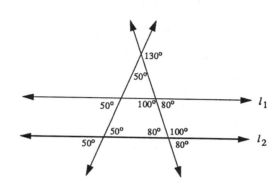

Solution

- A is true since interior angles on the same side of a transversal are supplementary.

    With this knowledge we can now insert the labeled angle measures (and others as needed) in the given figure. This is shown in the figure labeled [2] above.

- B is true ($e = 50°$); vertical angles have the same measure.

- C is true ($b = e = 50°$); alternate interior angles have the same measure.

- D is not true. $b + a = 180°$, but $c + e = 150°$.

Choice D is correct.

Example 12

Given that $\overline{AB} \parallel \overline{CD}$, $\overline{AB} \perp \overline{BF}$, $\overline{EF} \perp \overline{BF}$, and $\overline{CD} \cong \overline{DF}$, one of the statements given below is not true. Which statement is *not* true? (The measure of angle BAC is represented by x.)

A. $\overline{CD} \perp \overline{BF}$

B. $\angle BAC$ and $\angle ACD$ are supplementary angles

C. $y = 150°$

D. $z = u$

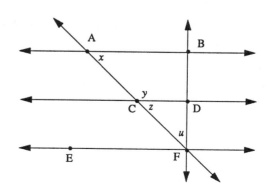

Continued on next page ...

344 CLAST REVIEW

Solution

- A is true. Since $\overline{AB} \parallel \overline{CD}$ and $\overline{AB} \perp \overline{BF}$, $\overline{CD}$ is also perpendicular to $\overline{BF}$.
- B is true. Since $\overline{AB} \parallel \overline{CD}$ and line $\overleftrightarrow{AF}$ is a transversal, $\angle BAC$ and $\angle ACD$ are interior angles on the same side of the transversal and are thus supplementary angles.
- C is not true. Since $\overline{CD} \cong \overline{DF}$, $z = u$. However, $\angle CDF$ is a right angle. Thus $z + u + m\angle CDF = 180°$ and $z + u + 90° = 180°$. Therefore, $z + u = 90°$ and $z = 45°$. As a result, $y = 180° - z = 180° - 45° = 135°$.
- D is true. This was shown in the explanation for C.

Choice C is correct.

Skill IIB 2. Classify simple plane figures by recognizing their properties

Figures to be classified can include:

1. types of angle, such as acute, right, straight, and obtuse;
2. types of angle pairs such as supplementary, complementary, adjacent, alternate interior angles, and corresponding angles (with parallel lines);
3. types of triangles such as acute, scalene, right, obtuse, isosceles, and equilateral; and
4. types of convex quadrilaterals such as square, rectangle, rhombus, parallelogram, and trapezoid.

    (**Note:** To understand the meaning of convex, study the figures below.)

Convex Polygons   Non-Convex Polygons

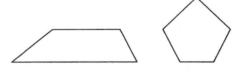 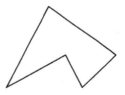

Example 13

Select the quadrilateral that possesses all of the following characteristics.

    i. Diagonals are equal in length

    ii. Opposite angles are congruent

    iii. Adjacent sides are equal in length

    A. parallelogram   B. rhombus   C. rectangle   D. square

Continued on next page ...

GEOMETRY 345

Solution

- Choice A is incorrect. The properties of a parallelogram are

    — Both pairs of opposite sides are parallel

    — Both pairs of opposite sides are congruent

    — Both pairs of opposite angles are congruent

    — Each pair of adjacent angles is supplementary

    — The diagonals bisect each other

    As a result, all parallelograms do not possess properties i and iii.

- Choice B is incorrect. A rhombus is a parallelogram with adjacent sides congruent. Thus the properties of a rhombus include those of a parallelogram and two additional properties.

    — All four sides are equal in length

    — The diagonals bisect the angles

    As a result, all rhombuses have properties ii and iii, but not property i.

- Choice C is incorrect. A rectangle is a parallelogram with four right angles. Thus a rectangle possesses all the properties of a parallelogram and two additional properties.

    — All four angles are right angles and hence are congruent

    — The diagonals are congruent

    All rectangles have properties i and ii, but not property iii.

- Choice D is correct. A square is a rectangle with all four sides congruent. It is also a rhombus with four right angles. Thus a square possesses all of the properties of a rectangle and a rhombus. Therefore, all squares have properties i, ii, and iii.

Choice D is correct.

Example 14: Which of the following triangles is both acute and scalene?

A.   B.   C.   D.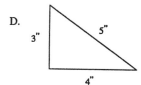

Solution

An acute triangle has three acute angles; that is, each of its angles has a measure less than 90°. A scalene triangle is a triangle in which no two sides have the same length.

- Choice A is incorrect. The triangle shown is indeed acute. Two of its angles however, have the same measure. Thus the triangle has two sides equal in length. As a result the triangle is acute, but it is not scalene.

- Choice B is incorrect. The figure shown is an obtuse triangle since one angle has a measure greater than 90°. Thus the triangle is not acute.

Continued on next page ...

346 CLAST REVIEW

- Choice C is correct. The third angle of the figure shown is equal to 180° − (70° + 50°), which is equal to 180° − 120° = 60°. Since all three angles have a measure less than 90° the triangle is acute. Furthermore, since all three angles have different measures, all three sides have different lengths. Thus the triangle is also scalene.
- Choice D is incorrect. The triangle shown is a right triangle, which can be confirmed by the Pythagorean Theorem. Since one angle has a measure of 90°, the triangle is not acute.

Choice C is correct.

Example 15: Which of the following pairs of angles is complementary?

A. 1 and 3

B. 1 and 4

C. 3 and 5

D. 2 and 5

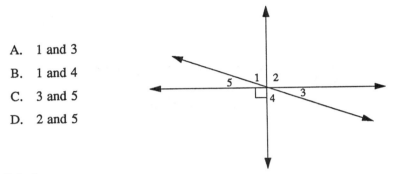

Solution

- Choice A is correct. The two lines are perpendicular as shown. Hence, ∠2 is a right angle. Further, ∠1 and ∠5 are complementary. However, ∠5 and ∠3 are vertical angles and have the same measure. Thus ∠1 and ∠3 are complementary.
- Choice B is incorrect. ∠1 and ∠4 are vertical and have the same measure, but the sum of their measures cannot be concluded to be 90° with the information given.
- Choice C is incorrect by the same argument given in B.
- Choice D is incorrect. Since ∠2 is a right angle, the measures of ∠2 and ∠5 cannot add to be 90°.

As stated, choice A is correct.

Skill IIB 3. Recognize similar triangles and their properties

The questions on this skill will take one of the following two formats:

1. Given a diagram of similar triangles, you will be asked to identify true statements about the diagram.

2. Given several groups of triangles, you will be asked to identify the group showing similar triangles.

Similar triangles (i.e. triangles with the same shape) have the following two properties. (See the figure below.) For the figure, △ ABC ~ △ DEF. (**Note:** The symbol ~ is read *is similar to*.)

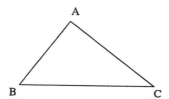
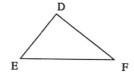

Corresponding angles are congruent	Corresponding sides are proportional.
$\angle A \cong \angle D$	$\dfrac{AB}{DE} = \dfrac{AC}{DF} = \dfrac{BC}{EF}$
$\angle B \cong \angle E$	
$\angle C \cong \angle F$	

There are three ways in which two triangles can be proven similar.

[1] AA Theorem

Show two pairs of corresponding angles to be congruent. For example, in the figure above, show $\angle A \cong \angle D$ and $\angle B \cong \angle E$.

[2] SSS Theorem

Show the corresponding sides are in proportion. That is, show $\dfrac{AB}{DE} = \dfrac{AC}{DF} = \dfrac{BC}{EF}$.

[3] SAS Theorem

Show two pairs of corresponding sides proportional and show the included angles to be congruent. For example, show $\dfrac{AB}{DE} = \dfrac{AC}{DF}$ and $\angle A \cong \angle D$.

A wide variety of information can be given in these problems for you to use to determine whether triangles are similar. Analysis of this information requires you to have an understanding of:

1. the sum of the measures of interior angles of a triangle;
2. the relationships among angles formed when parallel lines are cut by a transversal;
3. the Pythagorean Theorem;
4. the relationship between congruent angles of a triangle and the lengths of sides opposite these angles;
5. the three theorems discussed in [1], [2], and [3] above;
6. the properties of corresponding angles and sides of similar triangles; and
7. other relationships referred to in Skills IIB1 and IIB2.

348   CLAST REVIEW

Example 16

Which of the statements A-D is true for the pictured triangles? (The measure of angle C is represented by x.)

  A. $x = w$ B. $y = z$

  C. $\dfrac{AB}{AC} = \dfrac{DE}{BC}$ D. None of the above statements is true.

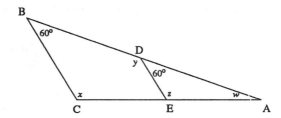

Solution

- Choice A is not correct. Since $\angle ADE \cong \angle ABC$, $\overline{DE} \parallel \overline{BC}$. This implies that $x = z$. By the AA theorem, triangle ADE is similar to triangle ABC. There is no evidence, however, to indicate that BC = BA. This would be necessary for choice A to hold.
- Choice B is incorrect. If $y = z$, then $\overleftrightarrow{BD} \parallel \overleftrightarrow{CE}$. However, these two lines are not parallel since they intersect at point A.
- Choice C is not true. Since triangle ADE is similar to triangle ABC, corresponding sides are proportional. Thus, $\dfrac{AB}{AD} = \dfrac{BC}{DE}$. This is not what choice C states.

As a result, choice D is correct.

Example 17: Which of the statements A-D is true for the pictured triangles?

  A. $\dfrac{AD}{DE} = \dfrac{BC}{AB}$ B. $\triangle AED \sim \triangle ABC$

  C. $\dfrac{AD}{DB} = \dfrac{AE}{EC}$ D. $\dfrac{AE}{AC} = \dfrac{AB}{AD}$

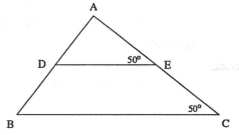

Solution

Note that $m\angle AED = m\angle C$, and $m\angle A = m\angle A$. As a result, $\triangle AED \sim \triangle ACB$ by the AA Theorem. Since the two triangles are similar, their corresponding sides must be proportional. That is, $\dfrac{AD}{AB} = \dfrac{AE}{AC} = \dfrac{DE}{BC}$.

Continued on next page ...

- Choice A is false; there is an inverted fraction in the given proportion.
- Choice B is false; the vertices are improperly matched. (The second triangle should be ACB, not ABC.)
- Choice C is true. Since $\frac{AD}{AB} = \frac{AE}{AC}$, a property of proportion can be used to show $\frac{AD}{AB - AD} = \frac{AE}{AC - AE}$. That is, $\frac{AD}{DB} = \frac{AE}{EC}$.
- Choice D is false; there is an inverted fraction in the proportion.

Choice C is correct.

Example 18

Which of the statements A-D is *not* true for the pictured triangles? (The measure of angle E is represented by x.)

A. $x = y$ B. $\triangle ABC \sim \triangle EDC$

C. $BC = 6$ cm   D. $\frac{AC}{CE} = \frac{DC}{CB}$

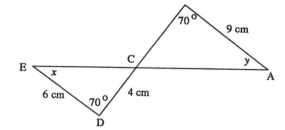

Solution

$m\angle B = m\angle D$. Also, since vertical angles are congruent, $\angle BCA \cong \angle DCE$. Therefore, $\triangle ABC \sim \triangle EDC$. Corresponding angles of these two triangles are congruent and corresponding sides are proportional.

- Choice A is true. $\angle A$ and $\angle E$ are corresponding angles in the two triangles. Hence their measures are equal, that is $x = y$.
- Choice B is true (from the above discussion).
- Choice C is true. $\frac{BC}{CD} = \frac{BA}{ED}$. If we let $BC = x$, then $\frac{x}{4} = \frac{9}{6}$. $6x = 36$. $x = 6$. Therefore $BC = 6$ cm.
- Choice D is false. The correct proportion is $\frac{AC}{CE} = \frac{CB}{DC}$. Choice D has an inverted fraction in the proportion.

Choice D is correct.

350 CLAST REVIEW

Example 19

Study figures A-D and then select the figure in which all the triangles are similar.

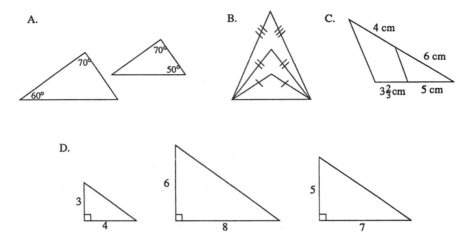

Solution

- Choice A is correct. The first triangle has angle measures of 60°, 70°, and 180° − (60° + 70°) = 50°. The second triangle has measures 70°, 50°, and 180° − (70° + 50°) = 60°. Since three pairs of angles for these two triangles are congruent, two pairs of angles are also congruent, and therefore the triangles are similar by the AA theorem.

- Choice B is incorrect. Note that each triangle is isosceles. Since this is a plane figure, corresponding angles at the common base of the three triangles are not congruent. Therefore the triangles are not similar.

- Choice C is incorrect. If the triangles were similar then $\frac{6}{6+4}$ would equal $\frac{5}{5+3\frac{2}{3}}$. However, $\frac{6}{6+4} = \frac{6}{10} = \frac{3}{5}$, and $\frac{5}{5+3\frac{2}{3}} = \frac{5}{5+\frac{11}{3}} = \frac{5}{\frac{26}{3}} = \frac{15}{26}$. $\frac{3}{5} \neq \frac{15}{26}$. Therefore the triangles are not similar.

- Choice D is incorrect. Although each triangle is a right triangle, corresponding sides are not proportional in all three triangles. $\frac{3}{4} = \frac{6}{8}$, but $\frac{6}{8} \neq \frac{5}{7}$. The first two triangles are similar, but they are not similar to the third triangle.

Choice A is correct.

Skill IIB 4. Identify appropriate units of measurement for geometric objects

The following three facts were mentioned in the discussion of Skill IB2, and will be needed for problems here.

1. Distances, lengths of segments, perimeters, and the circumference circles are all measured in **linear** units such as feet, miles, centimeters, and meters.

2. Areas, including the surface areas of solids, are measured in **square** units such as square inches, square miles, and square meters.

3. Volumes of solids are measured in **cubic** units such as cubic feet, cubic yards, and cubic centimeters. Units such as liters and gallons may be used.

To illustrate the type of questions that could be asked we will consider the figure shown below.

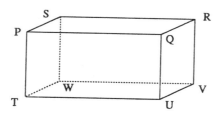

From this figure note that

- $\overline{TU}$ would be measured using linear units since $\overline{TU}$ involves length.
- The perimeter of rectangle PQRS would be measured using linear units since a perimeter is a sum of lengths.
- The interior of the solid would be measured using cubic units since volume would be involved.
- The area of rectangle QRVU would be measured using square units since area is measured in square units.

Example 20: The perimeter of the figure could be given in what type of measure?

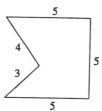

A. cubic inches   B. square inches

C. inches   D. degrees

Solution

A perimeter is a sum of lengths. A linear measure would be used. Choice C is correct.

Example 21

Which of the following would *not* be used to report the amount of oil in a tank of oil?

A. Gallons   B. Square yards   C. Liters   D. Cubic meters

Solution

Square yards is a unit of *area* measure. Gallons, liters, and cubic meters could all be used to describe the amount of oil (*capacity* or *volume*) in a tank.

Choice B is correct.

Skill IIIB 1. Infer formulas for measuring geometric figures

In these problems you will be given several geometric figures that illustrate some particular type of measurement. You will then be asked to solve a similar problem by generalizing from the figures. The figures can be two-dimensional or three-dimensional. Two-dimensional polygons may be regular or non-regular. Also, the geometric measure can be area, volume, surface area, angle measure, perimeter, circumference, or other distances. You should review the formulas from Skill IB2 since they might be used in these problems. (**Note**: A polygon is regular if all sides are congruent and all angles are congruent; otherwise it is non-regular. Also, a pentagon has five sides; a hexagon has six sides; a heptagon has seven sides; an octagon has eight sides; etc.)

Example 22: Study the information given for the convex polygons.

- 4 sides
- 1 diagonal at each vertex
- 2 total diagonals

- 5 sides
- 2 diagonals at each vertex
- 5 total diagonals

- 6 sides
- 3 diagonals at each vertex
- 9 total diagonals

Determine the number of total diagonals for an eight-sided convex polygon.

    A. 40   B. 16   C. 20   D. 12

Solution

Note that a diagonal joins two vertices. We need only to count the number of diagonals at one vertex to determine the total number of diagonals for a convex polygon.

# of sides	# of diagonals at each vertex	total # of diagonals
4	$1 = 4 - 3$	$\dfrac{4(1)}{2} = 2$
5	$2 = 5 - 3$	$\dfrac{5(2)}{2} = 5$
6	$3 = 6 - 3$	$\dfrac{6(3)}{2} = 9$
n	$n - 3$	$\dfrac{n(n-3)}{2}$
8	$5 = 8 - 3$	$\dfrac{8(5)}{2} = 20$

The correct choice is C.

Example 23: Study the information given for the right solids. (Each solid has altitude h.)

Volume = $\dfrac{25\pi h}{3}$

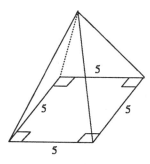
Volume = $\dfrac{25h}{3}$

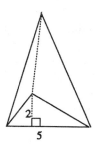
Volume = $\dfrac{5h}{3}$

Calculate the volume of the right solid shown with altitude h and a trapezoidal base. The lengths of the bases of the trapezoid are three and five, with a height of 5.

A. $\dfrac{10h}{3}$ B. $\dfrac{40h}{3}$ C. $10h$ D. $\dfrac{20h}{3}$

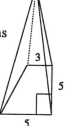
Volume = ?

Solution

- The first solid is a cylinder with a circular base. The area of the circular base is πr^2. This is equal to $\pi(5)^2 = 25\pi$. If we multiply the area by the altitude we get $25\pi h$. Moreover, this product is divided by three. It appears that the volume of this solid is equal to $\dfrac{(base\ area) \times (altitude)}{3}$, which can be expressed as $\dfrac{Bh}{3}$.

- The second solid is a square pyramid. The area of the square base is $s^2 = 5^2 = 25$. If we multiply the area by the altitude we get $25h$. Moreover, this product is divided by three. It appears that the general formula, $V = \dfrac{Bh}{3}$ holds.

- The third solid is a triangular pyramid. The area of the triangular base is $\dfrac{1}{2}(base)(height)$. This is equal to $\dfrac{1}{2}(5)(2) = 5$. Also, $5h = (base\ area) \times (altitude)$. This product is divided by three. Once again, the formula $V = \dfrac{Bh}{3}$ holds.

Using the development above, we now can determine the volume of the solid with a trapezoidal base.

The area of the trapezoidal base is $\dfrac{1}{2}h(b_1 + b_2) = \dfrac{1}{2}(5)(3+5) = \dfrac{(5)(8)}{2} = (5)(4) = 20$. Thus, $B = 20$.

The altitude of the solid is h. As a result, $V = \dfrac{Bh}{3} = \dfrac{20h}{3}$. Choice D is correct.

Skill IIIB 2. Identify applicable formulas for computing measures of geometric figures

The figures will be two-dimensional and/or three-dimensional. The two-dimensional figures can include parallelograms (squares, rectangles, rhombuses), circles, semicircles, triangles, and regular polygons. The three-dimensional figures can include rectangular solids, cylinders, and right circular cones. The measures given in the problems and calculated by the formulas in the problems can include length, perimeter, area of two-dimensional or three-dimensional figures (surface area), and volumes for three-dimensional figures. You should review the formulas for perimeter, area, and volume given in the discussion of Skill IB2 since these formulas might be used in these problems.

Example 24

Study the figure below, which shows a right isosceles triangle and one of the two triangles formed by using the altitude of the triangle as a side. Select the formula for computing the area of an isosceles triangle with base b.

A. $\dfrac{b}{2}$ B. $\dfrac{b^2}{4}$

C. $\dfrac{b^2}{2}$ D. $\dfrac{b^2}{8}$

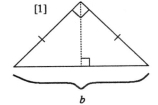

Solution

Since triangle [1] is isosceles, its base angles are congruent. Since it is a right triangle, each of these base angles has a measure of 45°. Thus triangle [2] also has angle measures of 90°, 45°, and 45°. Moreover, one of its legs has length $\dfrac{b}{2}$, and its other leg (i.e., the height) also has length $\dfrac{b}{2}$. Thus, $\dfrac{b}{2}$ is the height of the triangle [1]. Therefore the area of the isosceles right triangle with base b is $A = \dfrac{1}{2}bh = \dfrac{1}{2}b\left(\dfrac{b}{2}\right) = \dfrac{b^2}{4}$. Choice B is correct.

Example 25

Study the figure below, which shows a rectangular solid with a square base. Select the formula for computing the total surface area of the solid.

A. Surface area = $b^2 h$ B. Surface area = $4bh$

C. Surface area = $4bh + b^2$ D. Surface area = $4bh + 2b^2$

Solution

1. Read the problem carefully and restate it in your own words.

   A rectangular solid is a box. This box has a square base. We want to find the total area of the six faces. Four of the six faces have the same area and the other two (i.e., the bases) have the same area.

2. Select a strategy.

   Surface area = (area of the two bases) + (area of the four vertical faces)

   = 2(area of a square) + 4(area of a rectangle)

3. Carry out the strategy and solve the problem.

   - Area of each square base = b^2
   - Area of each vertical rectangular face = $lw = hb$ or bh
   - Total surface area = $SA = 2(b^2) + 4(bh) = 2b^2 + 4bh = 4bh + 2b^2$

Continued on next page ...

4. Check and reflect.

Have we found the areas of all surfaces involved? (yes) Is our algebra correct? (yes)

Choice D is correct.

Skill IVB 1. Solve real-world problems involving perimeters, areas, and volumes of geometric figures

These problems will be taken from a business, social studies, industry, education, economics, environmental studies, arts, physical science, sports, or a consumer-related context. The formulas given in the discussion of Skill IB2 will be involved. The problem, however, will involve more than calculating perimeter, area, or volume. Conversions (English-English or metric-metric) will be necessary. Also, the problems might contain irrelevant information.

Example 26

A toy airplane is attached to a string four yards in length. The string is held so that the airplane flies in a circular path. What is the area in square feet of the circular region determined by the path in which the airplane flies?

    A. 144π sq. ft.   B. 16π sq. ft.   C. 36π sq. ft.   D. 48π sq. ft.

Solution

1. Read the problem carefully and restate it in your own words.

    We want to find the area of a circle with a radius of four yards. Since the area is to be given in square feet, we will need to convert yards to feet.

2. Select a strategy.

    a. Convert four yards to feet.

    b. Find the area of the circle in square feet.

3. Carry out the strategy and solve the problem.

    a. 1 yard = 3 feet

      4 yards = 4(3) = 12 feet

    b. $A_{circle} = \pi r^2 = \pi(12)^2 = 144\pi$ sq. ft.

The correct choice is A.

4. Check and reflect.

    Solve the problem in a slightly different way. Specifically, find the area in square yards and then convert the answer to square feet.

    - $A_{circle} = \pi r^2 = \pi(4)^2 = 16\pi$ sq. yd.
    - 1 sq. yd. = (3 ft.)(3 ft.) = 9 sq. ft.
    - 16π sq. yd. = $(16\pi)(9)$ sq. ft. = 144π sq. ft.

Example 27

The cost of building a swimming pool is $30.00 per running foot along the edge of the pool plus $2.00 for each cubic foot of dirt removed. What would be the total cost of building a pool 26 feet long, 14 feet wide, and 120 inches deep along its entire length?

    A. $18,200   B. $9,440   C. $9,680   D. $11,136

Solution

1. Read the problem carefully and restate it in your own words.

   We want to find the total cost of building a rectangular pool that is 26 ft. × 14 ft. The depth of the pool along its entire length is 120 inches. The cost of building the pool will depend upon the perimeter of the pool and the volume of the pool. Conversion of units is needed.

2. Select a strategy.

   a. Find the perimeter of the pool in feet.

   b. Multiply the perimeter by $30.00.

   c. Convert 120 inches to feet and then find the volume in cubic feet.

   d. Multiply the volume by $2.00.

   e. Add the two amounts of money to obtain the total cost of building the pool.

3. Carry out the strategy and solve the problem.

   a. Perimeter of pool = $2l + 2w$
   = $2(26) + 2(14)$
   = $52 + 28$
   = 80 feet

   b. Cost based upon perimeter = ($30.00 per running foot) × (perimeter in feet)
   = $30.00 × 80
   = $2400.00

   c. Volume of pool = lwh
   = $(26)(14)(10)$
   = 3640 cubic feet

   (Note: 120 inches = 10 feet.)

   d. Cost based upon volume = ($2.00 per cubic feet of dirt) × (volume in cubic feet)
   = $2.00 × 3640
   = $7280

   e. Total cost of building pool = $2400 + $7280 = $9680

4. Check and reflect.

   Is the answer reasonable for the cost of a pool? (yes) Recheck the arithmetic.

Choice C is correct.

GEOMETRY

Skill IVB 2. Solve real-world problems involving the Pythagorean property

The problems for this skill will be taken from the same context areas as those of Skill IVB 1. The problems will involve more than just direct application of the Pythagorean Theorem. Conversions within the English and metric systems will be required. Also, the problems might contain irrelevant information. You should review the statement of the Pythagorean Theorem given in the discussion of Skill IB2.

It might be convenient for you to become aware of some common Pythagorean triples; that is, sets of three numbers that satisfy the Pythagorean Theorem. For example, {3, 4, 5} is a Pythagorean triple since $5^2 = 4^2 + 3^2$ (25 = 16 + 9). Other triples are {6, 8, 10}, {9, 12, 15}, {5, 12, 13}, {10, 24, 26}, {7, 24, 25}, and {8, 15, 17}. Try to look for others.

Example 28

A series of seven poles will be equally spaced along a block 100 yards long. Each pole is 30 feet tall and will be supported by three guywires that reach from halfway up each pole to a point on the ground eight feet from the base of the pole. If each pole costs $200 and the cable for the guywires is $0.80 per foot, what will be the total cost of the poles and guywires?

    A. $6255.20   B. $1440.80   C. $1495.20   D. $1685.60

Solution

1. Read the problem carefully and restate it in your own words.

    We want to find the total cost of seven poles and the guywires needed to support them. Each pole costs $200. Each pole has three guywires that reach from a point 15 feet up on the pole to a point on the ground eight feet from the pole. Cable costs 80¢ per foot. The 100 yards is irrelevant information.

2. Select a strategy.

    a. Use the Pythagorean Theorem to find the length of each guywire.

    b. Find the total cost of the guywires.

    c. Find the total cost of the poles.

    d. Add the two results to obtain the total cost requested.

3. Carry out the strategy and solve the problem.

    a. Let c = the length of one guywire in feet.

    $$\begin{aligned} c^2 &= a^2 + b^2 \\ &= 8^2 + 15^2 \\ &= 64 + 225 \\ &= 289 \\ c &= \sqrt{289} \\ &= 17 \text{ ft.} \end{aligned}$$

    (Note that knowledge of Pythagorean triples would yield an answer of 17 without any calculations.)

Continued on next page ...

b. There will be 7 × 3 or 21 guywires. Moreover, the total length of the guywires is equal to 21 × 17 feet, which is equal to 357 feet. The total cost of the guywires will be 357 × $0.80, which is $285.60

c. The total cost of the poles will be 7 × $200, which is $1400.

d. The total cost of the poles and the guywires is $1400 + $285.60, which is $1685.60.

4. Check and reflect.

   Was the Pythagorean Theorem used properly? (yes)
   Are the computations correct? (yes)

The correct choice is D.

Example 29

The Parks Department wants to construct a concrete pathway diagonally across a small park that is 400 meters long and 300 meters wide. Working together, three men can construct 20 running meters of pathway in two hours. If two of the men earn $8.00 per hour and the third man earns $10.00 per hour, what will be the labor costs for constructing the pathway?

A. $450   B. $1300   C. $3250   D. $650

Solution

- Find the length of the pathway.

   Let c = the length of the pathway.

$$\begin{aligned} c^2 &= a^2 + b^2 \\ &= (300)^2 + (400)^2 \\ &= 90000 + 160000 \\ &= 250000 \\ c &= \sqrt{250000} \\ &= 500 \text{ m} \end{aligned}$$

   (Recall that {3, 4, 5} is a Pythagorean triple.)

- The men construct 20 meters of pathway in two hours, which implies a rate of 10 meters in one hour.

- The men must work $\frac{500}{10}$ = 50 hours each.

- Two men earn a total of 50 × $8.00 × 2, which is $800 in 50 hours.

- The third man earns 50 × $10.00, which is $500 in 50 hours.

- The total labor cost is $800 + $500, or $1300.

The correct choice is B.

5. ALGEBRA

In this section we discuss the CLAST subject area Algebra. Each skill within this subject will be addressed separately, and examples of the type of problems that could be asked on the CLAST for each skill will be presented.

Skill IC 1. Add, subtract, multiply, and divide real numbers

The numbers you will be asked to work with will include irrational numbers in the form of square root radicals or π. To add or subtract square root radicals or other irrational numbers we apply the distributive property. Also, square root radicals usually must be simplified before they can be added or subtracted. The following two Laws of Radicals can be applied to simplify square root radicals.

I. $\sqrt{ab} = \sqrt{a}\sqrt{b}$ for $a > 0$, $b > 0$

II. $\sqrt{\dfrac{a}{b}} = \dfrac{\sqrt{a}}{\sqrt{b}}$ for $a > 0$, $b > 0$

Example 1: $\sqrt{3} + \sqrt{12} =$

    A. $\sqrt{15}$ B. $3\sqrt{3}$ C. $5\sqrt{3}$ D. 153

Solution

$$\sqrt{3} + \sqrt{12}$$
$$= \sqrt{3} + \sqrt{4}\sqrt{3} \quad \text{(Law I)}$$
$$= \sqrt{3} + 2\sqrt{3} \quad \text{(4 is a \emph{perfect square} and } \sqrt{4} = 2.\text{)}$$
$$= (1 + 2)\sqrt{3} \quad \text{(Factor out } \sqrt{3}.\text{)}$$
$$= 3\sqrt{3} \quad \text{(Combine like terms.)}$$

The correct choice is B.

(**Note:** $\sqrt{3} + \sqrt{12} \neq \sqrt{15}$.)

Laws I and II are also used to multiply and divide square root radicals. In general, leave radical results in simplified form. Additionally, do not leave a radical in the denominator of any fraction. (Examples 3 and 4 demonstrate this concept.)

Example 2: $\sqrt{5} \times \sqrt{10} =$

    A. 25×100 B. $25\sqrt{2}$ C. $5\sqrt{2}$ D. 50

Continued on next page ...

Solution

$$\sqrt{5} \times \sqrt{10} = \sqrt{50} \quad \text{(By Law I)}$$
$$= \sqrt{25}\sqrt{2} \quad \text{(By Law I)}$$
$$= 5\sqrt{2} \quad \text{(25 is a perfect square and } \sqrt{25} = 5.)$$

Choice C is correct.

Example 3: Simplify $\dfrac{\sqrt{72}}{\sqrt{3}}$.

    A. $\sqrt{24}$ B. $2\sqrt{6}$ C. $4\sqrt{6}$ D. $2\sqrt{3}$

Solution

$$\frac{\sqrt{72}}{\sqrt{3}} = \sqrt{\frac{72}{3}} \quad \text{(By Law II)}$$
$$= \sqrt{24} \quad (\tfrac{72}{3} = 24.)$$
$$= \sqrt{4}\sqrt{6} \quad \text{(By Law I)}$$
$$= 2\sqrt{6} \quad \text{(4 is a perfect square and } \sqrt{4} = 2.)$$

Choice B is correct.

Example 4: $\dfrac{5}{\sqrt{3}} =$

    A. $\dfrac{\sqrt{5}}{3}$ B. $\dfrac{5\sqrt{3}}{9}$ C. $\dfrac{5\sqrt{3}}{3}$ D. $5\sqrt{3}$

Solution

$$\frac{5}{\sqrt{3}} = \frac{5}{\sqrt{3}} \cdot \frac{\sqrt{3}}{\sqrt{3}} \quad \text{(Multiply by } \tfrac{\sqrt{3}}{\sqrt{3}}.)$$
$$= \frac{5\sqrt{3}}{\sqrt{9}} \quad \text{(By Law I)}$$
$$= \frac{5\sqrt{3}}{3} \quad \text{(9 is a perfect square and } \sqrt{9} = 3.)$$

Choice C is correct.

This skill can also involve adding, subtracting, multiplying, and dividing algebraic expressions.

Example 5: $3w + 5w - 4 =$

    A. $15w^2 - 4$ B. $8w - 4$ C. $4w$ D. $3w + 1$

Solution

$$
\begin{aligned}
&\ 3w + 5w - 4 \\
&= (3 + 5)w - 4 \quad \text{(Factor } w \text{ from the first two terms.)} \\
&= 8w - 4 \quad \text{(Combine like terms.)}
\end{aligned}
$$

The correct choice is B.

(**Note:** The terms $8w$ and -4 are *unlike terms* and hence cannot be combined.)

Skill IC 2. Apply the order of operations agreement to computations involving numbers and variables

The order of operations agreement is as follows:

[1] If parentheses or other grouping symbols are present

    a. work separately above and below any fraction bar using the rules in [2] below;

    b. use the rules in [2] below within each set of grouping symbols. (Start with the innermost and work outward.)

[2] If no grouping symbols are present

    a. simplify all powers and roots, working from left to right;

    b. do any multiplication or division in the order in which they occur, working from left to right;

    c. do any additions or subtractions in the order in which they occur, working from left to right.

Other rules of arithmetic and algebra allow this agreement to be changed in certain circumstances, but use of this agreement will always produce correct results.

Example 6: $36 - 24 \div 4 \times 3 - 1 =$

    A. 33   B. 17   C. 8   D. 0

Solution

$$
\begin{aligned}
&\ 36 - 24 \div 4 \times 3 - 1 \quad \text{(Divide first.)} \\
&= 36 - 6 \times 3 - 1 \quad \text{(Now multiply.)} \\
&= 36 - 18 - 1 \quad \text{(Finally, subtract in order from left to right.)} \\
&= 18 - 1 \\
&= 17
\end{aligned}
$$

The correct choice is B.

Example 7: $6y - 18y \div 6 + 12y^2 \div 3 \times 2 =$

    A. $8y^2 - 2y$ B. $2y^2 + 3y$ C. $2y^2 - 2y$ D. $8y^2 + 3y$

Solution

$$\begin{aligned}
& 6y - 18y \div 6 + 12y^2 \div 3 \times 2 && \text{(First compute the divisions.)} \\
= & \; 6y - 3y + 4y^2 \times 2 && \text{(Now multiply.)} \\
= & \; 6y - 3y + 8y^2 && \text{(Finally, subtract.)} \\
= & \; 3y + 8y^2 && \text{(These are unlike terms.)}
\end{aligned}$$

The correct choice is D, even though the terms are in a different order.

Example 8: $\dfrac{7}{8} - \dfrac{1}{4}\left(\dfrac{1}{2}\right)^2 + \dfrac{3}{16} =$

    A. $\dfrac{11}{32}$ B. $\dfrac{67}{64}$ C. 1 D. $\dfrac{15}{16}$

Solution

$$\begin{aligned}
& \dfrac{7}{8} - \dfrac{1}{4}\left(\dfrac{1}{2}\right)^2 + \dfrac{3}{16} && \text{(Perform the power first.)} \\
= & \; \dfrac{7}{8} - \dfrac{1}{4}\left(\dfrac{1}{4}\right) + \dfrac{3}{16} && \text{(Now multiply.)} \\
= & \; \dfrac{7}{8} - \dfrac{1}{16} + \dfrac{3}{16} && \text{(Next write each fraction with denominator 16.)} \\
= & \; \dfrac{14}{16} - \dfrac{1}{16} + \dfrac{3}{16} && \text{(Now add.)} \\
= & \; \dfrac{14 - 1 + 3}{16} \\
= & \; \dfrac{13 + 3}{16} \\
= & \; \dfrac{16}{16} && \text{(Now reduce.)} \\
= & \; 1
\end{aligned}$$

The correct choice is C.

ALGEBRA 363

Skill IC 3. Use scientific notation in calculations involving very large or very small measurements

To multiply or divide using scientific notation, first write each number in scientific notation. As an illustration, study the following conversions.

Standard Decimal Notation	Scientific Notation
3960000	3.96×10^6
7200	7.2×10^3
0.000045	4.5×10^{-5}
0.000123	1.23×10^{-4}

Example 9: $.00548 \div 274,000 =$

    A. 2.00×10^{-2} B. 2.00×10^2 C. 2.00×10^{-9} D. 2.00×10^{-8}

Solution

$$\frac{.00548}{274000} = \frac{5.48 \times 10^{-3}}{2.74 \times 10^5} \quad \text{(First express each number in scientific notation.)}$$

$$= \frac{5.48}{2.74} \times \frac{10^{-3}}{10^5} \quad \text{(Rewrite as two separate fractions.)}$$

$$= 2.00 \times 10^{-3-5} \quad \text{(Use the law } \frac{a^m}{a^n} = a^{m-n}.\text{)}$$

$$= 2.00 \times 10^{-8}$$

Note that all the answer choices are given in scientific notation. Choice D is correct.

Example 10: $(4.8 \times 10^5) \times (1.4 \times 10^{-9}) =$

    A. -67200 B. 67200 C. 0.000672 D. 0.00672

Solution

Note that multiplication is the only operation involved in this problem. Consequently, we can use the Commutative and Associative Properties of Multiplication to reexpress the problem in a more convenient form.

$$(4.8 \times 10^5) \times (1.4 \times 10^{-9})$$
$$= (4.8 \times 1.4) \times (10^5 \times 10^{-9})$$
$$= 6.72 \times (10^{5 + -9}) \quad \text{(Use the law } a^m a^n = a^{m+n}.\text{)}$$
$$= 6.72 \times 10^{-4}$$

Although our computation is complete, the form of the given answer choices is in standard notation. Converting 6.72×10^{-4} to standard notation, we get 0.000672, which is choice C. (**Note:** Since all the answer choices involve 672, we did not have to compute 4.8×1.4.)

Skill IC 4. Solve linear equations and inequalities

When solving equations or inequalities with grouping symbols try to avoid errors in sign when you remove grouping symbols. Also, recall that

$$-(a + b) = -a - b \quad \text{and} \quad -(a - b) = -a + b$$

Example 11: If $4b + 3 = 3(b - 1)$, then

    A. $b = -4$ B. $b = -6$ C. $b = 0$ D. $b = \dfrac{-6}{7}$

Solution

$4b + 3$	$=$	$3(b - 1)$	(First distribute on the right side.)
$4b + 3$	$=$	$3b - 3$	(Next subtract $3b$ from each side.)
$b + 3$	$=$	-3	(Now subtract 3 from each side.)
b	$=$	-6	

Choice B is correct.

Example 12: If $2(b - 4) = 4[b - (1 - 2b)]$ then

    A. $b = \dfrac{-7}{10}$ B. $b = \dfrac{2}{3}$ C. $b = \dfrac{2}{5}$ D. $b = \dfrac{-2}{5}$

Solution

$2(b - 4)$	$=$	$4[b - (1 - 2b)]$	(First distribute on the left side and remove the parentheses on the right side.)
$2b - 8$	$=$	$4[b - 1 + 2b]$	(Next combine like terms in the brackets.)
$2b - 8$	$=$	$4[3b - 1]$	(Now distribute on the right side.)
$2b - 8$	$=$	$12b - 4$	(Subtract $2b$ from each side.)
-8	$=$	$10b - 4$	(Add 4 to each side.)
-4	$=$	$10b$	(Divide both sides by 10.)
$\dfrac{-4}{10}$	$=$	b	(Finally, reduce the fraction to lowest terms.)
$\dfrac{-2}{5}$	$=$	b	

Thus choice D is correct.

ALGEBRA 365

When solving inequalities keep in mind the following Multiplication Property of Inequality.

[1] If $a < b$ and $c > 0$ then $ac < bc$

[2] If $a < b$ and $c < 0$ then $ac > bc$

This property implies that if we multiply (or divide) both sides of an inequality by the same *positive* number, then the direction of the inequality symbol remains the same (Property [1]). However, if we multiply (or divide) both sides of an inequality by a negative number then the direction of the inequality symbol must be reversed (Property [2]).

Example 13: If $4x + 7 < 7x - 8$ then

    A. $x < 5$ B. $x > -5$ C. $x < -5$ D. $x > 5$

Solution

$4x + 7$	$<$	$7x - 8$	(First subtract $7x$ from each side.)
$-3x + 7$	$<$	-8	(Now subtract 7 from each side.)
$-3x$	$<$	-15	(Finally, divide both sides by -3; reverse the inequality symbol.)
x	$>$	5	

The correct choice is D.

Example 14: If $4x + 6 > -3(x - 2)$, then

    A. $x < 0$ B. $x > 0$ C. $x < 12$ D. $x > 12$

Solution

$-4x + 6$	$>$	$-3(x - 2)$	(First distribute on the right side.)
$-4x + 6$	$>$	$-3x + 6$	(Next add $3x$ to both sides.)
$-x + 6$	$>$	6	(Now subtract 6 from both sides.)
$-x$	$>$	0	(Finally, multiply both sides by -1; reverse the inequality symbol.)
x	$<$	0	

The correct choice is A.

Skill IC 5. Use given formulas to compute results when geometric measurements are not involved

The formulas can be algebraic in nature or they can come from science, business, or everyday life. The formulas will be limited to first or second degree expressions involving rational numbers.

Example 15: Given $y = (t - 3)^2$, if $t = 8$ then $y =$

    A. 5   B. 25   C. 55   D. $\sqrt{5}$

Solution

Substitute 8 for t in the expression $(t - 3)^2$ and evaluate the expression.

$$y = (t - 3)^2$$
$$y = (8 - 3)^2$$
$$y = 5^2 = 25$$

The correct choice is B.

Example 16

If p dollars are deposited into a savings account at r percent simple interest for t years, the amount of money s in the account is given by $s = p(1 + rt)$. Find the amount of money in the account at the end of 5 years if $15,000 is deposited at 8% simple interest.

    A. $18,000   B. $21,000   C. $210,000   D. $75,000

Solution

Substitute the following values for r, p, and t into the given formula and then evaluate it using the correct order of operations.

$$r = 8\% = 0.08; \quad p = \$15{,}000; \quad \text{and} \quad t = 5$$

$$\begin{aligned} s &= p(1 + rt) \\ &= 15000(1 + (0.08)(5)) \\ &= 15000(1 + 0.40) \\ &= 15000(1.40) \\ &= 21000 \end{aligned}$$

Thus the amount of money is $21,000, which is choice B.

Skill IC 6. Find particular values of a function

Given a polynomial function in $f(x)$ notation, you will be asked to find the value of the function at some particular value of the independent variable x. For example, you might be asked to find $f(-1)$. Recall that $f(-1)$ is the value of the function when x has the value of -1. As a result, $f(-1)$ is found by substituting -1 for x and evaluating.

Example 17: $f(x) = x^3 + x^2 - 2x + 1$, find $f(-2)$.

    A. 17   B. -7 C. 1   D. -15

Continued on next page ...

Solution

Substitute -2 for x and evaluate using the correct order of operations.

$$f(x) = x^3 + x^2 - 2x + 1$$
$$f(-2) = (-2)^3 + (-2)^2 - 2(-2) + 1$$
$$= -8 + 4 + 4 + 1$$
$$= 1$$

The correct choice is C.

Example 18: Find $f(-3)$ given $f(x) = 3x^2 + 2x - 1$

    A. 20   B. 32   C. 74   D. -34

Solution

$$f(x) = 3x^2 + 2x - 1$$
$$f(-3) = 3(-3)^2 + 2(-3) - 1$$
$$= 3(9) - 6 - 1$$
$$= 27 - 6 - 1$$
$$= 21 - 1$$
$$= 20$$

The correct choice is A.

Skill IC 7. Factor a quadratic expression

You will be asked to identify linear factors of quadratic expressions of the form $ax^2 + bx + c$ where $a \neq 1$. In addition to reviewing your skills for factoring trinomials and for factoring out a common factor from a polynomial, you might need to recall these patterns.

$$[1] \quad x^2 + 2ax + a^2 = (x + a)^2$$
$$[2] \quad x^2 - 2ax + a^2 = (x - a)^2$$
$$[3] \quad x^2 - a^2 = (x - a)(x + a)$$

Example 19: Which is a linear factor of the expression $6x^2 - 11x - 7$?

    A. $3x + 7$ B. $3x - 7$ C. $2x - 1$ D. $6x - 1$

Continued on next page ...

Solution

$$6x^2 - 11x - 7 = (3x - 7)(2x + 1)$$

Both $3x - 7$ and $2x + 1$ are factors of $6x^2 - 11x - 7$. Only $3x - 7$ is listed. Choice B is correct.

Skill IC 8. Find the roots of a quadratic equation

You will be asked to solve quadratic equations that are solvable by factoring or that require the use of the quadratic formula. The factoring process requires the use of the Zero Factor Law.

$$\text{If } A \cdot B = 0 \text{ then } A = 0 \text{ or } B = 0$$

If the quadratic expression is not factorable then use the quadratic formula.

$$\text{If } ax^2 + bx + c = 0 \text{ then } x = \frac{-b \pm \sqrt{b^2 - 4ac}}{2a}.$$

Notice that this formula requires that the equation be in standard form; that is, in the form $ax^2 + bx + c = 0$.

To solve a quadratic equation we suggest that you employ the following strategy:

1. Write the equation in standard form.
2. Try to factor the quadratic expression.
3. If the expression does not factor readily with integer coefficients, then use the quadratic formula.

Example 20: Find the correct solutions to this equation: $2x^2 - 3x - 4 = 0$

A. $\dfrac{-3 + \sqrt{41}}{4}$ and $\dfrac{-3 - \sqrt{41}}{4}$ B. $\dfrac{-3 + \sqrt{17}}{4}$ and $\dfrac{-3 - \sqrt{17}}{4}$

C. $\dfrac{3 + \sqrt{17}}{4}$ and $\dfrac{3 - \sqrt{17}}{4}$ D. $\dfrac{3 + \sqrt{41}}{4}$ and $\dfrac{3 - \sqrt{41}}{4}$

Solution

First note that the quadratic equation $2x^2 - 3x - 4 = 0$ is in standard form and does not factor with integer coefficients. As a result we must use the quadratic formula. (This information can be derived from the nature of the answer choices.)

$$2x^2 \quad - \quad 3x \quad - \quad 4 \quad = \quad 0$$
$$a = 2 \qquad b = -3 \qquad c = -4$$

Continued on next page ...

We now substitute 2 for a, -3 for b, and -4 for c in the formula and evaluate it.

$$x = \frac{-b \pm \sqrt{b^2 - 4ac}}{2a}$$

$$= \frac{-(-3) \pm \sqrt{(-3)^2 - 4(2)(-4)}}{2(2)}$$

$$= \frac{3 \pm \sqrt{9 + 32}}{4}$$

$$= \frac{3 \pm \sqrt{41}}{4}$$

The correct choice is D.

Example 21: Find the correct solutions to this equation: $3x^2 - 4 = 6x$

A. $1 + 2\sqrt{21}$ and $1 - 2\sqrt{21}$ B. $\dfrac{3 + 2\sqrt{3}}{3}$ and $\dfrac{3 - 2\sqrt{3}}{3}$

C. $\dfrac{3 + \sqrt{21}}{3}$ and $\dfrac{3 - \sqrt{21}}{3}$ D. $\dfrac{-3 + \sqrt{21}}{3}$ and $\dfrac{-3 - \sqrt{21}}{3}$

Solution

First write the quadratic equation in standard form, $3x^2 - 6x - 4 = 0$. Since the equation does not factor with integer coefficients, we must use the quadratic formula.

$$3x^2 - 6x - 4 = 0$$
$$a = 3 \quad b = -6 \quad c = -4$$

We now substitute 3 for a, -6 for b, and -4 for c in the formula and evaluate it.

$$x = \frac{-b \pm \sqrt{b^2 - 4ac}}{2a}$$

$$= \frac{-(-6) \pm \sqrt{(-6)^2 - 4(3)(-4)}}{2(3)}$$

$$= \frac{6 \pm \sqrt{36 + 48}}{6}$$

$$= \frac{6 \pm \sqrt{84}}{6} \quad \text{(Now simplify the radical } \sqrt{84}.\text{)}$$

$$= \frac{6 \pm \sqrt{4}\sqrt{21}}{6} \quad \text{(By Law I for radicals.)}$$

Continued on next page ...

$$= \frac{6 \pm 2\sqrt{21}}{6}$$

$$= \frac{2(3 \pm \sqrt{21})}{6}$$

$$= \frac{3 \pm \sqrt{21}}{3}$$

The correct choice is C.

Example 22: Find the correct solution to this equation: $12x^2 - 6 = x$

A. $\frac{-3}{4}$ and $\frac{2}{3}$ B. $\frac{3}{4}$ and $\frac{-2}{3}$

C. 1 and $\frac{-1}{2}$ D. $\frac{1 + \sqrt{17}}{4}$ and $\frac{1 - \sqrt{17}}{4}$

Solution

$12x^2 - 6 = x$ (First write the equation in standard form.)

$12x^2 - x - 6 = 0$ (Next factor the quadratic expression.)

$(4x - 3)(3x + 2) = 0$ (Now use the Zero Factor Law to set the factors equal to zero.)

$4x - 3 = 0$ or $3x + 2 = 0$ (Finally, solve the simple equations.)

$4x = 3$ $3x = -2$

$x = \frac{3}{4}$ $x = \frac{-2}{3}$

The correct choice is B.

Skill IC 9. Solve a system of two linear equations in two unknowns.

A system of two linear equations in two unknowns has the form

$$\begin{cases} ax + by = c \\ cx + dy = e \end{cases}$$

where a, b, c, d, e, and f are real numbers. a and b cannot both be zero. c and d cannot both be zero. To solve such a system means to find the solution set of the system. The solution set is the set of all ordered pairs (x, y) of real numbers which make both equations true simultaneously. The solution set may contain exactly one ordered pair of real numbers, infinitely many ordered pairs of real numbers, or no ordered pairs of real numbers. In this last case, the solution set is the empty set, which may sometimes be symbolized as $\emptyset$.

There are many ways to solve such a system. We will concentrate on two methods, the Substitution Method (shown in Example 23), and the Addition/Subtraction Method, also referred to as the Elimination Method. The Addition/Subtraction Method is illustrated in Examples 24 and 25.

ALGEBRA 371

Example 23: Choose the correct solution set for the system of linear equations.

$$\begin{cases} 3x - 6y = 3 \\ x + y = 3 \end{cases}$$

A. $\{(-7, -4)\}$ B. $\{(\frac{7}{3}, \frac{2}{3})\}$ C. $\{(\frac{15}{7}, \frac{6}{7})\}$ D. $\{(\frac{2}{3}, \frac{7}{3})\}$

Solution (The Substitution Method)

$$\begin{cases} 3x - 6y = 3 \\ x + y = 3 \end{cases}$$

Select one of the equations, usually the more simple one, and solve for one of the variables. If a variable has coefficient 1 or −1, we usually solve for that variable.

$x + y$	=	3	(We select the second equation.)
y	=	$3 - x$	(We select y and solve for it in terms of x.)
$3x - 6(3 - x)$	=	3	(Substitute this expression for y into the *other* equation.)
$3x - 18 + 6x$	=	3	(Solve for x.)
$9x - 18$	=	3	
$9x$	=	21	
x	=	$\frac{21}{9}$	
x	=	$\frac{7}{3}$	

Now substitute the value $\frac{7}{3}$ for x into the equation $y = 3 - x$ and compute to solve for y.

$$y = 3 - x$$
$$y = 3 - \frac{7}{3}$$
$$y = \frac{9}{3} - \frac{7}{3}$$
$$y = \frac{2}{3}$$

The ordered pair $(\frac{7}{3}, \frac{2}{3})$ appears to be correct. However, this pair needs to be checked in both equations.

Continued on next page ...

Check in $3x - 6y = 3$			Check in $x + y = 3$		
$3(\frac{7}{3}) - 6(\frac{2}{3})$	$\stackrel{?}{=}$	3	$(\frac{7}{3}) + (\frac{2}{3})$	$\stackrel{?}{=}$	3
$7 - 4$	$\stackrel{?}{=}$	3	$\frac{9}{3}$	$\stackrel{?}{=}$	3
3	=	3 True	3	=	3 True

Since $(\frac{7}{3}, \frac{2}{3})$ satisfies both equations of the system, the solution set is $\{(\frac{7}{3}, \frac{2}{3})\}$. Choice B is correct.

Example 24: Choose the correct solution set for the system of linear equations.

$$\begin{cases} 3x - 2y = 10 \\ 5x + 3y = 23 \end{cases}$$

A. $\{(\frac{33}{19}, \frac{-91}{38})\}$ B. $\{(\frac{8}{3}, -1)\}$ C. $\{(4, -1)\}$ D. $\{(4, 1)\}$

Solution (The Addition/Subtraction Method)

$$\begin{cases} 3x - 2y = 10 \\ 5x + 3y = 23 \end{cases}$$

Select a variable to be eliminated. In this case, we select y. Multiply both sides of the first equation by 3 and both sides of the second equation by 2. The coefficients of y in the two equations will then be opposites.

$$\begin{cases} 9x - 6y = 30 \\ 10x + 6y = 46 \end{cases}$$ (Add the two equations term by term.)

$19x = 76$ (Now solve for x.)

$x = 4$

Now substitute the value 4 for x into one of the two original equations and solve for y. We substitute into the first equation.

$$3(4) - 2y = 10$$
$$12 - 2y = 10$$
$$-2y = -2$$
$$y = 1$$

The ordered pair (4, 1) needs to be checked in both equations.

Check in $3x - 2y = 10$			Check in $5x + 3y = 23$		
$3(4) - 2(1)$	$\stackrel{?}{=}$	10	$5(4) + 3(1)$	$\stackrel{?}{=}$	23
$12 - 2$	$\stackrel{?}{=}$	10	$20 + 3$	$\stackrel{?}{=}$	23
10	=	10 True	23	=	23 True

Since (4, 1) satisfies both equations of the system, the solution set is $\{(4, 1)\}$. Choice D is correct.

ALGEBRA 373

Example 25: Choose the correct solution set for the system of linear equations.

$$\begin{cases} 4x - 5y = 8 \\ -8x + 10y = -16 \end{cases}$$

A. $\{(0, -8)\}$ B. $\{(0, 0)\}$ C. $\{(x, y) \mid y = \frac{4}{5}x - \frac{8}{5}\}$ D. the empty set

Solution (The Addition/Subtraction Method)

$$\begin{cases} 4x - 5y = 8 \\ -8x + 10y = -16 \end{cases}$$

Select the variable x to be eliminated.

$$\begin{cases} 8x - 10y = 16 \\ -8x + 10y = -16 \end{cases}$$ (Multiply in the first equation by 2.)

$0 + 0 = 0$ (Add the two equations term by term.)

$0 = 0$

$0 = 0$ is a true statement. This indicates that the original system leads to a statement that is identically true. The original system has infinitely many ordered pairs in its solution set. To express that solution set, select the first equation and solve it for y in terms of x.

$$4x - 5y = 8$$
$$-5y = -4x + 8$$
$$y = \frac{4}{5}x - \frac{8}{5}$$

The solution set contains all ordered pairs (x, y) of real numbers such that $y = \frac{4}{5}x - \frac{8}{5}$. The solution set is expressed as $\{(x, y) \mid y = \frac{4}{5}x - \frac{8}{5}\}$. The correct choice is C. (**Note:** If we had arrived at a false statement rather than a true statement, the solution set of the system would have been the empty set.)

Skill IIC 1. Use properties of operations correctly

The properties involved can be any of the following:

[1] Commutative Property of Addition: $a + b = b + a$

[2] Commutative Property of Multiplication: $ab = ba$

[3] Associative Property of Addition: $a + (b + c) = (a + b) + c$

[4] Associative Property of Multiplication: $a(bc) = (ab)c$

[5] Identity Property of Addition: $a + 0 = a$ and $0 + a = a$

[6] Identity Property of Multiplication: $a \cdot 1 = a$ and $1 \cdot a = a$

[7] Inverse Property of Addition: $a + (-a) = 0$ and $(-a) + a = 0$

[8] Inverse Property of Multiplication: $a \cdot \frac{1}{a} = 1$ and $\frac{1}{a} \cdot a = 1$, for $a \neq 0$

[9] Distributive Property of Multiplication over Addition:
$a(b + c) = ab + ac$ and $(b + c)a = ba + ca$

374   CLAST REVIEW

You will be asked to perform one of the following two tasks:

1. Given an algebraic expression, identify an equivalent expression which illustrates the proper use of a property.
2. Given algebraic equations involving correct and incorrect uses of properties of operations, identify the incorrect use of a property.

Example 26

Choose the expression equivalent to the following:

$$3a^3(ab^2)$$

A. $3a^3a + 3a^3b^2$ B. $3a^3(b^2a)$

C. $(3a^3a)(3a^3b^2)$ D. $(3a^3a) + b^2$

Solution

- Choice A is incorrect. The given expression is a product. The expression in choice A is a sum.
- Choice B is correct. This shows the correct use of the Commutative Property of Multiplication.
- Choice C is incorrect. $3a^3$ does not "associate" with both of the factors a and b^2.
- Choice D is incorrect. The same argument used for choice A applies.

As stated, B is the correct choice.

Example 27

Choose the expression equivalent to the following:

$$(3x + 6y)(3x - 6y)$$

A. $3(x + 2y)(x - 2y)$ B. $(3x + 6y)(6y - 3x)$

C. $(6y + 3x)(-6y + 3x)$ D. $3xy(3x - 6y)$

Solution

- Choice A is incorrect. By the Distributive Property, 3 can be factored from both factors. The resulting expression would be $9(x + 2y)(x - 2y)$.
- Choice B is incorrect. Subtraction is not a commutative operation.
- Choice C is correct. The Commutative Property of Addition has been used properly in both factors.
- Choice D is incorrect. $3xy$ is not a factor of the original expression.

As stated, C is the correct choice.

Example 28

Choose the expression equivalent to the following:

$$4[x + (-x)]$$

A. $4x + 4x$ B. $4(0)$ C. $4x - x$ D. $-8x$

Solution

The correct choice is B since $x + (-x) = 0$ by the Inverse Property of Addition.

Example 29: Choose the statement which is *not* true for all real values of the variables.

A. $(a + b) + 4 = (b + a) + 4$ B. $(4a)b = 4(ab)$
C. $4(a + b) = 4a + b$ D. $(a + b) + 4 = a + (b + 4)$

Solution

- Choice A illustrates the proper use of the Commutative Property of Addition.
- Choice B illustrates the proper use of the Associative Property of Multiplication.
- Choice C improperly uses the Distributive Property of Multiplication over Addition. The statement should be $4(a + b) = 4a + 4b$.
- Choice D illustrates the proper use of the Associative Property of Addition.

The correct choice is C.

Skill IIC 2. Determine whether a particular number is among the solutions of a given equation or inequality

The equations or inequalities can be linear or quadratic and can involve absolute value.

Example 30

For each of the three statements below, determine if (-5) is a solution.

    i. $|x - 5| = 0$
    ii. $(t + 6)(t - 2) \leq 7$
    iii. $y^2 + 3y - 20 = 10$

Which option below satisfies every statement that has (-5) as a solution?

A. i and ii only B. iii only C. ii only D. ii and iii only

Solution

Notice the format of the choices. To answer correctly you must identify all of the sentences from i, ii, and iii for which (-5) is a solution. You must then select from A, B, C, and D the choice that correctly describes your selection. There are two ways to approach this problem.

Continued on next page ...

1. Solve each of the three equations/inequalities given in i, ii, and iii and note whether (−5) is a solution of each.
2. Simply check (−5) in each of the equations/inequalities in i, ii, and iii. If you obtain a true statement then (−5) is a solution; otherwise it is not a solution.

We choose to apply method 2.

i. $|x - 5| = 0 \quad |-5 - 5| = |-10| = 10$

(False; −5 is not a solution to i.)

ii. $(t + 6)(t - 2) \leq 7$
$(-5 + 6)(-5 - 2) \leq 7\ ?$
$(1)(-7) \leq 7\ ?$
$-7 \leq 7\ ?$
(True; −5 is a solution to ii.)

iii. $y^2 + 3y - 20 = 10$
$(-5)^2 + 3(-5) - 20 = 10\ ?$
$25 - 15 - 20 = 10\ ?$
$10 - 20 = 10\ ?$
$-10 = 10\ ?$
(False; −5 is not a solution to iii.)

As a result, (−5) is a solution to only statement ii, which is choice C.

Skill IIC 3. Recognize statements and conditions of proportionality and variation

Direct variation or inverse variation can be involved. Given a description of a problem involving variation, you will need to select an equation that mathematically represents the described conditions.

Example 31

In a housing division all of the grassy areas for the houses are the same size. It takes seven hours for a team of two men to mow the 13 grassy areas. Let y represent the number of hours for the two men to mow nine of the grassy areas. Select the correct statement of the given condition.

A. $\dfrac{7}{9} = \dfrac{y}{13}$ B. $\dfrac{y}{7} = \dfrac{13}{9}$ C. $\dfrac{7}{y} = \dfrac{13}{9}$ D. $\dfrac{7}{13} = \dfrac{9}{y}$

Solution

Using *units* only, the following proportion holds for this problem.

$$\frac{\text{hours to mow}}{\text{areas to be mowed}} = \frac{\text{hours to mow}}{\text{areas to be mowed}}$$

As a result, we have the proportion $\dfrac{7}{13} = \dfrac{y}{9}$. However, by properties of proportions, if $\dfrac{a}{b} = \dfrac{c}{d}$, then each of the following holds.

(i) $ad = bc$ (ii) $\dfrac{a}{c} = \dfrac{b}{d}$ (iii) $\dfrac{b}{a} = \dfrac{d}{c}$

Using (ii), $\dfrac{7}{13} = \dfrac{y}{9}$ results in $\dfrac{7}{y} = \dfrac{13}{9}$, which is choice C.

Example 32

The area of a set of rectangular pictures is held constant while the length and width can change. If the length is 16 units when the width is four units, select the statement of the condition when the width is five units. The variable l is used to represent the length.

A. $\dfrac{l}{5} = \dfrac{16}{4}$ B. $\dfrac{l}{5} = \dfrac{4}{16}$ C. $\dfrac{l}{4} = \dfrac{16}{5}$ D. $\dfrac{l}{4} = \dfrac{5}{16}$

Solution

Since the area stays the same,

$$\text{Area} = l_1 w_1 = l_2 w_2$$

or $\quad \text{Area} = 16 \cdot 4 = l \cdot 5 \quad$ (We replace l_2 with l.)

From this equation we can obtain several equivalent proportions.

$$\frac{l}{16} = \frac{4}{5}; \quad \frac{l}{4} = \frac{16}{5}; \quad \frac{5}{4} = \frac{16}{l}.$$

The second proportion is found in choice C.

Skill IIC 4. Recognize regions of the coordinate plane which correspond to specific conditions and vice versa

You can be given a graph and asked to identify the conditions (equations or inequalities) that describe it, or you can be given conditions and asked to identify the graph these conditions determine. Regions of the plane will be identified by shading. A dashed boundary indicates the boundary line is not included in the points determined by the conditions. A solid boundary indicates the boundary line is included in the points determined by the conditions. All conditions will be linear.

You will need to review these facts about lines.

1. The slope of a nonvertical line that passes through the two points P_1 and P_2, with coordinates (x_1, y_1) and (x_2, y_2), respectively, is given by $m = \dfrac{y_2 - y_1}{x_2 - x_1}$.

2. Forms of the equation of a line are

    a. **x = a** (Vertical line through the point (a,0))

    b. **y = b** (Horizontal line through the point (0,b))

    c. **ax + by = c** (Standard form)

    d. **y = mx + b** (Slope-intercept form; the slope is m and the y intercept is b.)

    e. **y − y₁ = m(x − x₁)** (Point-slope form; the slope is m and (x_1, y_1) is a specific point on the line.)

Example 33

Which option gives the conditions that correspond to the shaded region of the plane shown in the figure below?

Continued on next page ...

A. $2x - y \leq -2$ and $x \leq 2$
B. $2x - y \leq 4$ and $y \leq 2$
C. $2x - y \leq 4$ and $x \leq 2$
D. $2x - y \geq 4$ and $x \leq 2$

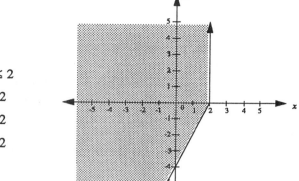

Solution

To identify the conditions for the region we must write the equation of the boundary lines. The vertical line is $x = 2$. The slanted line passes through the points $(0, -4)$ and $(2, 0)$. The slope is

$$\frac{y_2 - y_1}{x_2 - x_1} = \frac{-4-0}{0-2} = \frac{-4}{-2} = 2$$

Since the y intercept is -4, the equation of this line in slope-intercept form is $y = 2x - 4$. In the answer choices, however, the inequality is given in standard form. Re-expressing $y = 2x - 4$ in standard form we get $2x - y = 4$.

In the given figure, note that the shaded region is to the left of the vertical line $x = 2$, and above the slanted line $2x - y = 4$. Also, the boundaries are included. This information narrows our possible answers to choices C and D. To determine the correct choice, select a point in the region and check it in the inequalities given in choice C. (We choose the point $(0, 1)$ since it is in the region and is not on a boundary.)

Does $(0, 1)$ satisfy $2x - y \leq 4$?	Does $(0, 1)$ satisfy $x \leq 2$?
$2(0) - 1 \leq 4$?	$0 \leq 2$ (yes)
$0 - 1 \leq 4$?	
$-1 \leq 4$ (yes)	

Since $(0, 1)$ satisfies both inequalities in choice C, this choice correctly gives the conditions for the region of the plane.

Example 34

Which shaded region identifies the portion of the plane in which $x \geq -3$ and $y \leq 2$?

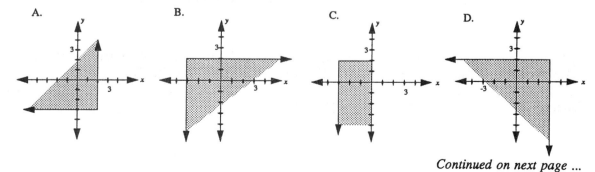

Continued on next page ...

Solution

- The line $x = -3$ is vertical, and the inequality $x > -3$ describes the region to the right of this line. As a result, $x \geq -3$ describes the vertical line and the points to its right.
- The line $y = 2$ is horizontal, and the inequality $y < 2$ describes the region below this line. As a result, $y \leq 2$ describes the horizontal line and the points below it.

$x \geq -3$ and $y \leq 2$ has as its graph the region in choice B.

Skill IIIC 1. This skill has been deleted.

Skill IIIC 2. Use applicable properties to select equivalent equations and inequalities

Given an equation or inequality (or inequalities), you will be asked to use an algebraic property or properties to identify an equivalent equation or inequality (inequalities). The equation or inequality can be first or second degree. The properties used will be from the following list:

$$[1] \quad a = b \text{ if and only if } a + c = b + c$$
$$[2] \quad a = b \text{ if and only if } ac = bc, c \neq 0$$
$$[3] \quad a > b \text{ if and only if } a + c > b + c$$
$$[4] \quad a > b \text{ if and only if } ac > bc \text{ and } c > 0$$
$$[5] \quad a > b \text{ if and only if } ac < bc \text{ and } c < 0$$
$$[6] \quad \text{If } a > b \text{ and } b > c \text{ then } a > c$$
$$[7] \quad \text{If } a = b \text{ and } b = c \text{ then } a = c$$
$$[8] \quad a = b \text{ if and only if } b = a$$

Note that an *if and only if* statement is two statements combined into a single statement. A problem can involve only one of the properties from such a combined property. Also, if a property is stated using the symbol >, it also holds for the symbol <.

Example 35: Choose the inequality equivalent to the following:

$$5 - 4x > 6$$

A. $-4x > 11$ B. $-4x < 11$ C. $-4x > 1$ D. $4x < 1$

Solution

$$5 - 4x > 6 \quad \text{(Subtract 5 from both sides of the inequality.)}$$
$$-4x > 1$$

By Property [3], from the list of properties, we can add (and thus subtract) the same number to both sides of an inequality to obtain an equivalent inequality. Choice C is correct.

Example 36: Choose the inequality equivalent to the following:

$$6 < -x < 9$$

A. $-6 < x < -9$ B. $-6 > x > -9$ C. $6 > x > -9$ D. $0 < -x < 3$

Continued on next page ...

Solution

- Choice A is incorrect. By Property [5], if both sides of an inequality are multiplied by a negative number, in this case −1, the direction of the inequality must be reversed.
- Choice B is correct by Property [5].
- Choice C is incorrect. 6 was not multiplied by −1.
- Choice D is incorrect. By Property [3], 6 must be subtracted from each member of the inequality. The correct inequality would be $0 < -x - 6 < 3$.

As stated, choice B is correct.

Example 37: Choose the equation equivalent to the following:

$$3y - 4 = 2y + 6$$

A. $3y = 2y + 2$ B. $5y - 4 = 6$
C. $3y - 2 = y + 3$ D. $2y + 6 = 3y - 4$

Solution

- Choice A is incorrect. By Property [1], the same number must be added to both sides. 4 was added to the left and subtracted from the right.
- Choice B is incorrect. 2y was added to the left side of the equation and subtracted from the right side.
- Choice C is incorrect. By Property [2], both sides must be multiplied (or divided) by the same nonzero number. Each term except 3y was divided by 2.
- Choice D is correct by Property [8].

Skill IVC 1. Solve real world problems involving the use of variables (aside from commonly used geometric formulas).

The problems will be taken from a business, social studies, industry, education, economics, environmental studies, arts, physical sciences, sports, or consumer-related context. The mathematical structure of the problem can involve linear or quadratic relationships, proportions, or any type of variation. Problems can also contain irrelevant information.

Example 38

The value of an $18,000 car depreciates linearly. After four years it is valued at $8,600. What is the value of the car after six years?

A. $2,350   B. $9,400   C. $3,900   D. $5,100

Continued on next page ...

Solution

Let y = value of car in dollars
and x = number of years after purchase

Since the value of the car depreciates linearly, $y = mx + b$ describes this depreciation. As a result, we must find m and b.

- When $x = 0$, $y = 18,000$. Thus $18000 = m(0) + b$, which implies that $b = 18000$.
- When $x = 4$, $y = 8,600$. Thus,

$$m = \frac{y_2 - y_1}{x_2 - x_1} = \frac{18000 - 8600}{0 - 4} = \frac{9400}{-4} = -2350$$

- As a result of the above, $b = 18000$ and $m = -2350$. Therefore, the equation is

$$y = -2350x + 18000.$$

- Finally, we find y when $x = 6$.

$$\begin{aligned} y &= -2350x + 18000 \\ &= -2350(6) + 18000 \\ &= -14100 + 18000 \\ &= 3900 \end{aligned}$$

The value after six years is $3,900, which is choice C.

Example 39

The kinetic energy (*KE*) of a moving object varies with the mass (*m*) and velocity (*v*) according to the formula $KE = kmv^2$, where k is some constant. The kinetic energy of a mass of five kilograms moving at a velocity of two meters per second is 10 joules. What is the kinetic energy of this same mass when it moves at six meters per second?

    A. 360 joules   B. 90 joules   C. 180 joules   D. 15 joules

Solution

- We must first determine the value of the constant k.

$$KE = 10 \text{ when } m = 5 \text{ and } v = 2$$

$$\begin{aligned} KE &= kmv^2 \\ 10 &= k(5)(2)^2 \\ 10 &= 20k \\ \tfrac{1}{2} &= k \end{aligned}$$

As a result, $KE = \frac{1}{2}mv^2$

- We now can determine the value of KE when $m = 5$ and $v = 6$.

$$KE = \frac{1}{2}mv^2$$

Continued on next page ...

$$KE = \left[\frac{1}{2}\right](5)(6)^2$$
$$= \left[\frac{1}{2}\right](5)(36)$$
$$= \frac{1}{2}(180)$$
$$= 90$$

The kinetic energy is 90 joules, which is choice B.

Example 40

The formula $S = \frac{1}{2}gt^2$ gives the distance an object falls from rest during t seconds. In this formula S is measured in feet and g, the gravitational constant, is 32 ft/sec^2. What distance S will an object fall from rest in 8 seconds?

    A. 128 ft.   B. 512 ft.   C. 1024 ft.   D. 2048 ft.

Solution

Evaluate the formula $S = \frac{1}{2}gt^2$ when $g = 32$ and $t = 8$. The distance S will be in feet.

$$S = \frac{1}{2}gt^2$$
$$= (\frac{1}{2})(32)(8)^2$$
$$= (\frac{1}{2})(32)(64)$$
$$= (16)(64)$$
$$= 1024$$

The object will fall 1024 ft., which is choice C.

Skill IVC 2. Solve problems that involve the structure and logic of algebra

Given a verbal problem that involves numbers, symbols, or relations, you will be asked to select the symbol(s), expression, or statement that satisfies the requirements of the problem. The problem will involve one of the following:

1. A relationship between the digits of an integer with two digits
2. A relationship between two integers
3. The sum of the digits of an integer with two digits
4. The sum of two integers
5. Possible numerical values or algebraic properties of rational numbers of some given form
6. Reciprocal relationships
7. Consecutive integer relationships

Example 41

The tens digits of a two digit number is three more than twice the ones digit. If n represents the ones digit, select a representation for the number.

  A. $(2n + 3) + n$ B. $(2n + 3)n$ C. $10(3n + 2) + n$ D. $10(2n + 3) + n$

Solution

$$
\begin{aligned}
n &= \text{ones digit} \\
2n + 3 &= \text{tens digit} \\
\text{The number} &= (10 \times \text{tens digit}) + (\text{ones digit}) \\
&= 10(2n + 3) + n
\end{aligned}
$$

The correct choice is D.

Example 42

The sum of two integers is 27. Their product is 152. Let x represent the smaller integer. What equation would be used to find x, the smaller integer?

  A. $x + x = 27$ B. $x(27 - x) = 152$ C. $x(x - 27) = 152$ D. $27x = 152$

Solution

$$
\begin{aligned}
x &= \text{the smaller integer} \\
27 - x &= \text{the larger integer} \\
x(27 - x) &= 152 \text{ (since the product of the two numbers is 152)}
\end{aligned}
$$

The correct choice is B.

Example 43

The product of two whole numbers is odd. Which of the following statements is true about these two numbers?

  A. One of the numbers can be zero.   B. Both of the numbers can be even.
  C. Neither of the numbers can be even.   D. The sum of the two numbers can be odd.

Solution

- A is false since $n \cdot 0 = 0$ for all whole numbers n. Also, 0 is an even number.
- B is false; the product of two even numbers is even.
- C is true; the product of two even numbers is even and the product of an even number and an odd number is even.
- D is false; the numbers must both be odd and the sum of the two odd numbers is even.

The correct choice is C.

Example 44: Which statement is *false* for every integer x?

A. $x \cdot 0 = 0$ B. $2x > 0$ C. $\dfrac{1}{x} < 0$ D. $\dfrac{2}{x} = x$

Solution

- A is incorrect; $x \cdot 0 = 0$ is true for every integer x.
- B is incorrect; $2x > 0$ is true for all positive integers.
- C is incorrect; $\dfrac{1}{x} < 0$ is true for all negative integers.
- D is correct; $\dfrac{2}{x} = x$ is false for x = 0. If $x \neq 0$ then $2 = x^2$. Thus $x = \sqrt{2}$ or $x = -\sqrt{2}$. Note that $\sqrt{2}$ and $-\sqrt{2}$ are not integers. Thus $\dfrac{2}{x} = x$ is false for all integers.

The correct choice is D.

Example 45

Choose the algebraic description that is equivalent to the verbal description:

For a number x, the sum of the numbers and $\dfrac{1}{2}$ of its reciprocal is greater than the sum of the reciprocal of the number and 1.

A. $x + \dfrac{1}{2x} > \dfrac{1}{x} + 1$ B. $x + \dfrac{1}{2x} > \dfrac{1}{x+1}$

C. $x + \dfrac{2}{x} > \dfrac{1}{x} + 1$ D. $x + \dfrac{2}{x} > \dfrac{1}{x+1}$

Solution

$$x = \text{the number}$$

$$\dfrac{1}{x} = \text{the reciprocal of the number}$$

$$\dfrac{1}{2}\left(\dfrac{1}{x}\right) = \dfrac{1}{2x} = \dfrac{1}{2} \text{ of the reciprocal of the number}$$

The sum of x and $\dfrac{1}{2x}$ > the sum of $\dfrac{1}{x}$ and 1

$$x + \dfrac{1}{2x} > \dfrac{1}{x} + 1$$

The correct choice is A.

6. PROBABILITY AND STATISTICS

In this section we discuss the CLAST subject area Probability and Statistics. Each skill within this subject will be addressed separately, and examples of the type of problems that could be asked on the CLAST for each skill will be presented.

Skill ID 1. Identify information contained in bar, line, and circle graphs

For this skill you will be asked to read a bar, line, or circle graph. The information that you will be asked to extract from these graphs will include the frequency of a given category, the proportion or percent that reflects the frequency of a given category, or a specific category from a described frequency. In order to this you might need to do the following computations:

1. Obtain sums or differences of frequencies
2. Obtain a percent of the whole
3. Obtain a frequency from a percent

Focusing on the second list item above, a problem will typically involve a frequency that must be expressed as a proportion or percent. To do this conversion divide the given frequency by the total frequency; this will yield a proportion. If a percent is requested, then convert the proportion to a percent.

Focusing on the third list item above, the formula

$$percent \times base = amount$$

can be used to obtain a frequency from a percent. The total frequency would be used as the base and the amount would be the frequency requested.

Warning: Read the information given in the graphs carefully. Do not let misreading of the graph cause you to answer a question incorrectly.

Example 1

The number of hours an individual spends in leisure activities is represented by the circle graph below. What percent of these hours is in listening to music and watching television combined?

A. 37.5% B. 42.5%
C. 30%   D. 15%

Solution

The total number of hours spent in leisure activities is 40. The total number of hours spent in listening to music and watching television combined is 17. As a result, the ratio is $\frac{17}{40}$.

Continued on next page ...

Converting this to a percent we have

$$\frac{17}{40} = 0.425 = 42.5\%$$

The correct choice is B.

Example 2

The graph below represents the number of appliances sold in the last six months of 1991 by an appliance dealer in Miami. In what month did the number of appliances sold differ the most from the number of appliances sold in August?

A. July
B. October
C. November
D. December

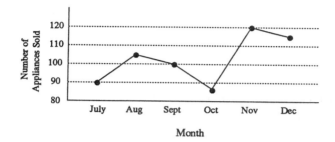

Solution

The number of appliances sold in August was 105. The number of appliances sold in the other months is listed below.

July	90
September	100
October	85
November	120
December	115

The number sold in October differs from 105 by 20. The correct choice is B.

Example 3

The graph below represents the yearly water level average of a lake for the years 1980 - 1985. Find the average water level for 1982.

A. 41 inches   B. 45 inches
C. 40 inches   D. 50 inches

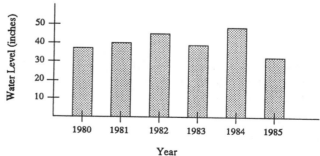

Solution

Read the graph for the year 1982. The average rainfall appears to be 45 inches. The correct choice is B.

Skill ID 2. Determine the mean, median, and mode of a set of numbers

The mean, median, and mode of a set of numbers (data) measure the tendency of the numbers to cluster about a middle number. They are called *measures of central tendency*.

The mean is the average of all the numbers. To find the mean, add all of the numbers and divide the sum by the total number of numbers that were listed. If a number is listed more than once it is added more than once and hence is counted more than once. To illustrate, consider the set of numbers {7,9,10,4,11,12,12,3,4}. The sum of the numbers is 72. Moreover, there are nine data values. Thus the mean is $72 \div 9 = 8$.

The median is the middle number. To determine the median, you will need to rank the data; that is arrange the data in order from smallest to largest or from largest to smallest. If there is an odd number of data values, the median is the middle number. If, on the other hand, there is an even number of data values, the median is found by averaging the middle two numbers. To illustrate, consider the set of numbers given above. This set of numbers, arranged from smallest to largest is {3,4,4,7,9,10,11,12,12}. The median is the fifth number, which is 9.

The mode is the most frequently occurring number. If no number is repeated, there is no mode. It is possible to have more than one mode if two or more numbers occur with greatest frequency. In this instance the distribution is said to be bimodal. For example, in the set of numbers above, both 4 and 12 are modes.

Example 4

What is the *mode* of the data in the following sample?

$$7, 9, 4, 8, 9, 6, 11, 2$$

A. 8   B. 7   C. 9   D. 11

Solution

The number that occurs with greatest frequency is 9 since it occurs twice and all the other numbers occur only once. The correct choice is C.

Example 5

What is the *median* of the data in the following sample?

$$3, 8, 4, 9, 11, 4, 6, 5$$

A. 6   B. 4   C. 10   D. 5.5

Solution

The data ranked from smallest to largest is {3,4,4,5,6,8,9,11}. There are eight data values. The median will be the average of the middle two numbers. Specifically, the median is the average of 5 and 6, which is $\frac{(5+6)}{2}$, which is 5.5. The correct choice is D.

Skill ID 3. Use the fundamental counting principle

You will be asked to identify the number of outcomes of a real-life experiment (situation), where the experiment consists of one of the following:

1. the selection of sequences of objects from a set of distinct objects
2. the selection of subsets of objects from a set of distinct objects
3. various stages with each stage resulting in one of several possible outcomes (choices).

The <u>Fundamental Counting Principle</u> (FTP) states that if one event, which can occur in k ways, is followed by a second event that can occur in m ways, then the total number of possible outcomes for the two events is $k \cdot m$. For example, let us assume that a couple wants to have two children. Since a child can be either a boy or a girl there are two possible outcomes for the first child (i.e., the first event), and two possible outcomes for the second child (the second event). By the FTP there are $2 \cdot 2$ or 4 possible outcomes. If the couple had wanted three children, then there would be $2 \cdot 2 \cdot 2$ or 8 possible outcomes, and so forth.

In addition to the FTP, you may find the following concepts useful:

1. <u>The number of subsets of a set</u>

    The total number of subsets of a set on n objects is 2^n. For example, a set with 4 elements has 2^4 or 16 total subsets.

2. <u>The number of permutations of n objects taken r at a time</u>

    A permutation is an ordered arrangement; that is, two different orders of r objects are counted as two different arrangements. For example, if there are seven books in a collection and four of them are to be arranged on a shelf, there will be seven choices for the first book, six choices for the second book, five choices for the third book, and four choices for the fourth book. The result is $7 \cdot 6 \cdot 5 \cdot 4 = 840$ permutations or possible arrangements of the four books. In general, a permutation of n total objects taken r at a time is the product of r factors.

    $$n(n-1)(n-2) \cdots (n-r+1)$$

    In the above illustration, $n = 7$, $r = 4$, $n - r + 1 = 7 - 4 + 1 = 4$. Therefore, the total number of permutations is $7 \times 6 \times 5 \times 4$.

3. <u>The number of combinations of n objects taken r at a time</u>

    A combination is an unordered arrangement; that is, two different orders of r objects are only counted as one arrangement. For example, consider the books of the previous illustration, except that the order on the shelf of the four books does not matter. You want to determine how many combinations of four books can be made from the seven books. The number of combinations will initially be a quotient in which the numerator consists of the product of r factors from a permutation perspective, and the denominator is the product of the integer factors from r to 1. Specifically, the number of combinations for the example is

    $$\frac{7 \cdot 6 \cdot 5 \cdot 4}{4 \cdot 3 \cdot 2 \cdot 1} = \frac{840}{24} = 35.$$

    In general, the number of combinations of n objects taken r at a time can be determined by the expression

    $$\frac{n(n-1)(n-2) \cdots (n-r+1)}{r(r-1)(r-2) \cdots 3 \cdot 2 \cdot 1}$$

The number of subsets containing r elements of a set of n elements is the same as the number of combinations of n objects taken r at a time. As a second example, consider a set with six objects. To find the number of subsets that each contain three elements, compute $\frac{6 \cdot 5 \cdot 4}{3 \cdot 2 \cdot 1} = 20$. There will be 20 subsets. Computing this number is much easier than listing all of the subsets and counting them.

Example 6

From a group of four boys and three girls, a boy and a girl will be selected to enter a contest. How many possible ways can the selection be made?

    A. 12   B. 7   C. 4   D. 6

Solution

This involves the FTP. One boy can be selected in four ways and one girl can be selected in three ways. The selection can thus be made in $4 \cdot 3 = 12$ ways. Choice A is correct.

Example 7

A perfume company wants to test six types of perfume, brands K, L, M, N, O, and P. The company will compare each type of perfume with the other types by pairing them. How many different pairs will result by selecting two different types at the same time?

    A. 360   B. 720   C. 15   D. 180

Solution

Since the pairings of perfume types will be unordered, combinations are involved in this problem. We want to know how many combinations there are of six objects taken two at a time. The number of different pairs is computed as

$$\frac{6 \cdot 5}{2 \cdot 1} = \frac{30}{2} = 15$$

Choice C is correct.

Example 8

Five people enter a pew in church where there is only space for three people. In how many ways can three of the five people be seated in the pew?

    A. 20   B. 60   C. 10   D. 15

Solution

Since order matters in the arrangement of three people in the pew, permutations are involved in this problem. We want to know how many permutations there are of five objects taken three at a time. The number of permutations is computed as $5 \cdot 4 \cdot 3 = 60$. The correct choice is B.

(**Note:** The FTP could also be used here. There are five choices for the first seat, four for the second seat, and three choices for the third seat. This produces a total of $5 \cdot 4 \cdot 3 = 60$ ways that three of the five people can be seated in the pew.)

Example 9

A car buyer can purchase a car with or without the following options: automatic transmission, aluminum wheels, deluxe interior, air conditioning. How many different combinations of these options are available?

      A. 4   B. 18   C. 16   D. 24

Solution

For each option, two choices are available. Therefore, by the FTP, there are $2 \cdot 2 \cdot 2 \cdot 2$ or 16 combinations of these options available. Choice C is correct.

Skill IID 1.   Recognize properties and interrelationships among the mean, median, and mode in a variety of distributions

You will be given a diagram and/or a description of a distribution of data and be asked to identify true statements about the mean, median, and mode. The data might be skewed left, skewed right, approximately normal, or symmetrical. In addition to the information about mean, median, and mode given in the discussion of skill ID2, you should know the following facts about normal distributions:

1. The mean, median, and mode are all equal.
2. Approximately 50% of the data falls above the mean and approximately 50% falls below the mean.

Example 10

On the first day of class, half the students scored 60 on a pretest. Most of the remaining students scored 45, except for a few students who scored 15. Which of the following statements is true about the distribution of scores?

      A. The mean and median are the same.   B. The mean and the mode are the same.
      C. The mean is greater than the mode   D. The mean is less than the median.

Solution

- The mode is 60, the most frequently occurring score.
- Since half the students scored 60, there is an even number of students. The median is thus $\frac{60 + 45}{2}$ or 52.5
- We cannot compute the mean, but since half the students scored 60 and the other half scored 45 or lower, the mean will be less than $\frac{60 + 45}{2}$, which is equal to 52.5.

The correct choice is D.

Example 11

The graph below represents the distribution of scores for students at one Florida Community College on the CLAST Essay Subtest. Select the statement that is true about the distribution of scores.

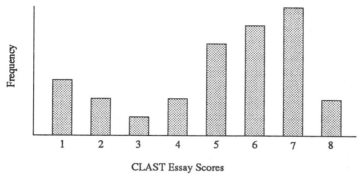

CLAST Essay Scores

A. The mode and the mean are the same.
B. The mode and the median are the same.
C. The mean is less than the mode.
D. The median is greater than the mode.

Solution

The mode is seven. The relationship between the mean and the median is unclear, but both the mean and the median are less than the mode. As a result, choices A, B, and D are false. The correct choice is C.

Skill IID 2. Choose the most appropriate procedure for selecting an unbiased sample from a target population

Given a situation where a sample is to be selected from a clearly specified target population, you will be asked to identify the most appropriate surveying method for obtaining a statistically unbiased sample. Remember that an *unbiased* sample results from a *random sample* of the total population involved. Common sense will help you rule out survey methods that use alphabetized lists, numerical lists, telephone directory lists, etc.

Example 12

The Florida Lottery Administrators want to determine the opinions of Florida voters about the lottery three years after the lottery has been in operation. They want to conduct a survey of a sample of Florida voters. Which of the following procedures would be the most appropriate for obtaining a statistically unbiased sample of the voters of Florida?

A. Send surveys to all registered voters and have voters voluntarily mail in the surveys.

B. Select a random sample of voters from the list of registered voters from the five largest cities of Florida.

C. Select a random sample from the list of all persons who had won any money in the lottery and survey these people.

D. Select a random sample from a compiled list of all registered voters in Florida and survey these people.

Continued on next page ...

Solution

- Choice A would produce a biased sample; only those persons choosing to mail in their survey would make up the sample.
- Choice B would bias the sample towards city-dwellers.
- Choice C would bias the sample towards lottery winners and might include nonvoters and non-Floridians.
- Choice D is correct.

Skill IID 3. Identify the probability of a specified outcome in an experiment

Since this skill can involve any or all of the various principles of probability listed below, it is difficult to predict how the skill will be tested. We do know, though, that no problems will involve dice or cards, and that real-world situations may be involved. The principles of probability that can be involved include the following:

[1] The probability of an event A is P(A), where

$$P(A) = \frac{\text{\# of ways that A can occur}}{\text{total \# of possible outcomes}}$$

Any of the concepts discussed after the statement of Skill ID3 might be necessary in determining the number of ways that an event A can occur or the total number of possible outcomes in an experiment.

[2] The probability that an event A does not occur is P(not A) where P(not A) = 1 − P(A). Note that if an event A is certain to occur then P(A) = 1 and P(not A) = 0. Also $0 \leq P(A) \leq 1$, for any event A.

[3] If two events A and B are independent (i.e., the occurrence of one event does not affect the probability of the other), then P(A and B) = P(A) · P(B).

(**Note:** P(A and B) means the probability that both events A and B occur.)

[4] If two events A and B are not independent (i.e., the occurrence of one event does affect the probability of the other), then P(A and B) = P(A) · P(B/A).

(**Note:** P(B/A) means the probability of event B given that event A has occurred.)

[5] If two events A and B are mutually exclusive (i.e., they cannot occur simultaneously since one prevents the other from occurring), then P(A or B) = P(A) + P(B).

[6] If two events A and B are not mutually exclusive then P(A or B) = P(A) + P(B) − P(A and B). In this case A and B can both occur.

Example 13

A box contains four red, three black, and six green marbles. Two marbles are drawn from the box at random without replacement. What is the probability that neither is red?

A. $\frac{72}{169}$ B. $\frac{81}{169}$ C. $\frac{6}{13}$ D. $\frac{1}{13}$

Continued on next page ...

Solution

Since the marbles are withdrawn without replacement the two events are not independent. Although the marbles are drawn simultaneously we can solve the problem by thinking of drawing first one marble and then the other.

- Let A be the event the first marble is not red. There are nine nonred marbles (out of a total of 13). Thus $P(A) = \frac{9}{13}$.

- Let B be the event the second marble is not red. After one nonred marble has been drawn there will be eight nonred marbles (out of a total of 12) remaining. Therefore the probability that the second marble is not red, given the fact that the first marble is not red (i.e., P(B/A)) is $\frac{8}{12}$.

As a result,

$$\begin{aligned} P(A \text{ and } B) &= P(A) \times P(B/A) \\ &= \frac{9}{13} \times \frac{8}{12} \\ &= \frac{9}{13} \times \frac{2}{3} \\ &= \frac{6}{13} \end{aligned}$$

The correct choice is C.

Example 14

Two children are born thirty minutes apart. What is the probability of both being girls or both being boys?

A. $\frac{1}{4}$ B. $\frac{1}{2}$ C. $\frac{3}{4}$ D. $\frac{1}{16}$

Solution

Let A = the event that both children are girls
Let B = the event that both children are boys

The events are mutually exclusive since both cannot occur simultaneously. Therefore, the P(A or B) = P(A) + P(B). There are four possible outcomes.

- BB (Both are boys)
- BG (First is a boy and second is a girl)
- GB (First is a girl and second is a boy)
- GG (Both are girls)

$P(A) = \frac{1}{4}$ and $P(B) = \frac{1}{4}$. As a result, $P(A \text{ or } B) = \frac{1}{4} + \frac{1}{4} = \frac{1}{2}$.

The correct choice is B.

Example 15

Two common sources of calcium for U.S. teenagers are ice cream and frozen yogurt. Sixty percent of U.S. teenagers eat ice cream but not yogurt, while 26% eat both. What is the probability that a randomly selected teenager does *not* eat ice cream?

    A. .86   B. .156   C. .74   D. .14

Solution

Let A be the event that a randomly selected U.S. teenager eats ice cream.

$$P(A) = .60 + .26 = .86$$

Thus, the probability that a randomly selected U.S. teenager does *not* eat ice cream is

$$P(\text{not } A) = 1 - P(A) = 1 - .86 = .14$$

Choice D is correct.

Example 16

Of the high school graduates in the state of Florida, 20% are participants in the Academic Scholars Program. Of these participants, 15% receive full scholarships to Florida colleges. What is the probability that a randomly selected high school graduate in Florida received a full scholarship through the Academic Scholars Program?

    A. .03   B. .35   C. .12   D. .65

Solution

Let $P(A)$ = the probability that a randomly selected Florida high school graduate was in the Academic Scholars Program and received a full scholarship

Then $P(A) = P(\text{the graduate was in the program}) \times P(\text{the program participant received a full scholarship})$

$$= (.20) \times (.15)$$
$$= .03$$

A is the correct choice

Example 17

Two representatives are chosen at random from a group of 50 students, consisting of 25 girls and 25 boys. What is the probability that the two representatives selected are either both boys or both girls?

    A. $\frac{1}{4}$ B. $\frac{12}{49}$ C. $\frac{24}{49}$ D. $\frac{97}{49}$

Continued on next page ...

Solution

$$\text{Let } A = \text{the event that both are girls}$$
$$\text{Let } B = \text{the event that both are boys}$$
$$\text{Also, } P(A \text{ or } B) = P(A) + P(B) - P(A \text{ and } B)$$

Note that the P(A and B) will be zero since both representatives cannot be both boys and girls; that is, the events are mutually exclusive.

$$P(A) = P(\text{first is a girl}) \times P(\text{second is a girl, given that the first was a girl})$$
$$= \frac{25}{50} \cdot \frac{24}{49}$$

In a similar fashion, $P(B) = \frac{25}{50} \cdot \frac{24}{49}$. As a result,

$$P(A \text{ or } B) = \frac{25}{50} \cdot \frac{24}{49} + \frac{25}{50} \cdot \frac{24}{49}$$
$$= 2\left[\frac{25}{50} \cdot \frac{24}{49}\right]$$
$$= 2\left[\frac{1}{2} \cdot \frac{24}{49}\right]$$
$$= \left[2 \cdot \frac{1}{2}\right] \cdot \frac{24}{49}$$
$$= 1 \cdot \frac{24}{49} = \frac{24}{49}$$

Choice C is correct.

Skill IIID 1.   Infer relations and make accurate predictions from studying statistical data

Given a table, histogram, broken line graph, scatter diagram, or line graph, you will be asked to infer relations and make predictions based on the data collected. The data will involve a single variable or two variables. The data may be from a business, economics, social science, environmental, physical science, or sports context. You may be asked to:

1. comment on the apparent relationship between two variables
2. comment on the trend in a single variable
3. estimate the value of a variable within the observed range
4. predict beyond the range of observations.

The incorrect options given as choices in the questions will involve one of the following:

1. stating an incorrect relationship between two variables (for example, stating that variables are positively related when, in fact, variables are negatively related)

2. not recognizing an obvious pattern or trend

3. inferring a cause and effect relationship between two variables

4. incorrectly reading, predicting, or estimating the value of a single variable at a particular time

5. giving a wrong value of a second variable based upon incorrectly reading the value of another variable, and

6. giving an incorrect value of a second variable based upon predicting the value of another variable well beyond the observed values to a point where it may make no sense physically.

As you study the following five examples, take care not to make the errors listed in 1 through 6.

Example 18

The graph below depicts the interest rates for an IRA offered by a life insurance company (···) and for Money Market funds (—) for the year 1991.

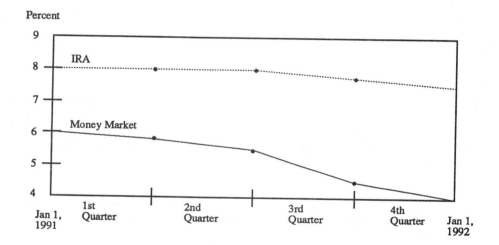

Which of the following *best* describes the relationship between time and interest rates (yield)?

A. Interest rates for both the IRA and the Money Market funds will decrease in 1992.

B. Interest rates showed a decrease in 1991; a greater decrease was shown in the interest rate of the IRA.

C. Interest rates showed a decrease in 1991; a greater decrease was shown in the interest rate of the Money Market funds.

D. Interest rates for both the IRA and the Money Market funds remained constant during the first two quarters of 1991.

Continued on next page ...

Solution

- Choice A is incorrect. Since many factors influence interest rates, and the data involve only the year 1991, we cannot predict from the data the trend in interest rates for 1992.

- Choice B is incorrect. Interest rates did show a decrease in 1991, but a greater decrease was shown in the rate for the Money Market funds, from 6% to 4%.

- Choice C is correct.

- Choice D is incorrect. Although the interest rate for the IRA remained constant during the first two quarters of 1991, the interest rate decreased for the Money Market funds.

As stated, C is the correct choice.

Example 19

Refer to the graph of Example 18. Which of the following is the best estimate of the decrease in the interest rate for Money Market funds during the third quarter of 1991?

A. 2% B. $\frac{1}{2}$% C. $\frac{1}{4}$% D. 1%

Solution

During the third quarter of 1991, the interest rate for the Money Market funds decreased from $5\frac{1}{2}$% to $4\frac{1}{2}$%, a net decrease of 1%. Choice D is correct.

Example 20

The graph below shows the percent of the total bond investments of a national insurance company in non-government bonds of various quality. The rating AAA is the highest quality of a bond while any rating below BBB is considered to be low quality. Which of the following *best* describes the relationship between bond quality and the percentage of the company's investments in non-government bonds?

A. The higher the quality of bond, the greater the percentage of investment in bonds of that quality.

B. The company tends to invest in bonds of quality rating BBB and above.

C. All of the company investments in non-government bonds are in high quality bonds.

D. There is no apparent relationship between bond quality and the percentage of investment in bonds of that quality.

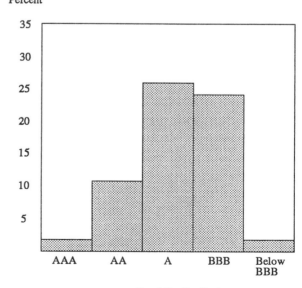

Continued on next page ...

Solution

- Choice A is incorrect. The company does not have its highest percentage of investments in AAA and AA rated bonds.
- Choice B is correct.
- Choice C is incorrect since a small percentage of the bond investments is in bonds with a rating below BBB.
- Choice D is incorrect since the company tends to invest in bonds with rating BBB or above.

As stated, choice B is correct.

Example 21

Refer to the graph of Example 20. Which of the following is the best estimate of the percent of investment of the company in bonds rated A or above?

    A. 39%   B. 32%   C. 63%   D. 37%

Solution

The approximate percent of investments is: AAA Bonds - 2%, AA Bonds - 11%, A Bonds - 26%. The sum is approximately 39%. Choice A is correct.

Example 22

The number of adjunct instructors and the number of students in the Arts and Sciences program at a community college is given below for eight consecutive years.

Year	Number of Adjunct Instructors	Number of Students
1984-85	88	4380
1985-86	94	4492
1986-87	97	4563
1987-88	98	4690
1988-89	103	5110
1989-90	113	5340
1990-91	116	5892
1991-92	120	6320

Which of the following *best* describes the relationship between the number of adjunct instructors and the number of students in the Arts and Sciences program?

  A. The increase in the number of students caused the increase in the number of adjunct instructors.

  B. There appears to be a positive association between the number of adjunct instructors and the number of students.

  C. The size of the adjunct instructor staff does not provide information to predict yearly student enrollment growth.

  D. The increase in the number of adjunct instructors caused an increase in the number of students.

Continued on next page ...

Solution

Choice B is correct. There is a positive association (correlation) between the number of adjunct instructors and the number of students, since both numbers have tended to increase over the eight year time range. Hence, the number of adjunct instructors could be used to predict the number of students. Cause and effect relationships cannot be concluded, however, since many factors operate to determine how many adjunct faculty are hired by a school and how many students choose to attend. In general, we avoid concluding that change in one variable causes change in another variable when interpreting statistical data. As stated, choice B is correct.

Skill IVD 1. **Interprets real-world data involving frequency and cumulative frequency tables.**

Given a table or tables specifying the distribution of a variable for one or two groups, you will be asked to select the correct response to a question about the table. The questions will be one of the following types:

1. What is the percent (proportion) above (below) ____?
2. What percent (proportion) lies between ____ and ____?
3. What is the mean?
4. What is the median?
5. What is the mode?
6. Identify the value so that ____% is above (below) it.

The distribution table will include a listing of values or intervals of possible values and the proportion (percentage) or cumulative proportion (percentage) of the group for each value or interval of values. Percentile rank may be used to specify proportions (percentages).

Example 23

The table below shows the distribution of the number of families living within the city limits of a Central American city and within a given distance of the center of the city in 1960 and the predictions for the year 2000.

Distance from the Center of the City in Miles	Proportion of Families (1960)	Proportion of Families (2000)
0-1, including 1	.05	.06
1-2, including 2	.08	.25
2-3, including 3	.10	.15
3-4, including 4	.12	.07
4-5, including 5	.10	.02
5-6, including 6	.15	.03
6-7, including 7	.20	.22
7-8, including 8	.20	.20

What is the median number of miles from the center of the city per family in 1960?

A. 4 - 5 miles   B. 5 - 6 miles   C. 6 - 7 miles   D. 3 - 4 miles

Continued on next page ...

Solution

The median is the number of miles, n, at which 50% of the families live between 0 and n miles from the center of the city. Add the proportions in the 1960 column of data until 50% is reached.

$$.05 + .08 + .10 + .12 + .10 + .15 = .60$$

Fifty percent (the proportion .50) is reached when the distance is five to six miles. The correct choice is B.

Example 24

Refer to the table of Example 23. What is the mode number of miles from the center of the city that is predicted for the location of families in the year 2000?

    A. 6 - 7 miles   B. 3 - 4 miles   C. 1 - 2 miles   D. 4 - 5 miles

Solution

The mode is the mile range within which the highest proportion (percentage) of families are predicted to live in the year 2000. The highest proportion shown is .25 for a distance of 1 - 2 miles. The correct choice is C.

Example 25

Refer to the table of Example 23. What is the mean number of miles from the center of the city that is predicted for the location of families in the year 2000?

    A. 3 - 4 miles   B. 4 - 5 miles   C. 5 - 6 miles   D. 1 - 2 miles

Solution

The mean is the average number of miles from the center of the city at which the families live. To find the mean do these computations:

# miles	×	%	=		
1	×	6	=	6	First multiply the number of miles by
2	×	25	=	50	the percentage (proportion) of families shown
3	×	15	=	45	in the column of the table for the year 2000.
4	×	7	=	28	
5	×	2	=	10	
6	×	3	=	18	
7	×	22	=	154	
8	×	20	=	160	
				471	Add the results.
471	+	100	=	4.71	Divide the sum by 100 (100%).

The result is the approximate mean number of miles. Select the choice that 4.71 miles would lie within.

The correct choice is B.

Example 26

Refer to the table of Example 23. What percentage of the families lived 5 or less miles from the center of the city in the year 1960?

    A. 35   B. 60   C. 55   D. 45

Solution

Add the proportions from the column for the year 1960 for each of the distance ranges 0-1, 1-2, 2-3, 3-4, and 4-5 miles. Since 5 miles is included in the range 4-5 miles, do not include the proportion for 5-6 miles. The sum is

$$.05 + .08 + .10 + .12 + .10 = .45$$

Convert the decimal result to a percent.

$$.45 = \frac{45}{100} = 45\%$$

The correct choice is D.

Example 27

Scores on a college entrance exam have been scaled so that the scores listed below correspond to the indicated percentile ranks.

Score	Percentile Rank
330	99
300	80
280	72
250	50
240	35
200	1

What percentage of the students who took the exam scored between 240 and 300?

    A. 60   B. 40   C. 45   D. 35

Solution

Recall that the percentile rank is the percentage of students that score below a given score. The percentile rank at score 300 is 80; the percentile rank at score 240 is 35. Subtract these percentile ranks. $80 - 35 = 45$. The percentage of students that scored between 240 and 300 is 45. The correct choice is D.

Skill IVD 2. Solve real-world problems involving probabilities

Given a real-world problem involving probability concepts, you will be asked to do one of the following:

1. attach a probability to an event composed of a single outcome or more than one outcome
2. compute a conditional probability by considering only a restricted sample space
3. find the expected number of occurrences of an event.

A table or graph (pie, bar, or line) describing relative frequency occurrence of all possible outcomes will be provided. The problems will be selected from a business, social studies, education, industry, economics, environmental studies, the arts, physical science, sports, or consumer-related context and will *not* be stated using odds.

Example 28

A county in Florida reported the following distribution of automobile accidents by category during 1991. The categories do not overlap.

Category	Percent of all Automobile Accidents
Alcohol-related	14
Equipment Malfunction	8
Speed-related	38
Due to Wrecklessness	22
Due to Carelessness	12
Other	6

What is the probability that an automobile accident in 1991 was either alcohol-related or speed-related?

A. .24 B. .52 C. .05 D. .38

Solution

Since the accident categories do not overlap, i.e., the events are mutually exclusive, we must use the law:

$$P(A \text{ or } B) = P(A) + P(B)$$

- Let A be the event that the accident is alcohol-related. Then

$$P(A) = \frac{14}{100} = .14$$

- Let B be the event that the accident is speed-related. Then

$$P(B) = \frac{38}{100} = .38$$

- Thus

$$P(A \text{ or } B) = P(A) + P(B) = .14 + .38 = .52$$

The correct choice is B.

Example 29

Refer to the table of Example 28. What is the probability that an automobile accident in 1991 was *not* due to either carelessness or wrecklessness?

A. .34 B. .78 C. .60 D. .66

Continued on next page ...

Solution

- Let A be the event that the accident was due to carelessness.

$$P(A) = \frac{12}{100} = .12$$

- Let B be the event that the accident was due to wrecklessness.

$$P(B) = \frac{22}{100} = .22$$

- Since the accident categories do not overlap, P(A or B) = P(A) + P(B). The probability that the accident was *not* due to either carlessness or wrecklessness is P[not(A or B)].

$$\begin{aligned} P(A \text{ or } B) &= P(A) + P(B) \\ &= .12 + .22 \\ &= .34 \end{aligned}$$

$$\begin{aligned} P[\text{not}(A \text{ or } B)] &= 1 - P(A \text{ or } B) \\ &= 1 - .34 \\ &= .66 \end{aligned}$$

The correct choice is D.

Example 30

Refer to the table of Example 28. If it is known that an accident in 1991 is *not* alcohol-related, find the probability that it is due to carelessness.

A. $\frac{6}{43}$ B. $\frac{3}{25}$ C. $\frac{6}{7}$ D. $\frac{37}{43}$

Solution

- Let B be the event that the accident is not alcohol-related.

$$P(B) = 1 - \frac{14}{100} = \frac{86}{100} = .86$$

- Let A be the event that the accident is due to carelessness.

$$P(A) = \frac{12}{100} = .12$$

- Now, P(A given B) = P(A|B) = the probability that the accident is due to carelessness given that it is not alcohol-related.

$$P(A|B) = \frac{P(A)}{P(B)} = \frac{.12}{.86} = \frac{12}{86} = \frac{6}{43}$$

Choice A is correct.

Example 31

The following is a distribution of all students at a community college in Florida by sex and student category.

	A.A. Degree Seeking	A.S. Degree Seeking	Certificate Program	Continuing Education
Male	20%	14%	12%	4%
Female	16%	18%	6%	10%

Find the probability that a randomly selected student at the community college is female or in a certificate program.

    A. .56   B. .68   C. .62   D. .50

Solution

- Let A be the event that the student is female.

$$P(A) = .16 + .18 + .06 + .10 = .50$$

- Let B be the event that the student is in a certificate program.

$$P(B) = .12 + .06 = .18$$

- Since the events A and B are not mutually exclusive, i.e., both can occur simultaneously,

$$P(A \text{ or } B) = P(A) + P(B) - P(A \text{ and } B)$$

In this problem, since P(A and B) = .06,

$$P(A \text{ or } B) = .50 + .18 - .06 = .62$$

Thus the probability that the student is female or in a certificate program is .62. C is the correct choice.

Example 32

Refer to the table of Example 31. Find the probability that a randomly chosen student is an A.S. degree seeking student.

    A. .36   B. .32   C. .14   D. .18

Solution

An A.S. degree seeking student can be either male or female. Thus

$$P(\text{Student is A.S. degree seeking}) = .14 + .18 = .32$$

The correct choice is B.

7. LOGIC AND SETS

In this section we discuss the CLAST subject area Logic and Sets. Each skill within this subject will be addressed separately, and examples of the type of problems that could be asked on the CLAST for each skill will be presented.

7.1 Sets

Skill IE 1. Deduce facts of set-inclusion or non-set-inclusion

In a diagram of sets, the universal set, U, is usually shown as a rectangle. Sets whose members lie entirely within U are shown by circles inside the rectangle. Overlapping circles indicate that sets might have elements in common. Circles that do not overlap indicate sets that have no elements in common. You might need the following symbols:

U   the universal set; the set containing all objects under discussion

$\in$ *is a member of* or *is an element of*; a $\in$ B means that object a is an element of set B

$\notin$ *is not a member (element) of*

$\cup$ *union*; A $\cup$ B represents the set of all elements found in set A or in set B, or in both sets A and B

$\cap$ *intersection*; A $\cap$ B represents the set of all elements found in both sets A and B (their common elements)

$\emptyset$ the empty set; the set containing no elements

Example 1

Sets A, B, and C are related as shown in the diagram below. Which of the following statements is *not* true, assuming none of the regions is empty?

A. Any element that is an element of set A is also an element of set B.

B. No element is a member of all three sets A, B, and C.

C. Any element that is a member of sets B and C is also a member of set A.

D. Sets B and C have some elements in common.

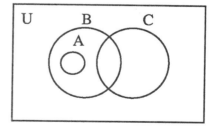

Solution

- Choice A is true since set A is entirely contained in set B.
- Choice B is true since set A does not intersect with set C.
- Choice C is not true since none of the common elements of sets B and C is found in set A. That is, set A does not intersect both sets B and C.
- Choice D is true since none of the regions is empty and sets B and C overlap.

The remaining skills concern logic, although sets might be employed to determine answers to questions involving logic. Some remarks on the basic principles of logic are appropriate. Study these remarks before proceeding to the skills and their respective examples.

7.2 Remarks on Logic

A simple statement is a statement that is either true or false, but not both. Simple statements are represented by the letters p, q, r, and s. The truth value of a simple statement can be altered by negating the statement. The negation of p is ~p and is read *not p*. When p is true, ~p is false. When p is false, ~p is true. Simple statements can be combined to form compound statements. These are defined below.

(**Note:** The numbers within brackets [ ] are for reference purposes.)

[1] **Disjunction** (p ∨ q) Read as *p or q*

The disjunction is true if either p is true or q is true, or if both p and q are true. The disjunction is false only when both p and q are false.

[2] **Conjunction** (p ∧ q) Read as *p and q*

The conjunction is true if both p and q are true. It is false if either p or q is false or if both p and q are false.

[3] **Implication** or **Conditional** (p → q) Read as *If p then q*

An implication is true in all cases except when p is true and q is false. (**Note:** p → q is sometimes read *q if p*.)

Each of the statements discussed above have a negation associated with them. The correct way to negate each of the statements is described below by giving an equivalent form of the negation. The symbol ≐ means *is equivalent to*. Two statements are equivalent if their truth values agree under identical conditions.

[4]   ~(~p) ≐ p

(This is a *double negation*.)

[5]   ~(p ∨ q) ≐ ~p ∧ ~q

(Note that ∨ is replaced by ∧ and both p and q are negated.)

[6]   ~(p ∧ q) ≐ ~p ∨ ~q

(Note that ∧ is replaced by ∨ and both p and q are negated.)

(**Note:** [5] and [6] are referred to as DeMorgan's Laws.)

[7]   ~(p → q) ≐ p ∧ ~q

(Note that the negation of an implication is a conjunction, not an implication.)

Consider the following examples of statements and their negations.

Statement (p): Christmas is December 25.

Negation (~p): Christmas is not December 25.

Statement (p V q): The roof caved in or the house burned down.

Negation (~p ∧ ~q): The roof did not cave in and the house did not burn down.

Statement (p ∧ q): The roof caved in and the house burned down.

Negation (~p V ~q): The roof did not cave in or the house did not burn down.

Statement (p → q): If the roof caved in, then the house burned down.

Negation (p ∧ ~q): The roof caved in and the house did not burn down.

Associated with an implication are three other statements.

Implication	p → q
Converse of the implication	q → p
Inverse of the implication	~p → ~q
Contrapositive of the implication	~q → ~p

These four statements are equivalent in pairs.

[8] $p \rightarrow q \doteq \sim q \rightarrow \sim p$

(The implication is equivalent to its contrapositive.)

[9] $q \rightarrow p \doteq \sim p \rightarrow \sim q$

(The converse of an implication is equivalent to the inverse of the implication.)

An implication also has an equivalent disjunctive form.

[10] $p \rightarrow q \doteq \sim p \vee q$

From [10] we obtain two equivalent forms for a disjunction.

[11] $p \vee q \doteq \sim p \rightarrow q$

[12] $p \vee q \doteq q \vee p \doteq \sim q \rightarrow p$

Study the following example of the various statements associated with an implication.

Implication (p → q): *If two angles are vertical then they are congruent.*

Converse (q → p): *If two angles are congruent then they are vertical.*

Inverse (~p → ~q): *If two angles are not vertical then they are not congruent.*

Contrapositive (~q → ~p): *If two angles are not congruent then they are not vertical.*

Equivalent disjunction (~p V q): *Two angles are not vertical or they are congruent.*

The words *some* or *all* are called **quantifiers**. Statements that involve the use of these words are called **quantified statements**. Listed on the next page are some quantified statements and their equivalent forms.

[13]   *All p are q* ≐ *There are no p that are not q*

[14]   *Some p are q* ≐ *There is a p that is a q*

[15]   *Some p are not q* ≐ *There is some p that is not q*
　　　　　　　　　　　≐ *Not all p are q*

[16]   *~(All p are q)* ≐ There is some p that is not q
　　　　　　　　　　　≐ Some p are not q
　　　　　　　　　　　≐ Not all p are q

[17]   *~(Some p are q)* ≐ There is no p that is q
　　　　　　　　　　　≐ No p is a q

[18]   *~(Some p are not q)* ≐ All p are q

Study the examples given below.

Statement:	All dogs have fleas.
Equivalent Form:	There is no dog that does not have fleas.
Negation:	Some dog does not have fleas. Not all dogs have fleas.
Statement:	Some cats kill rats.
Equivalent Form:	There is a cat that kills a rat.
Negation:	There is no cat that kills rats. No cats kill rats.
Statement:	Some birds cannot sing.
Equivalent Form:	There is a bird that cannot sing. Not all birds can sing.
Negation:	All birds can sing.

Statements can be combined to form **arguments**. An argument usually consists of two or more statements, called **hypotheses**, followed by a statement, called the **conclusion**, which logically follows from the hypotheses. An argument is either **valid** or **invalid**. A valid argument uses the principles of logic and the equivalence of statements to arrive at the conclusion. An invalid (**fallacious**) argument often results from the misuse of the principles of logic or from assuming that nonequivalent statements are equivalent. Note that the truth or falsity of the statements involved in an argument do not determine whether the argument is valid or invalid.

On the next page are listed some common forms of valid arguments and some common fallacies (invalid arguments). When an argument is symbolized, the hypotheses are listed one above the other. Below the hypotheses is a vertical line, and below this line is the conclusion. The conclusion is usually preceded by the symbol ∴, which means *therefore*.

[19]   **Valid Arguments**

Modus Ponens	Modus Tollens (Law of Contrapositive)	Hypothetical Syllogism	Disjunctive Syllogism	Disjunctive Syllogism
$p \to q$	$p \to q$	$p \to q$	$p \lor q$	$p \lor q$
p	$\sim q$	$q \to r$	$\sim p$	$\sim q$
$\therefore q$	$\therefore \sim p$	$\therefore p \to r$	$\therefore q$	$\therefore p$

[20]   **Common Fallacies (Invalid Arguments)**

Fallacy of Converse	Fallacy of Inverse	Fallacy of Disjunction	Fallacy of Disjunction
$p \to q$	$p \to q$	$p \lor q$	$p \lor q$
q	$\sim p$	p	q
$\therefore p$	$\therefore \sim q$	$\therefore \sim q$	$\therefore \sim p$

7.3 Logic

Skill IIE 1. Identify statements equivalent to the negations of simple and compound statements

The solution of this type of problem can be as simple as using DeMorgan's Laws (see remarks [5] and [6] in section 7.2) or much more complex. Statements to be negated can be simple quantified statements, conjunctions, disjunctions, conditionals (implications), or the negations of any of these four forms of statements.

Example 2

Select the statement that is the negation of the statement *John is a ball player and Mike is in the band.*

- A. John is not a ball player and Mike is not in the band.
- B. John is not a ball player or Mike is not in the band.
- C. John is not a ball player or Mike is in the band.
- D. If John is not a ball player then Mike is not in the band.

Solution

- If we let p = *John is a ball player* and q = *Mike is in the band* the given statement can be symbolized as the conjunction $p \land q$.
- Recall DeMorgan's Law $\sim(p \land q) \doteq \sim p \lor \sim q$. (The negation of a conjunction is logically equivalent to a disjunction in which each of the statements is negated.) As a result, the negation of the given conjunction is *John is not a ball player or Mike is not in the band*, which is choice B.

Example 3: Select the negation of the statement *Sue can dive or Bill cannot swim.*

    A. Sue cannot dive or Bill cannot swim   B. Sue cannot dive or Bill can swim.
    C. Sue cannot dive and Bill can swim   D. Sue can dive and Bill cannot swim.

Solution

- Symbolize the statements.

$$p = \text{Sue can dive.}$$
$$q = \text{Bill can swim.}$$
$$\sim q = \text{Bill cannot swim}$$
$$p \vee \sim q = \text{Sue can dive or Bill cannot swim.}$$

- By DeMorgan's Law $\sim(p \vee q) \doteq \sim p \wedge \sim q$. As a result,

$$\sim(p \vee \sim q) \doteq \sim p \wedge \sim(\sim q)$$
$$\doteq \sim p \wedge q$$

$$\sim p \wedge q = \textit{Sue cannot dive and Bill can swim.}$$

Choice C is correct.

Example 4: Select the negation of the statement *If the brakes fail then the car will crash.*

    A. If the brakes fail then the car will not crash.
    B. If the brakes do not fail then the car will not crash.
    C. The brakes do not fail or the car will not crash.
    D. The brakes fail and the car will not crash.

Solution

Symbolize the statements.

$$p = \text{The brakes fail.}$$
$$q = \text{The car will crash.}$$
$$p \rightarrow q = \text{If the brakes fail then the car will crash.}$$

Recall that $\sim(p \rightarrow q) \doteq p \wedge \sim q$, which is a conjunction. As a result, $p \wedge \sim q =$ *The brakes fail and the car will not crash.*

Choice D is correct.

Example 5:

Select the statement that is the negation of the statement *All insects have six legs.*

    A. Some insects have six legs.   B. There is no insect that has six legs.
    C. Some insects do not have six legs.   D. It is not true that some insects have six legs.

Continued on next page ...

Solution

- *All insects have six legs* can be symbolized as *All p are q*.

- ~(All p are q) $\doteq$ There is some p that is not q
  $\doteq$ Some p are not q
  $\doteq$ Not all p are q

- As a result,

  ~(All insects have six legs) $\doteq$ There is some insect that does not have six legs.
  $\doteq$ Some insects do not have six legs.
  $\doteq$ Not all insects have six legs.

  The second form of the negation is given in choice C.

Skill IIE 2. Determine equivalence or nonequivalence of statements

This skill can involve conjunctions, disjunctions, conditionals (implications), and negations of these three forms. The converse, inverse, or contrapositive of an implication can be involved as well as the equivalent disjunctive form of the implication.

Example 6

Select the statement that is logically equivalent to *If Robert is in Atlanta, then he is in Georgia.*

    A. If Robert is not in Atlanta, then he is not in Georgia.

    B. If Robert is in Georgia, then he is in Atlanta.

    C. Robert is in Atlanta or he is in Georgia.

    D. If Robert is not in Georgia, then he is not in Atlanta.

Solution

- Symbolize the statements.

  p = Robert is in Atlanta.
  q = Robert is in Georgia.
  $p \rightarrow q$ = If Robert is in Atlanta then he is in Georgia.

- The conditional $p \rightarrow q$ has two equivalent forms.

  Contrapositive: $\sim q \rightarrow \sim p$ If Robert is not in Georgia, then he is not in Atlanta.
  Disjunction: $\sim p \vee q$ Robert is not in Atlanta or he is in Georgia.

  The contrapositive, $\sim q \rightarrow \sim p$, is found in choice D.

  (**Note:** Choice A is the *inverse* of the implication and choice B is the *converse* of the implication. An implication is not equivalent to its inverse and it is not equivalent to its converse.)

Example 7

Select the statement that is logically equivalent to *If the play is canceled, Mary will do homework.*

    A. Mary will not do homework or the play was canceled.

    B. Mary will do homework or the play was not canceled.

    C. If Mary does homework, then the play is canceled.

    D. If the play is not canceled, then Mary will do homework.

Solution

- Symbolize the statements.

$$p = \text{The play is canceled.}$$
$$q = \text{Mary will do homework.}$$
$$p \to q = \text{If the play is canceled, then Mary will do homework.}$$

- Recall that $p \to q \equiv {\sim}q \to {\sim}p$ and $p \to q \equiv {\sim}p \vee q$.
- Symbolize the choices and compare with the equivalent forms of an implication.

    Choice A: ${\sim}q \vee p$ (This is not equivalent to $p \to q$)

    Choice B: $q \vee {\sim}p$ (This is equivalent to ${\sim}p \vee q$, which is equivalent to $p \to q$)

    Choice C: $q \to p$ (This is the converse, which is not equivalent to $p \to q$)

    Choice D: ${\sim}p \to q$ (This is not equivalent to $p \to q$)

Choice B is correct.

Example 8

Select the statement that is *not* logically equivalent to *It is not true that both Scat and Blackie are cats.*

    A. Scat is not a cat or Blackie is not a cat.

    B. If Scat is a cat then Blackie is not a cat.

    C. If Blackie is a cat then Scat is not a cat.

    D. Scat is not a cat and Blackie is not a cat.

Solution

- Symbolize the statements.

$$p = \text{Scat is a cat.}$$
$$q = \text{Blackie is a cat.}$$
$$p \wedge q = \text{Both Scat and Blackie are cats.}$$
$$= \text{Scat is a cat and Blackie is a cat.}$$
$${\sim}(p \wedge q) = \text{It is not true that both Scat and Blackie are cats.}$$

Continued on next page ...

- By DeMorgan's Law, $\sim(p \land q) \doteq \sim p \lor \sim q$. Moreover, recall that a disjunction has two equivalent forms that are implications.

$$[1] \quad p \lor q \doteq \sim p \to q$$

$$[2] \quad p \lor q \doteq \sim q \to p$$

As a result,

$$\sim p \lor \sim q \doteq p \to \sim q \quad \text{and} \quad \sim(p \land q) \doteq p \to \sim q$$

Also,

$$\sim p \lor \sim q \doteq q \to \sim p \quad \text{and} \quad \sim(p \land q) \doteq q \to \sim p$$

- Symbolize the choices.

    Choice A: $\sim p \lor \sim q$ (This is equivalent to $\sim(p \land q)$.)

    Choice B: $p \to \sim q$ (This form is equivalent to $\sim(p \land q)$.)

    Choice C: $q \to \sim p$ (This form is equivalent to $\sim(p \land q)$.)

    Choice D: $\sim p \land \sim q$ (This form is not equivalent to $\sim(p \land q)$.)

The correct choice is D.

Skill IIE 3. Draw logical conclusions from data

Data given in the form of negations, disjunctions, conjunctions, implications, and quantified statements might require computations or comparisons to determine conclusions. The forms of valid arguments from section 7.2 (Remarks On Logic) might also be necessary to determine conclusions. Two examples follow. Note that the solution to example 9 introduces a new process.

Example 9

Given that

i. *All dancers are athletic.*

ii. *Some polo players are dancers.*

determine which conclusion can be logically deduced.

    A. All polo players are athletic.   B. Some polo players are athletic.
    C. No polo players are athletic.   D. All athletic people are polo players.

Solution

This problem can be readily solved by set diagrams called Euler Circles. We first assign symbol names to the appropriate sets.

    D = the set of all dancers
    A = the set of all athletic people
    P = the set of all polo players

We next construct a diagram, using circular shapes for sets, that displays the statements made in the hypotheses. Several drawings can be possible. Draw all possibilities.

Continued on next page ...

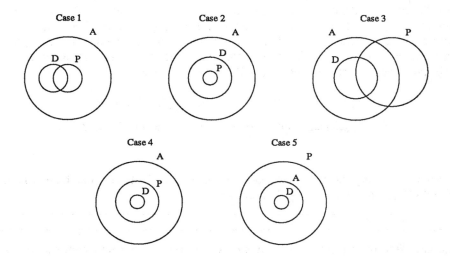

Finally, consider each choice of a conclusion and determine if it holds for all cases.. (A conclusion that can be deduced from the hypotheses is one that holds in all possible cases.)

- Choice A holds in cases 1, 2, and 4, but not in cases 3 and 5.
- Choice B holds in all five cases.
- Choice C does not hold in any of the five cases.
- Choice D holds only in case 5.

Choice B is correct.

(**Note:** When statements in an argument involve quantifiers (*all* or *some*), Euler circles can also be used to determine whether the argument is valid or invalid.)

Example 10

Read the following requirements for a job and each applicant's qualifications. Identify which of the applicants would qualify for the job.

Job Notice

Teacher of mathematics in the AA degree program at Seminole Community College. Applicants must have a masters degree with 18 semester hours of mathematics at the graduate level. Two years teaching experience at the high school level or higher and some experience with computers in education are required.

Applicant Information

- Applicant A has a masters degree with 24 hours in mathematics. She has taught seven years at Nova Middle School where she was instrumental in establishing a Computer Learning Center in mathematics.

- Applicant B holds a Ph.D. in mathematics with 40 postgraduate hours in mathematics. He has been teaching mathematics courses and the course *Computer Applications in Education* at Stetson University for 10 years.

Continued on next page ...

- Applicant C recently completed his masters degree at Stetson University with 22+ hours in mathematics and 16+ hours in computer science. He took the course *Computer Applications in Education* from applicant B. For a special project in this course, C designed a computer lab for community college mathematics students. While a graduate student he taught for two years as a full time instructor at Daytona Beach Community College.

    A. Applicant B only   B. Applicant C only

    C. Applicants B and C only   D. Applicants A, B, and C

Solution

Applicant A fails to qualify because her teaching experience is below the high school level. Applicants B and C both qualify. Choice C is correct.

Skill IIE 4. Recognize that an argument might not be valid even though its conclusion is true

Although these problems will be in verbal form, the valid arguments and fallacies of section 7.2 (Remarks On Logic) will be helpful.

Example 11

All of the arguments A-D have true conclusions, but one of the arguments is *not* valid. Select the argument that is not valid.

    A. All dogs bark and all poodles are dogs. Therefore, all poodles bark.

    B. All dogs bark and all poodles bark. Therefore, all poodles are dogs.

    C. All vertebrates have backbones and all snakes are vertebrates. Therefore, all snakes have backbones.

    D. Every county of Florida is in the U.S.A. Seminole is a county in Florida. Therefore, Seminole is in the U.S.A.

Solution

To solve this problem we construct Euler Circles (as in the solution to example 9) and determine if the conclusion holds in all cases. If so, then the argument is valid.

Continued on next page ...

Choice A

B = Set of barking animals
D = Set of dogs
P = Set of poodles

The conclusion,
All poodles bark holds;
the argument is valid.

Choice B

B = Set of barking animals
D = Set of dogs
P = Set of poodles

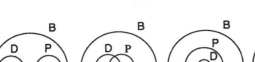

Case 1   Case 2   Case 3   Case 4

The conclusion, *All poodles are dogs* is forced to hold only in Case 4.
The argument is invalid.

Choice C

B = Set of animals with backbones
V = Set of vertebrates
S = Set of snakes

The conclusion,
All snakes have backbones holds;
the argument is valid.

Choice D

U = Set of counties in USA
F = Set of counties in Florida
x = Seminole county

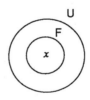

The conclusion, *Seminole is in the USA* holds;
the argument is valid.

The correct choice is B.

Example 12

All of the examples in A–D have true conclusions, but one of the arguments is not valid. Select the argument that is not valid.

A. If a lawyer practices law in Florida, then the lawyer has passed the Florida Bar exam. Attorney J. F. Fleming of Pensacola practices law in Florida. Therefore, attorney Fleming has passed the Florida Bar exam.

B. If $\triangle ABC \cong \triangle DEF$, then $\angle C \cong \angle F$. If $\angle C \cong \angle F$, then m $\angle C$ = m $\angle F$. Therefore, if $\triangle ABC \cong \triangle DEF$, then m $\angle C$ = m $\angle F$.

C. Either a parrot is a bird or a parrot is a fish. A parrot is not a fish. Therefore a parrot is a bird.

D. If two lines are parallel, then they do not intersect. Two horizontal lines do not intersect. Therefore, two horizontal lines are parallel.

Continued on next page ...

Solution

Symbolize each argument and determine whether the form of the argument is the form of a valid argument or of an invalid argument. (See items [19] and [20] in section 7.2 (Remarks On Logic).)

Choice A	Choice B	Choice C	Choice D
$p \to q$	$p \to q$	$p \lor q$	$p \to q$
p	$q \to r$	$\sim q$	q
$\therefore q$	$\therefore p \to r$	$\therefore p$	$\therefore p$
Modus ponens; a valid form	Hypothetical syllogism; a valid form	Disjunctive syllogism; a valid form	Fallacy of converse; an invalid form

Choices A, B, and C give valid arguments. The argument in Choice D is invalid and hence D is the correct choice.

Skill IIIE 1. Recognize valid reasoning patterns as illustrated by valid arguments in everyday language.

You will be given the premises of an argument, expressed in words, and will be asked to identify the conclusion that will make the argument a valid argument. The valid argument form will be chosen from those given in item [19] of section 7.2 (Remarks on Logic). Invalid conclusions may result from the fallacies listed above in item [20] of section 7.2 or from other misuses of the premises.

Example 13

Select the conclusion that will make the following argument valid.

If Andrea is to receive the makeup kit, her mother must pay the modeling agency before the first lesson. Andrea's mother will not pay the modeling agency until the third lesson.

A. Andrea will receive the makeup kit.

B. Andrea will not receive the makeup kit.

C. Andrea's mother will pay the modeling agency in time for Andrea to receive the makeup kit.

D. The modeling agency will give Andrea a free modeling lesson.

Continued on next page ...

Solution

Symbolize the premise statements. Then search for a conclusion that results in one of the valid forms of arguments given in item [19] of section 7.2.

    p = Andrea is to (will) receive the makeup kit.
    q = Andrea's mother must pay the modeling agency before the first lesson.
   ~q = Andrea's mother will not pay the modeling agency until the third lesson, (i.e., does not pay before the first lesson).

Symbolize the argument:

$$p \to q$$
$$\sim q$$
$$\therefore\ ?$$

The valid conclusion by the argument form Modus Tollens is ~p, which translates to *Andrea will not receive the makeup kit*. The correct conclusion is given in choice B.

Choice B is correct.

Example 14

Select the conclusion which will make the following argument valid.

If the Duke University team wins the final game, then the Duke team will be the national championship team. If the Duke team is the national championship team, then the Duke team will go on a world tour.

  A. If the Duke team goes on a world tour, then the Duke team is the national championship team.

  B. If the Duke team goes on a world tour, then the Duke team won the final game.

  C. If the Duke team does not win the final game, then the Duke team will not be the national championship team.

  D. If the Duke team wins the final game, then the team will go on a world tour.

Solution

Symbolize the premise statements. Then search for a conclusion that results in one of the valid forms of arguments discussed previously.

    p = The Duke team wins the final game.
    q = The Duke team is the national championship team.
    r = The Duke team will go on a world tour.

Symbolize the argument.

$$p \to q$$
$$q \to r$$
$$\therefore\ ?$$

Continued on next page ...

- Choice A. The symbolic form of this choice is r → q, the converse of the second statement, q → r. If an implication holds, its converse does not necessarily hold. This choice is incorrect.

- Choice B. The symbolic form of this choice is r → p, the converse of the statement p → r. Using the valid argument form Hypothetical Syllogism, p → r can be concluded from the given premises. Hence its converse r → p is incorrect.

- Choice C. The symbolic form of this choice is ~p → ~q, the inverse of the first premise statement p → q. If an implication holds, its inverse does not necessarily hold. This choice is incorrect.

- Choice D. The symbolic form of this choice is p → r, the desired conclusion to the given premises by the Law of Hypothetical Syllogism.

Choice D is correct.

Example 15

Select the conclusion that will make the following argument valid.

If all people are healthy, then no hospitals are needed. Some hospitals are needed.

A. Some people are healthy.

B. If no hospitals are needed, then all people are healthy.

C. Some people are not healthy.

D. All people are healthy.

Solution

Symbolize the premise statements. Then search for a conclusion that results in a valid argument form.

$$p \;=\; \text{All people are healthy.}$$
$$q \;=\; \text{No hospitals are needed.}$$

Symbolize the argument.

$p → q$ = If all people are healthy, then no hospitals are needed.
$~q$ = Some hospitals are needed.
∴ ?

(Note: By item [17] of section 7.2, ~(Some p are q) $\doteq$ No p is a q. From these equivalent forms, we see that ~(No p is a q) $\doteq$ Some p are q.)

The correct conclusion for a valid argument is ~p [Modus Tollens] or one of its equivalent forms. The negation of *All people are healthy* is *Some people are not healthy* by item [16] of section 7.2. Thus choice C is the correct choice.

420 CLAST REVIEW

Skill IIIE 2. Select applicable rules for transforming statements without affecting their meaning

Two equivalent statements will be given in verbal form or in mathematical form. You will be asked to select the rule that establishes the equivalence of the statements. The correct rule in the given choices will be in exactly the same form, but it may be in either symbolic or verbal form. These problems appear to be extremely complex, but it is actually quite simple to eliminate the wrong choices because they do not look like (i.e., are not in exactly the same form as) the original problem. The problems will involve the previously discussed equivalent forms of statements.

Example 16

Select the rule of logical equivalence that directly (in one step) *transforms* statement i into statement ii.

i. If x^2 is odd, then x is odd.

ii. x^2 is not odd or x is odd.

A. *If p, then q* is equivalent to *If not q, then not p*.

B. *If p, then q* is equivalent to *not p or q*.

C. *Not (p or q)* is equivalent to *not p and not q*.

D. *If p, then q* is equivalent to *all p are q*.

Solution

- Statement i in symbolized form is p → q and has the meaning *If p, then q*.
- Statement ii in symbolized form is ~p V q and has the meaning *Not p or q*.

Choice B is correct.

Example 17

Select the rule of logical equivalence that directly (in one step) *transforms* statement i into statement ii.

i. It is not true that all spiders are poisonous.

ii. Some spiders are not poisonous.

A. *Not (Some p are q)* is equivalent to *Not all p are q*.

B. *All p are q* is equivalent to *There is no p that is not q*.

C. *Some p are q* is equivalent to *There is a p that is a q*.

D. *Not (All p are q)* is equivalent to *Some p are not q*.

Solution

Use letters to symbolize statements i and ii.

- Statement i symbolized is ~*(All p are q)*
- Statement ii symbolized is *Some p are not q*

Statement i must be transformed directly to statement ii. Choice D is correct.

Skill IVE 1. Draw logical conclusions when the facts warrant them

This type of problem requires skill in arriving at valid conclusions using the same logical principles as in skills IIIE1, IIE3, and IIE4. As before, statements can involve disjunctions, conjunctions, implications (conditionals), quantified statements, negations, or any combinations of these. In this skill you are given an argument expressed in words and you must select the correct conclusion from the multiple choices, where one of the choices could be *none of the above*.

Example 18

Study the information given below. If a logical conclusion is given, select that conclusion. If none of the conclusions given is warranted, select the option expressing this condition.

If you study hard, then you will pass mathematics. If you pass mathematics, then you will be able to get an A.A. degree. You pass mathematics.

- A. You will not be able to get an A.A. degree.
- B. You will be able to get an A.A. degree.
- C. You studied hard.
- D. None of the above is warranted.

Solution

Symbolize the argument.

$$p \rightarrow q \quad \text{If you study hard, then you will pass mathematics.}$$
$$q \rightarrow r \quad \text{If you pass mathematics, then you will be able to get an A.A. degree.}$$
$$q \quad \text{You pass mathematics.}$$
$$\therefore ?$$

Using the second and third statements, the rule of valid arguments called Modus Ponens allows the conclusion r, *You will be able to get an A.A. degree*. As a result, the correct choice is B.

(**Note:** A common fallacy, Fallacy of the Converse, leads to the conclusion p, *You studied hard*. However, this is not a logical conclusion. With this conclusion, the argument will be invalid.)

Example 19

Study the information given below. If a logical conclusion is given, select that conclusion. If none of the conclusions given is warranted, select the option expressing this condition.

All race car drivers wear helmets. Some people who wear helmets will be protected from head injuries during an accident. Jack is a race car driver.

- A. Jack will be protected from head injuries during an accident.
- B. Jack will not be protected from head injuries during an accident.
- C. Jack does not wear a helmet.
- D. None of the above is warranted.

Continued on next page ...

Solution

Use Euler Circles.

Let R = the set of people who are race car drivers
     H = the set of people who wear helmets
     A = the set of people who are protected from head injuries during an accident
     x = Jack

Draw all possible cases that symbolize the hypotheses.

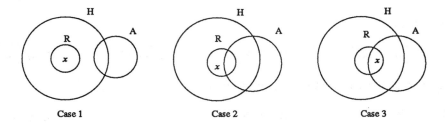

- Conclusion A holds only in Case 3.
- Conclusion B holds in Cases 1 and 2, but not Case 3.
- Conclusion C holds in none of the three cases.

Choice D is correct.

(**Note:** The conclusion *Jack wears a helmet* would hold in all three cases and would be a logical conclusion.)

8. PRACTICE CLAST: FORM A

Limit your time to 90 minutes. Choose the *one* correct answer.

1. $7\frac{5}{8} + 6\frac{2}{3} =$

    A. $13\frac{7}{24}$ B. $13\frac{7}{11}$ C. $14\frac{7}{24}$ D. $13\frac{1}{24}$

2. $\frac{2}{3} + (\frac{-1}{5}) =$

    A. $\frac{-3}{10}$ B. $\frac{-2}{15}$ C. $\frac{10}{3}$ D. $\frac{-10}{3}$

3. $5.8652 \div 68.2 =$

    A. 0.860   B. 0.0806   C. 11.63   D. 0.086

4. If 20 is increased to 25, what is the percent increase?

    A. 5%   B. 20%   C. 25%   D. 80%

5. 42 is what % of 840?

    A. 50%   B. 20%   C. 5%   D. 2%

6. Round off to the nearest meter: 48.67 m

    A. 48 m   B. 50 m   C. 49 m   D. 49.7 m

7. Find the total surface area of a cube of 8 cm.

    A. 64 sq. cm   B. 512 sq. cm   C. 384 sq cm   D. 256 sq cm

8. What is the volume of a right circular cylinder that is 8 inches high and has a base diameter of 6 inches?

    A. 288π cubic inches   B. 72π cubic inches
    C. 96π cubic inches   D. 48π cubic inches

9. $\sqrt{32} - \sqrt{2} =$

    A. $3\sqrt{2}$ B. $\sqrt{30}$ C. 4   D. $15\sqrt{2}$

10. $12w - 8w \cdot 2w + 12w + 4 =$

    A. $2w^2 + 3w$ B. $8w^2 + 3w$ C. $-4w^2 + 6w$ D. $-16w^2 + 15w$

11. $(5.2 \times 10^{-6}) \times (2.4 \times 10^4) =$

    A. 12.48   B. 1248   C. 0.1248   D. −12.48

12. Solve the inequality: $9 - 4x \leq 14$

    A. $x \leq \dfrac{-5}{4}$ B. $x \geq \dfrac{-5}{4}$ C. $x \leq \dfrac{5}{4}$ D. $x \geq \dfrac{-23}{4}$

13. Given $x = (y + 1)^2 - y$, if $y = -3$, then $x =$

    A. 13   B. 1   C. -1 D. 7

14. Find $f(-2)$ given $f(x) = 2x^3 + 3x - 1$

    A. 9   B. -23 C. -71 D. -11

15. Which is a linear factor of the expression $6x^2 - x - 15$?

    A. $2x + 3$ B. $3x + 5$ C. $2x - 3$ D. $2x - 5$

16. Find the real roots of this equation: $4x^2 + 3x - 2 = 0$

    A. $\dfrac{-3 + \sqrt{17}}{8}$ and $\dfrac{-3 - \sqrt{17}}{8}$ B. $\dfrac{3 + \sqrt{41}}{8}$ and $\dfrac{3 - \sqrt{41}}{8}$

    C. $\dfrac{-3 + \sqrt{41}}{8}$ and $\dfrac{-3 - \sqrt{41}}{8}$ D. $\dfrac{3 + \sqrt{17}}{8}$ and $\dfrac{3 - \sqrt{17}}{8}$

17. Find the real roots of this equation: $6x^2 - 35 = x$

    A. $\dfrac{-5}{3}$ and $\dfrac{7}{2}$ B. $\dfrac{-7}{3}$ and $\dfrac{5}{2}$ C. $\dfrac{-1 + \sqrt{841}}{12}$ and $\dfrac{-1 - \sqrt{841}}{12}$ D. $\dfrac{7}{3}$ and $\dfrac{-5}{2}$

18. Choose the correct solution set for the system of linear equations:

    $$\begin{cases} 7x - 3y = 8 \\ x - y = 2 \end{cases}$$

    A. $\{(\dfrac{5}{2}, \dfrac{1}{2})\}$ B. $\{(\dfrac{1}{2}, \dfrac{-3}{2})\}$ C. $\{(\dfrac{7}{2}, \dfrac{3}{2})\}$ D. $\{(\dfrac{9}{2}, \dfrac{5}{2})\}$

19. The number of hours Andrea spends each school day on various activities is represented by the circle chart below. What percent of these hours is spent on homework and miscellaneous tasks combined?

    A. 25%   B. $33\dfrac{1}{3}\%$

    C. 20%   D. 40%

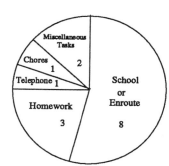

20. What is the median of the data in the following sample? 3, 12, 13, 18, 23, 12, 17

    A. 13 B. 12 C. 18 D. 14

21. A committee has five members. In how many different ways can a subcommittee of three people be selected from the committee?

    A. 60 B. 10 C. 20 D. 32

22. Sets A, B, C, and U are related as shown in the diagram. Which of the following statements is true assuming none of the eight regions is empty?

    A. All elements of sets A and B are also in set C.
    B. Some elements in set A are not in set U.
    C. All elements of set C are in set B.
    D. Some elements of set B are not in set A.

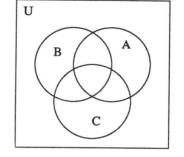

23. $(4^3)^2 =$

    A. 4^5 B. $4^3 \times 4^3$ C. $(4 \times 3)^2$ D. $4^3 \times 4^2$

24. Select the correct expanded notation for 3012.050.

    A. $(3 \times 10^4) + (1 \times 10^2) + (2 \times 10^1) + (5 \times \frac{1}{10^2})$
    B. $(3 \times 10^3) + (1 \times 10^1) + (2 \times 10^0) + (5 \times \frac{1}{10^2})$
    C. $(3 \times 10^3) + (1 \times 10^1) + (2 \times 10^0) + (5 \times \frac{1}{10^3})$
    D. $(3 \times 10^2) + (1 \times 10^1) + (2 \times 10^0) + (5 \times \frac{1}{10^2})$

25. $62.5\% =$

    A. 6250 B. $\frac{625}{100}$ C. $\frac{5}{8}$ D. 6.25

26. Select the correct relation: $\frac{-7}{12} \square \frac{-9}{16}$

    A. = B. > C. <

27. One hundred students took an achievement test. All of the students scored less than 94, but more than 64. Which of the following values is a reasonable estimate of the average score of the students?

    A. 64 B. 97 C. 55 D. 72

28. Which of the statements below is true for the figure shown, given that l_3 and l_4 are parallel lines? (The measure of $\angle BAC$ is represented by x.)

A. $z = 70°$

B. $x + y + z = 180°$

C. $x = z$

D. None of the above statements is true.

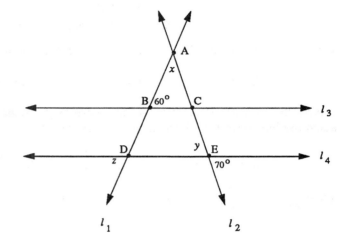

29. Study the figures in A, B, C, and D. Select the figure in which all the triangles are similar.

A.

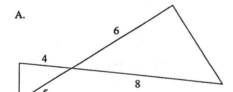

B.

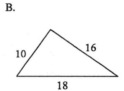 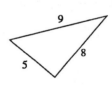

C.

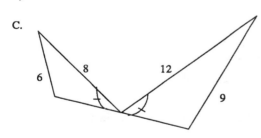

D.

30. Which of the following could *not* be used to report the distance between two cities?

    A. kilometers   B. meters   C. square meters   D. miles

31. Choose the statement which is *not* true for all real numbers x and y.

A. $4x - 4y = 4(x - y)$

B. $5xy(2x + y) = (2x + y)(5xy)$

C. $(x - y)(x + y) = (y - x)(y + x)$

D. $5(xy) = (5x)y$

32. For each of the statements below determine whether −6 is a solution.

   i. $t^2 - 36 = 0$
   ii. $(x - 5)(x + 6) \geq 0$
   iii. $2w + 12 = 2(w + 6) - 20$

   Which option below identifies every statement that has −6 as a solution and only those statements?

   A. i only   B. ii only   C. i and ii only   D. All of the statements

33. A man uses 14 buckets of paint in three days of painting. Let b = the number of buckets of paint he would use in seven days. Select the correct statement.

   A. $\dfrac{14}{3} = \dfrac{7}{b}$ B. $\dfrac{14}{3} = \dfrac{b}{7}$ C. $\dfrac{14}{7} = \dfrac{3}{b}$ D. $\dfrac{14}{b} = \dfrac{7}{3}$

34. Which option gives the conditions that correspond to the shaded region of the plane shown in the figure?

   A. $x - y > 2$ and $y < 2$
   B. $x - y > 2$ and $y > 2$
   C. $x - y < 2$ and $y < 2$
   D. $x - y < 2$ and $y > 2$

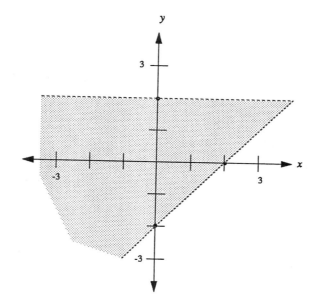

35. In a physics class, half of the students scored 60 on a test. Most of the remaining students scored 80 except a few who scored 90. Which of the following is true about the distribution of scores?

   A. The median and the mean are the same.   B. The mode and the median are the same.
   C. The mean is less than the mode.   D. The mean is more than the mode.

36. Management personnel of a grocery store want to estimate the average amount spent on groceries per month by the customers who shop there. To do so members of the store's marketing department will conduct a survey. Which of the following procedures would be most appropriate for selecting a statistically unbiased sample?

   A. Survey the first 100 customers from an alphabetical list of customers that hold a check cashing card.

   B. Randomly select two days of the week and randomly select three hours of the day. During these hours randomly survey customers who enter the store.

   C. Conduct a one week survey of all persons who shop from 8 A.M. to 10 A.M.

   D. Have customers voluntarily complete a questionnaire.

37. A box contains six portable radios, two of which are defective. If two portable radios are randomly selected from the box at the same time, find the probability that both are defective.

   A. $\frac{1}{15}$ B. $\frac{1}{3}$ C. $\frac{1}{9}$ D. $\frac{8}{15}$

38. Select the statement that is the negation of the statement *Mary is a secretary or Jane is a nurse.*

   A. Mary is not a secretary or Jane is not a nurse.

   B. Mary is not a secretary and Jane is a nurse.

   C. Mary is not a secretary and Jane is not a nurse.

   D. Mary is a secretary and Jane is not a nurse.

39. Determine which statement is logically equivalent to *If you do not rest then you will become tired.*

   A. If you became tired, then you did not rest.

   B. If you do not rest, then you will not become tired.

   C. If you rest, then you will not become tired.

   D. If you did not become tired, then you did rest.

40. Given that

   i. *If one studies hard, then one will make good grades.*

   ii. *If one makes good grades, then one will obtain a job.*

determine which conclusion can be logically deduced.

   A. If one does not obtain a job, then one did not study hard.

   B. If one obtains a job, then one studied hard.

   C. If one does not study hard, then one will not obtain a job.

   D. If one makes good grades, then one studies hard.

41. All of the following have true conclusions. Which of the following also has a valid argument?
    A. All insects have wings and all grasshoppers have wings. Therefore, all grasshoppers are insects.
    B. All insects are animals and all grasshoppers have wings. Therefore, all grasshoppers are animals.
    C. All insects have wings and all grasshoppers are insects. Therefore, all grasshoppers have wings.
    D. All insects are animals and all grasshoppers have wings. Therefore, all grasshoppers are insects.

42. Select the next term in the following geometric progression.

    $$81, -27, 9, -3, 1, \ldots$$

    A. 0 B. $\frac{1}{3}$ C. -1 D. $\frac{-1}{3}$

43. Study the given information.

    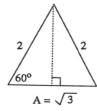 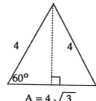 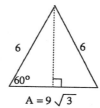 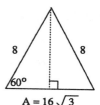

    Calculate the area of a triangle in which two sides have a measure of 10 and the angle opposite one of these sides has a measure of 60°.

    A. $20\sqrt{3}$ B. $100\sqrt{3}$ C. $10\sqrt{3}$ D. $25\sqrt{3}$

44. Select the formula for computing the area of a regular pentagon.

    A. Area = 5(h + b)   B. Area = 10bh

    C. Area = $\frac{5bh}{2}$ D. Area = 5bh

    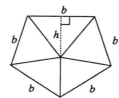

45. If $y < 0$, then $y^3 > xy + y^2$ is equivalent to which of the following?

    A. $y^2 > x + y$ B. $y^2 < x + y$ C. $y^4 < xy^2 + y^2$ D. $y^2 > -x - y$

46. Consider the following plot for Bob's time for running a mile.

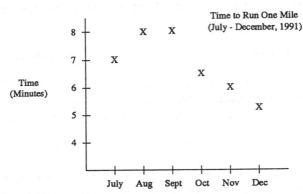

Which of the following best describes the trend in the time for Bob to run a mile?

A. Bob's time for running a mile is steadily decreasing into the months of 1992.

B. Cooler weather causes Bob's time for running a mile to decrease.

C. Bob's time for running a mile rapidly dropped off in the last months of 1991.

D. There is no trend in Bob's time for running a mile.

47. Select the conclusion that will make the following argument valid.

Sam bought a Toyota or Sam bought a Honda. Sam did not buy a Toyota.

A. If Sam did not buy a Toyota, then Sam bought a Honda.   B. Sam bought a Honda.
C. Sam did not buy a Honda.   D. Sam did not buy a car.

48. Select the rule of logical equivalence that directly (in one step) transforms statement "i" into statement "ii".

i. Some students do not take Physics.

ii. Not all students take Physics.

A. *If p then q* is equivalent to *If not q, then not p*.

B. *Some are not p* is equivalent to *Not all are p*.

C. *All are not p* is equivalent to *None are p*.

D. *Not (not p)* is equivalent to *p*.

49. An $1150 refrigerator is on sale for $1000. What fractional part of the original price is the savings?

A. $\frac{3}{23}$ B. $\frac{20}{23}$ C. $\frac{8}{20}$ D. $\frac{3}{20}$

50. How many whole numbers leave a remainder of zero when divided into 56 and a remainder of three when divided into 31?

A. 2   B. 3   C. 4   D. 5

51. The dimensions of a house are 25 feet by 15 yards. The dimensions of the lot on which the house is located are 55 feet by 27 yards. What is the area of the lawn, assuming that the area of the lot not occupied by the house is lawn?

    A. 1100 sq. yd.   B. 122 sq. yd.   C. 3330 sq. ft.   D. 1098 sq. ft.

52. Employees of Florida Power plan to erect a 40 ft. pole each block in downtown Sanford. Each pole will be supported by three cables, with each cable reaching from the top of the pole to a point on level ground 30 ft. from the base of the pole. If cable costs $2 per foot, find the total cost of the cable for one pole.

    A. $100   B. $400   C. $210   D. $300

53. The management of the Florida Game and Freshwater Fish Commission wanted to estimate the number of alligators in Everglades National Park. They caught and tagged 200 alligators from the park and released them back into the park. Later, the management team caught 300 alligators from sites in the park. Of those caught, 40 were tagged. Determine a reasonable estimate of the number of alligators in the park.

    A. 80,000   B. 4500   C. 3000   D. 1500

54. A Southern airport has collected data on all of its flights for twenty years. The following is a distribution of the causes of accidents occurring in the twenty year period. No accident has more than one cause. What is the probability that an accident was caused by neither pilot error nor traffic controller error?

Cause	Percent of all Accidents
Weather Conditions	42
Equipment Malfunction	38
Pilot Error	12
Traffic Controller Error	6
Other	2

    A. .18
    B. .92
    C. .72
    D. .82

55. If a logical conclusion is given, select that conclusion; otherwise select *none of the above is warranted*.

    All who practice music will perform well.
    All who perform well will be applauded.
    You practiced music.

    A. You are not applauded.   B. You did not perform well.
    C. You were applauded.   D. None of the above is warranted.

9. PRACTICE CLAST: FORM B

Limit your time to 90 minutes. Choose the *one* correct answer.

1. $6\frac{3}{4} - 4\frac{7}{8} =$

    A. $2\frac{7}{8}$ B. $1\frac{7}{8}$ C. $2\frac{1}{8}$ D. $2\frac{1}{2}$

2. $\frac{3}{8} \times (-6\frac{1}{3}) =$

    A. $2\frac{3}{8}$ B. $-6\frac{1}{8}$ C. $-2\frac{3}{8}$ D. $3\frac{1}{8}$

3. $5.709 - 7.358 =$

    A. 1.649 B. 2.649 C. −1.658 D. −1.649

4. If you increase 30 by 40% of itself, what is the result?

    A. 12 B. 52 C. 42 D. 30.12

5. 42 is 15% of what number?

    A. 630 B. 63 C. 28 D. 280

6. Round off to the nearest tenth of a meter: 52.74 m

    A. 52.7 m B. 52.8 m C. 52 m D. 52.75 m

7. Find the total surface area of a rectangular solid that is 12 inches long, 8 inches wide and 6 inches high.

    A. 576 cubic inches B. 432 cubic inches C. 216 square inches D. 432 square inches

8. What is the volume of a right circular cone that is 15 cm high and has a base diameter of 10 cm?

    A. 375π cubic cm B. 125π cubic cm C. 25π cubic cm D. 500π cubic cm

9. $\sqrt{3} \times \sqrt{18} =$

    A. $3\sqrt{2}$ B. $27\sqrt{2}$ C. $9\sqrt{6}$ D. $3\sqrt{6}$

10. $14y - 5y \cdot 2y + 16y + 8 =$

    A. $-10y^2 + 16y$ B. $18y^2 + 2y$ C. $6y$ D. $10y^2 - 16y$

11. $(7.1 \times 10^3) \times (1.4 \times 10^{-5}) =$

    A. 994 B. 0.0994 C. 0.994 D. −994

12. Solve the inequality: $4x - 3 \geq 7x + 2$

    A. $x \leq \frac{-5}{3}$ B. $x \geq \frac{-5}{3}$ C. $x \leq \frac{5}{3}$ D. $x \leq \frac{1}{3}$

13. Given $y = (x + 1)^2 - x$, if $x = -3$, then $y =$

    A. 13 B. 1 C. 7 D. 19

14. Find $f(-2)$ given $f(x) = -3x^3 - 2x^2 - 1$

    A. 31 B. 15 C. -33 D. 9

15. Which is a linear factor of the expression $8x^2 - 2x - 3$?

    A. $2x - 1$ B. $8x - 3$ C. $4x + 3$ D. $4x - 3$

16. Find the real roots of this equation: $5x^2 + 3x - 1 = 0$

    A. $\dfrac{-3 + \sqrt{14}}{10}$ and $\dfrac{-3 - \sqrt{14}}{10}$
    B. $\dfrac{3 + \sqrt{14}}{10}$ and $\dfrac{3 - \sqrt{14}}{10}$
    C. $\dfrac{-3 + \sqrt{29}}{10}$ and $\dfrac{-3 - \sqrt{29}}{10}$
    D. $\dfrac{3 + \sqrt{29}}{10}$ and $\dfrac{3 - \sqrt{29}}{10}$

17. Find the real roots of this equation: $6x^2 - 7 = 19x$

    A. $\dfrac{-7}{3}$ and $\dfrac{1}{2}$ B. $\dfrac{7}{2}$ and $\dfrac{-1}{3}$ C. $\dfrac{-7}{2}$ and $\dfrac{-1}{3}$ D. $\dfrac{7}{3}$ and $\dfrac{-1}{2}$

18. Choose the correct solution set for the system of linear equations:

    $$\begin{cases} 2x + y = 6 \\ 2y = 8 - 4x \end{cases}$$

    A. $\{(0, -4)\}$ B. $\{(-2, 10)\}$ C. $\{(5, -4)\}$ D. The empty set

19. The graph below represents the number of cars sold in the first six months of 1988 by a Chevrolet dealer in Orlando, Florida. In what month did the number of cars sold differ the most from the number of cars sold in March?

    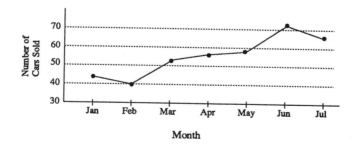

    A. February B. May C. June D. July

20. What is the median of the data in the following sample? 7, 13, 6, 19, 11, 13

    A. 11 B. 12 C. 13 D. 11.5

20. What is the median of the data in the following sample? 7, 13, 6, 19, 11, 13

    A. 11   B. 12   C. 13   D. 11.5

21. A shelf has 6 library books. In how many ways can a person select four books from the shelf?

    A. 360   B. 30   C. 24   D. 15

22. Sets X, Y, Z, and U are related as in the diagram. Which of the following statements is true assuming none of the eight regions is empty?

    A. Some elements of set X are not in set U.
    B. Every element that is in both sets X and Y is also in set Z.
    C. Some element of set X is in both sets Y and Z.
    D. No element lies outside of all three of the sets X, Y, and Z.

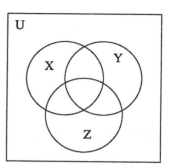

23. $(5^2)^3 =$

    A. 5^5 B. $5^2 \times 5^2 \times 5^2$ C. $(5 \times 2)^3$ D. $5^2 \times 5^3$

24. Select the numeral for $(4 \times 10^2) + (1 \times 10^1) + (4 \times \frac{1}{10^1}) + (5 \times \frac{1}{10^3})$

    A. 41.45   B. 41.405   C. 410.405   D. 401.4005

25. 6.25% =

    A. $\frac{625}{100}$ B. $\frac{1}{16}$ C. $\frac{5}{8}$ D. 0.625

26. Select the correct relation: $\frac{-7}{9} \square \frac{-3}{4}$

    A. <   B. >   C. =

27. Two hundred students took a history test. All of the students scored less than 88, but more than 60. Which of the following values is a reasonable estimate of the average score of the students?

    A. 76   B. 88   C. 58   D. 90

28. Which of the statements below is true for the figure shown, given that l_1 and l_2 are both perpendicular to l_3? (Angle measures are referred to by x, y, z, etc.)

A. $v + z = 240°$
B. $x = y$
C. $x = 140°$
D. $w + y = 140°$

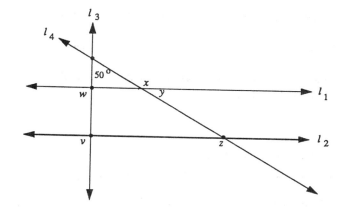

29. Study the figures in A, B, C, and D. Select the figure in which all the triangles are similar.

A.   B.    C. 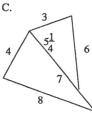   D.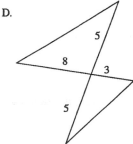

30. Which of the following could *not* be used to report the area of a circular lake?

    A. square miles   B. kilometers   C. acres   D. square meters

31. Choose the expression equivalent to the following: $6wz(4w + z^2)$

    A. $24w^2z + z^2$ B. $24wz + 6wz^2$
    C. $24wz + 6wz^3$ D. $24w^2z + 6wz^3$

32. For each of the statements below determine whether -7 is a solution.

    i. $|t + 9| = 2$
    ii. $(x - 7)(x + 2) \geq 0$
    iii. $2y - 6 = -2(y + 3) - 28$

Which option below identifies every statement that has -7 as a solution and only those statements?

    A. ii and iii only   B. i and iii only   C. i and ii only   D. All of the statements

33. A surveyor uses three rolls of string to mark off eight blocks of equal dimensions. Let w = the number of rolls of string to mark off 11 such blocks. Select the correct statement.

   A. $\dfrac{8}{3} = \dfrac{w}{11}$ B. $\dfrac{8}{w} = \dfrac{3}{11}$ C. $\dfrac{8}{3} = \dfrac{11}{w}$ D. $\dfrac{3}{w} = \dfrac{11}{8}$

34. Which option gives the conditions that correspond to the shaded regions of the plane shown in the figure?

   A. $x - y < 2$ and $|x| < 2$
   B. $x - y > 2$ and $|x| < 2$
   C. $x - y < 2$ and $|x| > 2$
   D. $x - y > 2$ and $|x| > 2$

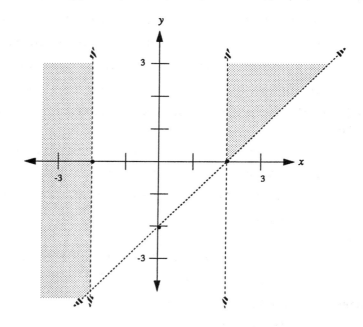

35. On a science test, 40% of the students scored 70, 20% of the students scored 90, and the remaining students scored less than 70 with no two of these students scoring the same. Which of the following is true about the distribution of scores?

   A. The mode is less than the median   B. The mode is equal to the median
   C. The mean is greater than the median   D. The mean is greater than the mode

36. K-Mart wants to determine the average expenditure per Christmas gift of the households in Sanford, Florida. To do so members of the store's marketing department will conduct a survey. Which of the following procedures would be most appropriate for selecting a statistically unbiased sample?

   A. Randomly select a week in November and survey all persons who shop K-Mart during that week.

   B. Ask customers in December to voluntarily complete a questionnaire.

   C. Randomly select several letters of the alphabet and survey by telephone all persons in the telephone directory whose last name begins with one of the letters.

   D. Randomly select households from the post office listing of addresses and survey members of the selected households.

37. Ten percent of the baseballs used in a professional baseball game are defective. If two balls are randomly selected with replacement from the balls used in the game, find the probability that both of them are defective.

    A. $\dfrac{1}{5}$ B. $\dfrac{1}{100}$ C. $\dfrac{1}{10}$ D. $\dfrac{1}{90}$

38. Select the statement that is the negation of the statement *Bob is a lawyer and Bob is a real estate broker.*

    A. Bob is not a lawyer or Bob is not a real estate broker.
    B. Bob is not a lawyer and Bob is not a real estate broker.
    C. Bob is not a lawyer and Bob is a real estate broker.
    D. Bob is a lawyer and Bob is not a real estate broker.

39. Determine which statement is *not* logically equivalent to *If you study hard then you will not have a difficult time.*

    A. If you have a difficult time, then you did not study hard.
    B. If you do not have a difficult time, then you studied hard.
    C. You do not study hard or you do not have a difficult time.
    D. You do not have a difficult time or you do not study hard.

40. Given that
    i. *Mary obtained the job or John was fired.*
    ii. *John was not fired.*

    determine which conclusion can be logically deduced.

    A. Mary did not obtain the job.
    B. Mary obtained the job and John was fired.
    C. Mary obtained the job.
    D. None of the conclusions can be logically deduced.

41. All of the following have true conclusions. Which of the following also has a valid argument?

    A. Some cats have hair and all mammals have hair. Therefore, some cats are mammals.
    B. All mammals have hair and all people have hair. Therefore, all people are mammals.
    C. All mammals have hair and all cats are mammals. Therefore, all cats have hair.
    D. All reptiles are cold blooded and all snakes are cold blooded. Therefore, all snakes are reptiles.

42. Select the next term in the following geometric progression.

    $$\dfrac{-1}{25}, \dfrac{1}{5}, -1, 5, -25, \ldots$$

    A. -5 B. -125 C. 125 D. 625

43. Let c = the length of the hypotenuse.

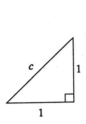
First Triangle
$c = \sqrt{2}$

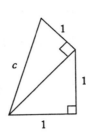
Second Triangle
$c = \sqrt{3}$

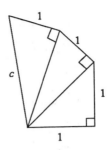
Third Triangle
$c = 2$

Give the length of the hypotenuse of the tenth triangle in the above construction.

A. 3 B. $\sqrt{11}$ C. 11 D. $\sqrt{10}$

44. Select the formula for calculating the total surface area (SA) of a regular triangular pyramid.

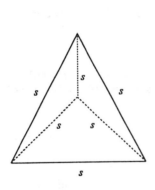
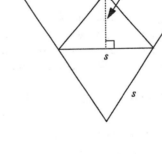

A. $SA = \dfrac{s^2}{4}\sqrt{3}$ B. $SA = s^2\sqrt{3}$ C. $SA = 2s^2\sqrt{3}$ D. $SA = 4s^2$

45. Choose the inequality equivalent to the following: $-8 < -2x + 2 < 5$

A. $-10 < -2x < 7$ B. $4 > x - 1 > \dfrac{-5}{2}$

C. $4 < x - 1 < \dfrac{-5}{2}$ D. $-10 > -2x > 3$

46. The size of a shoe store's sales force and its yearly sales revenue are given below for five consecutive years.

Year	Number of Salespersons	Sales (Thousands of Dollars)
1987	2	18.4
1988	3	19.6
1989	4	20.1
1990	6	24.4
1991	8	32.9

Which of the following best describes the relationship between the sales force and yearly sales?

A. Increasing the number of salespersons caused the increased sales.

B. Increased sales caused the shoe store to hire more salespersons.

C. There appears to be a positive association between the size of the sales force and yearly sales revenue.

D. The size of the sales force does not provide information needed to predict yearly sales revenue.

47. Select the conclusion that will make the following argument valid.

If Andrea washes the cars, then she will be rewarded.
Andrea was not rewarded.

A. Andrea washed the cars.
B. Andrea did not wash the cars because she was not rewarded.
C. If Andrea does not wash the cars, then she will not be rewarded.
D. Andrea did not wash the cars.

48. Select the rule of logical equivalence that directly (in one step) transforms statement "i" into statement "ii".

  i. *It is not true that all snakes have rattles.*

  ii. *Some snakes do not have rattles.*

A. *Not (not P)* is equivalent to *P*.

B. *Not (Some are P)* is equivalent to *All are not P*.

C. *Not (All are P)* is equivalent to *Some are not P*.

D. *All are not P* is equivalent to *None are P*.

49. A tour bus has 60 seats at $20 per seat. It costs $6 per seat to operate the bus. On a given tour there were 18 empty seats. What were the company's profits on this tour?

A. $360   B. $480   C. $840 $D. $1092

50. Find the smallest positive multiple of five that leaves a remainder of three when divided by four, and a remainder of one when divided by three.

A. 35   B. 45   C. 55   D. 75

51. Find the cost of painting the four inside walls of a gymnasium if the inside of the gymnasium is 60 feet long, 40 feet wide, and 20 feet high, and the cost of painting one square foot is 15 cents.

    A. $600   B. $300   C. $7200   D. $1320

52. Mary follows the jogging path (shown in the figure below) from A to B to C to D to E to F and then usually reverses the path to get back to A. One day she was tired so on her return trip she walked directly from point F to point A. How many feet did she save?

    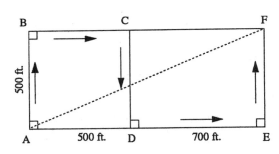

    A. 1300 ft.   B. 2700 ft.   C. 2600 ft.   D. 1400 ft.

53. An equation for centripetal force is $F = \dfrac{mv^2}{r}$. A car of mass $M = 1200$ kilograms, traveling at a constant speed of $v = 8$ meters per second, travels around a circular track of radius $r = 30$ meters. What is the force (F) of the car as it travels around the track? (Force is measured in Newtons.)

    A. 320 Newtons   B. $562\dfrac{1}{2}$ Newtons   C. 2560 Newtons   D. $10\dfrac{2}{3}$ Newtons

54. A Southern airport has collected data on all of its flights for twenty years. The following is a distribution of the causes of accidents occurring in the twenty year period. No accident has more than one cause. What is the probability that an accident was caused by either weather conditions or pilot error?

Cause	Percent of all Accidents
Weather Conditions	42
Equipment Malfunction	38
Pilot Error	12
Traffic Controller Error	6
Other	2

    A. .504
    B. .54
    C. .46
    D. .58

55. If a logical conclusion is given, select that conclusion; otherwise select *none of the above is warranted*.

    If you drink carbonated soda, you drink Pepsi.
    If you drink Pepsi, then you get indigestion.
    You get indigestion.

    A. You drink Pepsi and you drink carbonated soda.
    B. You do not drink Pepsi.
    C. You do not drink carbonated soda.
    D. None of the above is warranted.

10. PRACTICE CLAST ANSWERS

Answers to Practice CLAST/Form A

1. C	2. D	3. D	4. C	5. C	6. C	7. C	8. B	9. A	10. D
11. C	12. B	13. D	14. B	15. A	16. C	17. B	18. B	19. B	20. A
21. B	22. D	23. B	24. B	25. C	26. C	27. D	28. B	29. B	30. C
31. C	32. C	33. B	34. C	35. D	36. B	37. A	38. C	39. D	40. A
41. C	42. D	43. D	44. C	45. B	46. C	47. B	48. B	49. A	50. C
51. C	52. D	53. D	54. D	55. C					

Answers to Practice CLAST/Form B

1. B	2. C	3. D	4. C	5. D	6. A	7. D	8. B	9. D	10. A
11. B	12. A	13. C	14. B	15. D	16. C	17. B	18. D	19. C	20. B
21. D	22. C	23. B	24. C	25. B	26. A	27. A	28. C	29. C	30. B
31. D	32. D	33. C	34. C	35. B	36. D	37. B	38. A	39. B	40. C
41. C	42. C	43. B	44. B	45. B	46. C	47. D	48. C	49. B	50. C
51. A	52. D	53. C	54. B	55. D					

APPENDIX

A. KEY TO NOTATION AND SYMBOLS

{ }	braces, used to enclose members of a set	iff	if and only if
$\in$	"is an element of"	$\equiv$	is the same as
$\notin$	"is not an element of"	P(A)	the probability that event A will occur
$\cdots$	proceed in the indicated pattern	$\left\{\begin{array}{c} A:B \\ A \text{ to } B \\ \dfrac{A}{B} \end{array}\right\}$	ratio of A and B
$\emptyset$	the empty set; also denoted by { }	n!	n factorial, which is equal to $n(n-1)(n-2)\ldots(3)(2)(1)$
$\subset$	"is a proper subset of"	$_nP_r$	the number of permutations of n things taken r at a time
$\subseteq$	"is a subset of"	$_nC_r$	the number of combinations of n things taken r at a time
$\not\subseteq$	"is not a subset of"	$\bar{x}$	mean (read "x bar")
U	the universal set	a^2	a squared
A'	the complement of A	$\sqrt{a}$	the positive square root of a
$\cap$	intersection	σ	standard deviation
$\cup$	union	$\bullet\, C$	point C
n(A)	the cardinal number of set A	$\overleftrightarrow{AB}$	line AB
$A \times B$	the Cartesian product of sets A and B	$\overrightarrow{AB}$	ray AB
(a,b)	the ordered pair a and b	$\overline{AB}$	line segment AB
=	"is equal to"	$\overset{\rightarrow}{A\overset{\circ}{B}}$	half line AB
$P \wedge Q$	P and Q	$\angle RST$	angle RST
$P \vee Q$	P or Q	$\overset{\frown}{TR}$	open line segment TR
$P \underline{\vee} Q$	P or Q, but not both	$\overleftrightarrow{MN} \perp \overleftrightarrow{BT}$	line MN is perpendicular to line BT
$P \to Q$	if P, then Q	$m(\overline{AB})$	measure of line segment AB
$P \leftrightarrow Q$	P if and only if Q		
$\sim P$	$\left\{\begin{array}{l}\text{not P}\\ \text{it is false that P}\\ \text{it is not the case that P}\end{array}\right.$		

B. TABLES

B.1 FACTORIALS

n	n!
0	1
1	1
2	2
3	6
4	24
5	120
6	720
7	5,040
8	40,320
9	362,880
10	3,628,800
11	39,916,800
12	479,001,600
13	6,227,020,800
14	87,178,291,200
15	1,307,674,368,000

B.2 AREAS OF THE STANDARD NORMAL DISTRIBUTION

Amount of Standard Deviation from Mean (one direction only)	Percentage of Data	Amount of Standard Deviation from Mean (one direction only)	Percentage of Data
0.1	4.0	1.6	44.5
0.2	7.9	1.7	45.5
0.3	11.8	1.8	46.4
0.4	15.5	1.9	47.1
0.5	19.2	2.0	47.7
0.6	22.6	2.1	48.2
0.7	25.8	2.2	48.6
0.8	28.8	2.3	48.9
0.9	31.6	2.4	49.2
1.0	34.1	2.5	49.4
1.1	36.4	2.6	49.5
1.2	38.5	2.7	49.65
1.3	40.3	2.8	49.7
1.4	41.9	2.9	49.8
1.5	43.3	3.0	49.87

B.3 SQUARES, SQUARE ROOTS, AND PRIME FACTORIZATIONS

n	n²	√n	Prime factoriztaion	n	n²	√n	Prime factoriztaion
1	1	1.000	—	51	2,601	7.141	3·17
2	4	1.414	prime	52	2,704	7.211	2·2·13
3	9	1.732	prime	53	2,809	7.280	prime
4	16	2.000	2·2	54	2,916	7.348	2·3·3·3
5	25	2.236	prime	55	3,025	7.416	5·11
6	36	2.449	2·3	56	3,136	7.483	2·2·2·7
7	49	2.646	prime	57	3,249	7.550	3·19
8	64	2.828	2·2·2	58	3,364	7.616	2·29
9	81	3.000	3·3	59	3,481	7.681	prime
10	100	3.162	2·5	60	3,600	7.746	2·2·3·5
11	121	3.317	prime	61	3,721	7.810	prime
12	144	3.464	2·2·3	62	3,844	7.874	2·31
13	169	3.606	prime	63	3,969	7.937	3·3·7
14	196	3.742	2·7	64	4,096	8.000	2·2·2·2·2·2
15	225	3.873	3·5	65	4,225	8.062	5·13
16	256	4.000	2·2·2·2	66	4,356	8.124	2·3·11
17	289	4.123	prime	67	4,489	8.185	prime
18	324	4.243	2·3·3	68	4,624	8.246	2·2·17
19	361	4.359	prime	69	4,761	8.307	3·23
20	400	4.472	2·2·5	70	4,900	8.367	2·5·7
21	441	4.583	3·7	71	5,041	8.426	prime
22	484	4.690	2·11	72	5,184	8.485	2·2·2·3·3
23	529	4.796	prime	73	5,329	8.544	prime
24	576	4.899	2·2·2·3	74	5,476	8.602	2·37
25	625	5.000	5·5	75	5,625	8.660	3·5·5
26	676	5.099	2·13	76	5,776	8.718	2·2·19
27	729	5.196	3·3·3	77	5,929	8.755	7·11
28	784	5.292	2·2·7	78	6,084	8.832	2·3·13
29	841	5.385	prime	79	6,241	8.888	prime
30	900	5.477	2·3·5	80	6,400	8.944	2·2·2·2·5
31	961	5.568	prime	81	6,561	9.000	3·3·3·3
32	1,024	5.657	2·2·2·2·2	82	6,724	9.055	2·41
33	1,089	5.745	3·11	83	6,889	9.110	prime
34	1,156	5.831	2·17	84	7,056	9.165	2·2·3·7
35	1,225	5.916	5·7	85	7,225	9.220	5·17
36	1,296	6.000	2·2·3·3	86	7,396	9.274	2·43
37	1,369	6.083	prime	87	7,569	9.327	3·29
38	1,444	6.164	2·19	88	7,744	9.381	2·2·2·11
39	1,521	6.245	3·13	89	7,921	9.434	prime
40	1,600	6.325	2·2·2·5	90	8,100	9.487	2·3·3·5
41	1,681	6.403	prime	91	8,281	9.539	7·13
42	1,764	6.481	2·3·7	92	8,464	9.592	2·2·23
43	1,849	6.557	prime	93	8,649	9.644	3·31
44	1,936	6.633	2·2·11	94	8,836	9.695	2·47
45	2,025	6.708	3·3·5	95	9,025	9.747	5·19
46	2,116	6.782	2·23	96	9,216	9.798	2·2·2·2·2·3
47	2,209	6.856	prime	97	9,409	9.849	prime
48	2,304	6.928	2·2·2·2·3	98	9,604	9.899	2·7·7
49	2,401	7.000	7·7	99	9,801	9.950	3·3·11
50	2,500	7.071	2·5·5	100	10,000	10.000	2·2·5·5

B.4 ENGLISH-METRIC CONVERSION FACTORS

B.4.1 Length and Area English-Metric Conversions

ENGLISH TO METRIC		
When you know	**Multiply by**	**To find**
inches (in.)	2.56	centimeters (cm)
feet (ft.)	30.0	centimeters (cm)
yards (yd.)	0.9	meters (m)
miles (mi.)	1.6	kilometers (km)
square inches ($in.^2$)	6.5	square centimeters (cm^2)
square feet ($ft.^2$)	0.09	square meters (m^2)
square yards ($yd.^2$)	0.8	square meters (m^2)
acres	0.4	hectares (ha)
METRIC TO ENGLISH		
When you know	**Multiply by**	**To find**
millimeters (mm)	0.04	inches (in.)
centimeters (cm)	0.4	inches (in.)
meters (m)	3.3	feet (ft.)
meters (m)	1.1	yards (yd.)
kilometers (km)	0.6	miles (mi.)
square centimeters (cm^2)	0.16	square inches ($in.^2$)
square meters (m^2)	1.2	square yards ($yd.^2$)
hectares (ha)	2.5	acres

B.4.2 Capacity English-Metric Conversions

ENGLISH TO METRIC		
When you know	**Multiply by**	**To find**
teaspoons (tsp.)	5	milliliters (ml)
tablespoons (tbsp.)	15	milliliters (ml)
fluid ounces (fl. oz.)	30	milliliters (ml)
cups (c.)	0.24	liters (l)
pints (pt.)	0.47	liters (l)
quarts (qt.)	0.95	liters (l)
gallons (gal.)	3.8	liters (l)
cubic feet (cu. ft.)	0.03	cubic meters (cu. m)
cubic yards (cu. yd.)	0.76	cubic meters (cu. m)
METRIC TO ENGLISH		
When you know	**Multiply by**	**To find**
milliliters (ml)	0.03	fluid ounces (fl. oz.)
liters (l)	2.1	pints (pt.)
liters (l)	1.06	quarts (qt.)
liters (l)	0.26	gallons (gal.)
cubic meters (cu. m)	35	cubic feet (cu. ft.)
cubic meters (cu. m)	1.3	cubic yards (cu. yd.)

B.4.3 Mass English-Metric Conversions

ENGLISH TO METRIC		
When you know	Multiply by	To find
ounces (oz.)	28	grams (g)
pounds (lb.)	0.45	kilograms (kg)
tons (T.)	0.9	tonnes (t)
METRIC TO ENGLISH		
When you know	Multiply by	To find
grams (g)	0.035	ounces (oz.)
kilograms (kg)	2.2	pounds (lb.)
tonnes (t)	1.1	tons (T.)

B.4.4 Fahrenheit to Celsius Conversion

When you know	Subtract	Then multiply by	To find
°F	32	$\frac{5}{9}$	°C

Note: This is written as $C = \frac{5}{9}(F - 32)$

B.4.5 Celsius to Fahrenheit Conversion

When you know	Multiply	Then add by	To find
°C	$\frac{9}{5}$	32	°F

Note: This is written as $F = (\frac{9}{5} \times C) + 32$

B.5 MONTHLY MORTGAGE PAYMENTS

	Length of Mortgage		
Rate (%)	10 Years	20 Years	30 Years
8	$12.13	$8.36	$7.34
9	12.67	9.00	8.05
10	13.22	9.65	8.78
10½	13.49	9.98	9.15
12	14.35	11.01	10.29
12½	14.64	11.36	10.67
14	15.53	12.44	11.85
15	16.13	13.17	12.64
16	16.75	13.91	13.45
18	18.02	15.43	15.07
20	19.33	16.99	16.71

B.6 COMPOUND AMOUNT OF $1.00

Periods	\multicolumn{9}{c}{Interest Rate per Period}
	2%	3%	4%	6%	8%	10%	12%	14%	16%
1	1.020	1.030	1.040	1.060	1.080	1.100	1.120	1.140	1.160
2	1.040	1.061	1.082	1.124	1.166	1.210	1.254	1.300	1.346
4	1.082	1.126	1.170	1.262	1.360	1.464	1.574	1.689	1.811
6	1.126	1.194	1.265	1.419	1.587	1.772	1.974	2.195	2.436
8	1.172	1.267	1.369	1.594	1.851	2.144	2.476	2.853	3.278
10	1.219	1.344	1.480	1.791	2.159	2.594	3.106	3.707	4.411
12	1.268	1.426	1.601	2.012	2.518	3.138	3.896	4.818	5.936
14	1.319	1.513	1.732	2.261	2.937	3.797	4.887	6.261	7.988
16	1.373	1.605	1.873	2.540	3.426	4.595	6.130	8.137	10.748
20	1.486	1.806	2.191	3.207	4.661	6.728	9.646	13.743	19.461
24	1.608	2.033	2.563	4.049	6.341	9.850	15.179	23.212	35.236
28	1.741	2.288	2.999	5.112	8.627	14.421	23.884	39.204	63.800

B.7 ANNUAL PREMIUM RATE PER $1,000 OF LIFE INSURANCE

Age	5-Year Term	Straight Life	Limited Payment (20-year)	Endowment (20-year)
20	$6.99	$15.48	$25.96	49.32
25	7.57	17.50	28.83	49.73
30	8.60	20.15	32.10	50.17
35	9.98	23.84	35.92	51.13
40	12.44	28.37	40.74	52.69
45	15.59	33.90	46.57	55.33
50	19.89	41.23	53.01	59.00
55	27.09	50.88	59.73	64.55
60	—	63.95	70.00	—

B.8 PREMIUM RATES

Payment Period	Percentage of Annual Premium
Semiannually	51%
Quarterly	26%
Monthly	9%

C. GEOMETRY SUMMARY

C.1 TRIANGLE CLASSIFICATION CHART

Type of Classification	Name of Triangle	Description	Sample Figure
Triangles classified according to the characteristics of their sides. (**Note:** Hash marks are used in the sample figures to represent sides that are of equal length.)	Scalene	No sides are equal	
	Isosceles	Two sides are equal	
	Equilateral	All three sides are equal	
Triangles classified according to the characteristics of their angles. (**Note:** The sum of the measures of the interior angles of a triangle is equal to 180°.)	Acute	All three angles are acute	
	Obtuse	One angle is obtuse	
	Right	One angle is a right angle	
	Equiangular	All three angles are of equal measure	

C.2 QUADRILATERAL CLASSIFICATION CHART

Name of Quadrilateral	Description	Sample Figure
Trapezoid	• Two sides are parallel In the sample figure $\overline{AB} \parallel \overline{CD}$.	
Isosceles Trapezoid	• Two sides are parallel • Two nonparallel sides are equal in length In the sample figure $\overline{AB} \parallel \overline{CD}$. Also, $m(\overline{AC}) = m(\overline{BD})$.	
Parallelogram	• Both pairs of opposite sides are parallel In the sample figure $\overline{AB} \parallel \overline{CD}$, and $\overline{AC} \parallel \overline{BD}$.	
Rhombus	• Both pairs of opposite sides are parallel • The adjacent sides are of equal length, which implies that all four sides are of equal length. In the sample figure $\overline{AB} \parallel \overline{CD}$, and $\overline{AC} \parallel \overline{BD}$. Also, $m(\overline{AB}) = m(\overline{BD}) = m(\overline{CD}) = m(\overline{AC})$.	
Rectangle	• Both pairs of opposite sides are parallel • One right angle, which implies that all the angles are right angles In the sample figure $\overline{AB} \parallel \overline{CD}$, and $\overline{AC} \parallel \overline{BD}$. Also, the measure of all angles is 90°.	
Square	• All four sides are of equal length • All four interior angles are of equal measure, which implies that all angles are right angles and that the opposite sides are parallel.	

C.3 COMMON PERIMETER AND AREA FORMULAS

Name of Figure	Formulas (P = Perimeter) (A = Area)	Sample Figure
Triangle	$P = a + b + c$ $A = \frac{1}{2}bh$	
Square	$P = 4s$ $A = s^2$	
Trapezoid	$A = \frac{1}{2}h(b_1 + b_2)$	
Rectangle	$P = 2l + 2w$ $A = lw$ or bh	
Circle	Circumference $= 2\pi r = \pi d$ $A = \pi r^2$	
Parallelogram	$P = 2a + 2b$ $A = bh$	

C.4 COMMON VOLUME FORMULAS

Name of Figure	Volume Formula	Sample Figure
Rectangular Solid	$V = lwh$	
Right Circular Cylinder	$V = \pi r^2 h$	
Right Circular Cone	$V = \frac{1}{3}\pi r^2 h$	
Sphere	$V = \frac{4}{3}\pi r^3$	

ANSWERS

CHAPTER 1

Answers to Supplementary Exercise 1.2

1. {m,u,s,k}   2. {Tuesday, Thursday}   3. {2,3,5,8}   4. {1,3,9}   5. {April, June, September, November}

6. {1st, 2nd, 3rd}   7. {0,1,2,3,4,5,6,7,8,9}   8. {2}   9. $\{x \mid x$ is a multiple of 5 from 5 to 100$\}$

10. $\{x \mid x$ is a day of the week that begins with the letter S$\}$ 11. $\{x \mid x$ is one of the three rs$\}$

12. True   13. False   14. True   15. True   16. True   17. True   18. True
19. True   20. False   21. True   22. True   23. False   24. True   25. False

Answers to Supplementary Exercise 1.3

1. False   2. True   3. True   4. True   5. True   6. True   7. True
8. False   9. False   10. False   11. True   12. True   13. False   14. False
15. True   16. False   17. True   18. True   19. True   20. False
21. {2,4,6,8,10}   22. {1,3,5,7,9}   23. {1,2,3,10}   24. {8,9,10}

25. { }, {2}, {8}, {9}, {2,8}, {2,9}, {8,9}, {2,8,9}

Answers to Supplementary Exercise 1.4

1. {1}   2. {4,8}   3. {3,7}   4. { }   5. {1,2,3,4,5,6,7,8}
6. {1,3,4,5,6,7,8,9,10}   7. {3,5,6,7,9,10}   8. {1,2,3,5,7,9,10}   9. {4,5,6,8}   10. {1,2}
11. {1,2,3,5,6,7,9,10}   12. U   13. { }   14. A'   15. {1,4,8}
16. {1,2,4,8}   17. {1,3,4,5,6,7,8,9,10}   18. {1,3,7}   19. {5,6}   20. {3,7}

Answers to Supplementary Exercise 1.5

1. I, II, IV   2. II, III, IV   3. I, II, III, V, VIII   4. I, II, III, IV, VI, VII, VIII
5. VIII   6. I, II, III, VII, VIII   7. IV, VI, VII   8. IV, V, VI, VII, VIII
9. ∅   10. I, II, IV, V, VII, VIII   11. 16   12. 60
13. 41   14. 37   15. 3   16. 18   17. 47   18. 1   19. 1
20. 0   21. True   22. True   23. False   24. False   25. False   26. False

Answers to Supplementary Exercise 1.6

1		2		3		4		5		6		7		8	
a.	3	a.	48	a.	12	a.	8	a.	311	a.	0	a.	47	a.	3
b.	5	b.	1	b.	18	b.	3	b.	5,630	b.	7	b.	4	b.	3
c.	9	c.	5	c.	4	c.	18	c.	6,738	c.	4	c.	10	c.	4
d.	65	d.	4			d.	58	d.	5,395	d.	9			d.	9

Answers to Supplementary Exercise 1.7

1. A × B = {(Pepper,Kit), (Pepper,Claudia), (Spike,Kit), (Spike,Claudia)}
   B × A = {(Kit,Pepper), (Kit,Spike), (Claudia,Pepper), (Claudia,Spike)}
   $n(A \times B) = 4$

2. A × B = {(he,see),(he,be),(he,E),(me,see),(me,be),(me,E),(fee,see),(fee,be),(fee,E)}
   B × A = {(see,he),(see,me),(see,fee),(be,he),(be,me),(be,fee),(E,he),(E,me),(E,fee)}
   $n(A \times B) = 9$

3. A × B = {(AB,x),(AB,y),(AB,z),(C,x),(C,y),(C,z)}
   B × A = {(x,AB),(x,C),(y,AB),(y,C),(z,AB),(z,C)}
   $n(A \times B) = 6$

4. A × B = {(R,yes),(U,yes),(E,yes),(Z,yes),(?,yes),(R,no),(U,no),(E,no),(Z,no),(?,no)}
   B × A = {(yes,R),(yes,U),(yes,E),(yes,Z),(yes,?),(no,R),(no,U),(no,E),(no,Z),(no,?)}
   $n(A \times B) = 10$

5. A = {a,e,i,o,u} and B = {8,9,10}

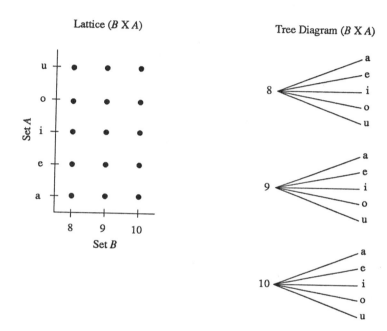

6. B ∩ C = {3,8} and A × (B ∩ C) = {(2,3),(2,8),(4,3),(4,8)}
7. (B ∪ C) = {1,3,8} and (B ∪ C) × A = {(1,2),(1,4),(3,2),(3,4),(8,2),(8,4)}
8. A ∩ B = ∅ and (A ∩ B) × C = ∅
9. (A ∪ C) = {1,2,3,4,8}, (B ∩ C) = {3,8}, and
   (A ∪ C) × (B ∩ C) = {(1,3)(1,8),(2,3),(2,8),(3,3),(3,8),(4,3),(4,8),(8,3),(8,8)}
10. $n(A \times B \times C) = 12$

Answers to Chapter Test

1. True   2. False   3. False   4. False   5. True   6. True
7. False   8. True   9. True   10. False   11. False   12. False

13. $2^3 = 8$; ∅,{e},{f},{g},{ef},{eg},{fg},{efg}   14. Finite   15. Infinite

16. Region III of a two-set Venn diagram should be shaded
17. Region VII of a three-set Venn diagram should be shaded

18. {1,2,3,4,7,8,9,10} 19. {5} 20. {2,3,4,5,6,7,8}
21. {1,4,5,9} 22. 7 23 {(5,4),(5,5),(6,4),(6,5),(7,4)(7,5),(8,4),(8,5)}

24 a.

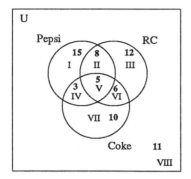

b. 70   c. 28   d. 39

CHAPTER 2

Answers to Supplementary Exercise 2.2

1. P → ~Q   2. P V Q   3. ~P ∧ ~Q   4. ~(P V Q)
5. Q ∧ ~ P   6. P ↔ Q   7. ~P → Q   8. P V ~Q

9. Shannon loves chicken and Pepper does not need to lose weight.
10. It is not the case that if Shannon loves chicken then Pepper needs to lose weight.
11. Shannon does not love chicken iff Pepper needs to lose weight.
12. Shannon does not love chicken or Pepper needs to lose weight.
13. If Pepper does not need to lose weight then Shannon loves chicken.
14. It is false that Shannon loves chicken and Pepper needs to lose weight.
15. Either Pepper needs to lose weight or Shannon loves chicken.
16. Pepper needs to lose weight iff Shannon loves chicken.

Answers to Supplementary Exercise 2.3

1. (P → ~Q) ∧ R   2. none needed   3. ~P V (Q ∧ R)   4. Q V (R → ~P)
5. P V Q ↔ R   6. ~(P ∧ R)   7. ~Q ∧ R → ~P   8. (P V Q) ∧ R
9. ~(Q ↔ ~P)   10. R → P V Q   11. ~P ∧ ~Q

12. It rained today or July is not a hot month, iff I got wet.
13. It is false that it rained today or I got wet.
14. If it did not rain today or I got wet, then July is not a hot month.
15. It rained today or I got wet, and July is a hot month.
16. It did not rain today,, or if I got wet then July is a not a hot month.

Answers to Supplementary Exercise 2.4-2.5

1.

P	Q	~Q	~Q → P
T	T	F	T
T	F	T	T
F	T	F	T
F	F	T	F
1	2	3	4

2.

P	Q	P V Q	~(P V Q)
T	T	T	F
T	F	T	F
F	T	T	F
F	F	F	T
1	2	3	4

3.

P	Q	~Q	~Q ∧ P	P ↔ ~Q ∧ P
T	T	F	F	F
T	F	T	T	T
F	T	F	F	T
F	F	T	F	T
1	2	3	4	5

4.

P	Q	~P	~P → Q	(~P → Q) V P
T	T	F	T	T
T	F	F	T	T
F	T	T	T	T
F	F	T	F	F
1	2	3	4	5

5.

P	Q	~P	Q ∧ ~P	P V (Q ∧ ~P)
T	T	F	F	T
T	F	F	F	T
F	T	T	T	T
F	F	T	F	F
1	2	3	4	5

6.

P	Q	~Q	~P	P ∧ ~Q	~P V Q	P ∧ ~Q → ~P V Q
T	T	F	F	F	T	T
T	F	T	F	T	F	F
F	T	F	T	F	T	T
F	F	T	T	F	T	T
1	2	3	4	5	6	7

7.

P	Q	P ↔ Q	~P	(P ↔ Q) V ~P
T	T	T	F	T
T	F	F	F	F
F	T	F	T	T
F	F	T	T	T
1	2	3	4	5

8.

P	Q	~P	~P → Q	~(~P → Q)
T	T	F	T	F
T	F	F	T	F
F	T	T	T	F
F	F	T	F	T

9.

P	Q	P V Q	~(P V Q)	~(P V Q) ∧ P
T	T	T	F	F
T	F	T	F	F
F	T	T	F	F
F	F	F	T	F
1	2	3	4	5

10.

P	Q	P ∧ Q	~Q	P ∧ Q ↔ ~Q
T	T	T	F	F
T	F	F	T	F
F	T	F	F	T
F	F	F	T	F
1	2	3	4	5

11.

P	Q	R	~P	~P V Q	~P V Q → R
T	T	T	F	T	T
T	T	F	F	T	F
T	F	T	F	F	T
T	F	F	F	F	T
F	T	T	T	T	T
F	T	F	T	T	F
F	F	T	T	T	T
F	F	F	T	T	F
1	2	3	4	5	6

12.

P	Q	R	~Q	R V P	P ∧ ~Q	~(R V P)	P ∧ ~Q ↔ ~(R V P)
T	T	T	F	T	F	F	T
T	T	F	F	T	F	F	T
T	F	T	T	T	T	F	F
T	F	F	T	T	T	F	F
F	T	T	F	T	F	F	T
F	T	F	F	F	F	T	F
F	F	T	T	T	F	F	T
F	F	F	T	F	F	T	F
1	2	3	4	5	6	7	8

Answers to Supplementary Exercise 2.6

1. ~(P V ~Q) 2. ~P V Q 3. ~(~P ∧ ~Q)
4. ~(P ∧ ~Q) 5. (~P V ~Q) ∧ R

6. Ann didn't go to the beach or Sue didn't go to the beach.
7. It is not the case that I went to the store and I didn't play tennis.
8. Fred cooks dinner and John doesn't wash the dishes.
9. It is false that Margaret or Rita took the test.
10. It is not the case that I don't wash my car and I don't bake peach muffins.

Answers to Supplementary Exercise 2.7

1. Converse: B → S
   Inverse: ~S → ~B
   Contrapositive: ~B → ~S
2. Converse: ~S → M
   Inverse: ~M → S
   Contrapositive: S → ~M
3. Converse: ~R → ~S
   Inverse: S → R
   Contrapositive: R → S
4. Q → P
5. P → Q
6. P ↔ Q
7. ~P → ~Q
8. P → Q
9. P ↔ Q
10. ~P → ~Q
11. P → Q
12. P → Q
13. Q → P

Answers to Supplementary Exercise 2.8

1. Valid

W	B	W V B	~B	(W V B) ∧ ~B	(W V B) ∧ ~B → W
T	T	T	F	F	T
T	F	T	T	T	T
F	T	T	F	F	T
F	F	F	T	F	T
1	2	3	4	5	6

2. Invalid

C	M	C → M	(C → M) ∧ M	(C → M) ∧ M → C
T	T	T	T	T
T	F	F	F	T
F	T	T	T	F
F	F	T	F	T
1	2	3	4	5

3. Invalid

R	H	R ↔ H	~R	(R ↔ H) ∧ ~R	(R ↔ H) ∧ ~R → H
T	T	T	F	F	T
T	F	F	F	F	T
F	T	F	T	F	T
F	F	T	T	T	F
1	2	3	4	5	6

4. Invalid

H	L	~H	~H → L	(~H → L) ∧ H	~L	(~H → L) ∧ H → ~L
T	T	F	T	T	F	F
T	F	F	T	T	T	T
F	T	T	T	F	F	T
F	F	T	F	F	T	T
1	2	3	4	5	6	7

5. Invalid

N	E	~N	~N → E	(~N → E) ∧ E	(~N → E) ∧ E → N
T	T	F	T	T	T
T	F	F	T	F	T
F	T	T	T	T	F
F	F	T	F	F	T
1	2	3	4	5	6

6. Valid

P	J	P ∧ J	(P ∧ J) ∧ J	(P ∧ J) ∧ J → P
T	T	T	T	T
T	F	F	F	T
F	T	F	F	T
F	F	F	F	T
1	2	3	4	5

7. Valid

C	W	~C	~C → W	~W	(~C → W) ∧ ~W	(~C → W) ∧ ~W → C
T	T	F	T	F	F	T
T	F	F	T	T	T	T
F	T	T	T	F	F	T
F	F	T	F	T	F	T
1	2	3	4	5	6	7

8. Invalid

S	P	S V P	~S	(S V P) ∧ ~S	~P	(S V P) ∧ ~S → ~P
T	T	T	F	F	F	T
T	F	T	F	F	T	T
F	T	T	T	T	F	F
F	F	F	T	F	T	T
1	2	3	4	5	6	7

9. Valid

P	D	~D	P → ~D	(P → ~D) ∧ P	(P → ~D) ∧ P → ~D
T	T	F	F	F	T
T	F	T	T	T	T
F	T	F	T	F	T
F	F	T	T	F	T
1	2	3	4	5	6

10. Valid

O	W	~O	~W	O ↔ W	(O ↔ W) ∧ ~O	~W
T	T	F	F	T	F	T
T	F	F	T	F	F	T
F	T	T	F	F	F	T
F	F	T	T	T	T	T
1	2	3	4	5	6	7

462 STUDENT STUDY GUIDE

Answers to Supplementary Exercise 2.9

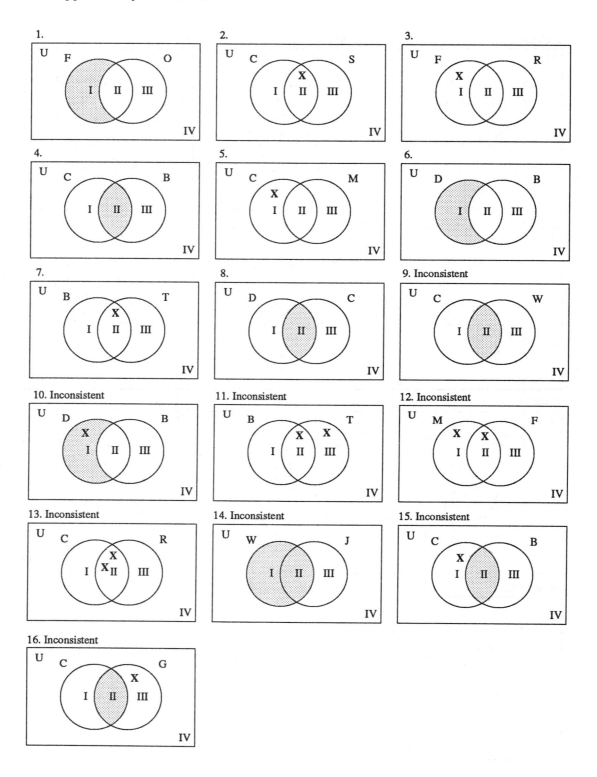

Answers to Supplementary Exercise 2.10

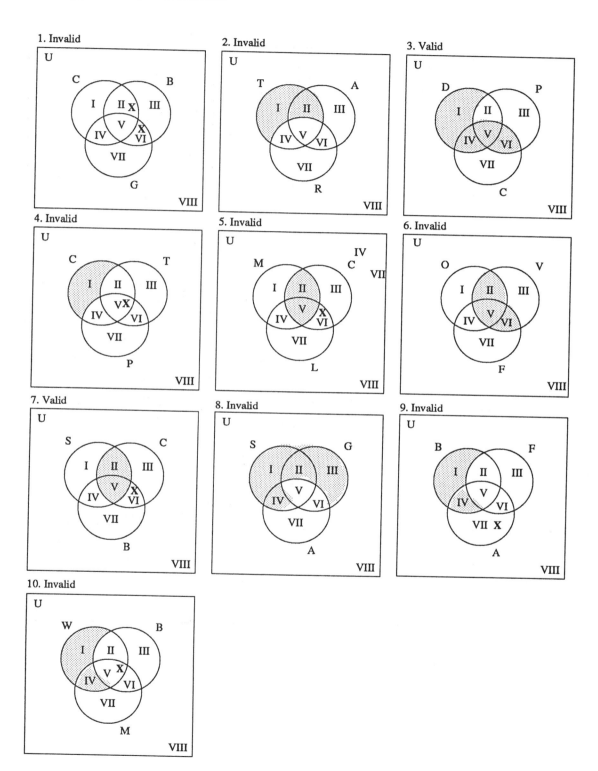

464 STUDENT STUDY GUIDE

Answers to Supplementary Exercise 2.11

1. P ∧ [(Q ∧ R) V P]
2. P ∧ Q ∧ (~P V Q)
3. [P V (~Q ∧ P)] ∧ R
4. (P ∧ R) V (~Q ∧ ~P) V (~R ∧ P)
5. [~P V (Q ∧ P)] ∧ [Q V (R ∧ ~P)]

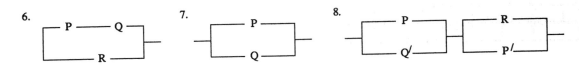

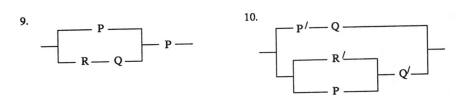

Answers to Chapter Test

1. F 2. T 3. F 4. F 5. T 6. F 7. T 8. T
9. Q → P 10. ~Q → ~P 11. P ∧ ~Q 12. P V ~Q 13. ~(P → Q)

14. If you use sunscreen, then you don't get a sunburn.
15. You don't use sunscreen and you get a sunburn.
16. It is not the case that if you get a sunburn then you don't use sunscreen.

17.

P	Q	~Q	P ↔ ~Q
T	T	F	F
T	F	T	T
F	T	F	T
F	F	T	F
1	2	3	4

18.

P	Q	~P	~P V Q	~(~P V Q)
T	T	F	T	F
T	F	F	F	T
F	T	T	T	F
F	F	T	T	F
1	2	3	4	5

19. No

P	Q	~Q	P V ~Q	P	Q	~P	Q → ~P
T	T	F	T	T	T	F	F
T	F	T	T	T	F	F	T
F	T	F	F	F	T	T	T
F	F	T	T	F	F	T	T
1	2	3	4	1	2	3	4

20. Valid

L	M	L V M	~M	(L V M) ∧ ~M	(L V M) ~M → L
T	T	T	F	F	T
T	F	T	T	T	T
F	T	T	F	F	T
F	F	F	T	F	T
1	2	3	4	5	6

21. Invalid 22. Valid 23. Valid
24. Converse: ~Q → P Inverse: ~P → Q Contrapositive: Q → ~P
25. Inconsistent 26. Inconsistent

CHAPTER 3

Answers to Supplementary Exercise 3.2

1. $\frac{4}{6} = \frac{2}{3}$ 2. $\frac{2}{6} = \frac{1}{3}$ 3. $\frac{1}{6}$ 4. $\frac{0}{6} = 0$ 5. $\frac{2}{6} = \frac{1}{3}$
6. $\frac{1}{52}$ 7. $\frac{4}{52} = \frac{1}{13}$ 8. $\frac{6}{52} = \frac{3}{26}$ 9. $\frac{13}{52} = \frac{1}{4}$ 10. $\frac{26}{52} = \frac{1}{2}$
11. $\frac{16}{52} = \frac{4}{13}$ 12. $\frac{1}{52}$ 13. $\frac{22}{52} = \frac{11}{26}$ 14. $\frac{0}{52} = 0$ 15. $\frac{2}{52} = \frac{1}{26}$
16. $\frac{4}{18} = \frac{2}{9}$ 17. $\frac{5}{18}$ 18. $\frac{9}{18} = \frac{1}{2}$ 19. $\frac{14}{18} = \frac{7}{9}$ 20. $\frac{6}{18} = \frac{1}{3}$

Answers to Supplementary Exercise 3.3-3.4

1. P (at least one head) = $\frac{3}{4}$

Nickel	Dime	Total Outcomes
H	H	HH
	T	HT
T	H	TH
	T	TT

2. P (all tails) = $\frac{1}{8}$

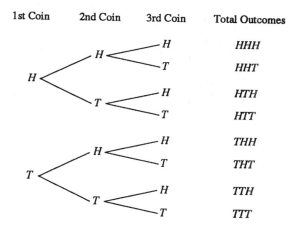

466   STUDENT STUDY GUIDE

3. $\frac{3}{8}$ 4. $\frac{7}{8}$ 5. $\frac{1}{3}$ 6. $\frac{1}{24}$ 7. $\frac{1}{12}$ 8. $\frac{1}{12}$

9 a. none   b. $\frac{3}{20}$ c. $\frac{2}{20} = \frac{1}{10}$ d. $\frac{1}{10}$ e. $\frac{19}{20}$ f. $\frac{1}{20}$

10 a. $\frac{7}{50}$ b. $\frac{2}{50} = \frac{1}{25}$ c. $\frac{10}{50} = \frac{1}{5}$ d. $\frac{1}{50}$ e. $\frac{3}{50}$

Answers to Supplementary Exercise 3.5

1. 4 to 32, or 1 to 8   2. 9 to 27, or 1 to 3   3. 33 to 3, or 11 to 1
4. 33 to 3, or 11 to 1   5. 5 to 31   6. 2 to 50, or 1 to 25
7. 4 to 48, or 1 to 12   8. 8 to 44, or 2 to 11   9. 46 to 6, or 23 to 3
10. 24 to 48, or 6 to 7   11. $\frac{3}{7}$ 12. $\frac{6}{11}$ 13. $\frac{1}{6}$
14. $\frac{7}{11}$ 15. 3:10 in favor; 10:3 against   16. $\frac{2}{7}$ 17. $4
18. 5 to 31; $1.39   19. $1,667   20. $\frac{1}{10000}$; 20¢

Answers to Supplementary Exercise 3.6

1. $\frac{25}{102}$ 2. $\frac{1}{221}$ 3. $\frac{1}{204}$ 4. $\frac{13}{102}$ 5. $\frac{32}{529}$ 6. $\frac{36}{529}$
7. $\frac{40}{529}$ 8. $\frac{16}{529}$ 9. $\frac{1}{8}$ 10. $\frac{5}{8}$ 11. $\frac{1}{52}$ 12. $\frac{7}{13}$
13. $\frac{120}{5814}$ 14. $\frac{280}{5814}$ 15. $\frac{240}{5814}$ 16. $\frac{160}{5814}$

Answers to Supplementary Exercise 3.7

1. 35   2. 720; 7776   3. 72   4. 160   5. 3450
6. 56   7. 72   8. 5040   9. 5040   10. 60480
11. 210   12. 90   13. 8   14. 907,200   15. 15,818,400
16. 2,002   17. 60,480   18. 120   19. 870   20. 14,406

Answers to Supplementary Exercise 3.8

1. 84   2. 126   3. 6   4. 19,600   5. 624
6. 35   7. 369,600   8. 1,440,000   9. 1680   10. 9900

Answers to Supplementary Exercise 3.9

1. $\frac{16}{1265}$ 2. $\frac{36}{270,725}$ 3. $\frac{4}{35}$ 4. $\frac{3}{10}$ 5. $\frac{5}{14}$
6. $\frac{16}{270,725}$ 7. $\frac{87}{136}$ 8. $\frac{70}{1287}$ 9. $\frac{12}{55}$ 10. $\frac{4}{17}$

Answers to Chapter 3 Test

1. $\frac{8}{20} = \frac{2}{5}$ 2. $\frac{7}{20}$ 3. $\frac{5}{20} = \frac{1}{4}$ 4. 0   5. $\frac{12}{20} = \frac{3}{5}$
6. $\frac{13}{20}$ 7. $\frac{12}{2652} = \frac{1}{221}$ 8. $\frac{156}{2652} = \frac{13}{221}$ 9. $\frac{8}{36} = \frac{2}{9}$ 10. $\frac{1}{4}$
11. $\frac{3}{4}$ 12. $\frac{28}{37}$ 13. $\frac{4}{312} = \frac{1}{78}$ 14. $\frac{72}{312} = \frac{3}{13}$ 15. $\frac{26}{312} = \frac{1}{12}$

16. $\frac{182}{312} = \frac{7}{12}$ 17. $\frac{7}{3}$ 18. $\frac{3}{7}$ 19. $\frac{5}{8}$ 20. $8
21. $3025   22. $36   23. 260,000   24. $\frac{8!}{2!3!} = 3360$ 25. $\frac{810}{21,924}$
26. $\frac{720}{21,924}$

CHAPTER 4

Answers to Supplementary Exercise 4.2

1. mean: 5
   median: 5
   mode: 5
   midrange: 5.5

2. mean: 5
   median: 5
   mode: none
   midrange: 5

3. mean 8.4
   median: 8
   mode: 8
   midrange: 9

4. mean: 4.8
   median: 4.5
   mode: 1 and 8
   midrange: 5

5. mean: 6.4
   median: 6.5
   mode: 9
   midrange: 6

6. mean: 60.8
   median: 61.5
   mode: none
   midrange: 62.5

7. mean 39.4
   median: 32
   mode: none
   midrange: 49.5

8. mean: 411.5
   median: 395
   mode: none
   midrange: 411.5

9. mean: 636.6
   median: 202
   mode: 101
   midrange: 1505.5

10. mean: 3090.3
    median: 2784
    mode: none
    midrange: 3216.5

11. midrange: 57.5
    median: 80

12a. $44,200
12b. $33,000
12c. $55,000
12d. $133,000

Answers to Supplementary Exercise 4.3

1. 1.4   2. 2.3   3. 1.5   4. 1.4   5. 24.5   6. 4.9
7. 2.8   8. 1.9   9. 9.2   10. 13.0
11a. $66   11b. $62   11c. no mode   11d. $54   11e. $70   11f. $18.20
12a. 54   12b. 55   12c. 34   12d. 45   12e. 53.5   12f. 15.8

Answers to Supplementary Exercise 4.4

1. 93rd   2. 85th   3. 75th   4. 375th   5. 24th
6. 24th   7. 24   8. 8th   9. 50th   10. 87 or 88th
11. 62   12. 90   13. 85   14. none   15. Claudio

Answers to Supplementary Exercise 4.5

1. See the graph below.

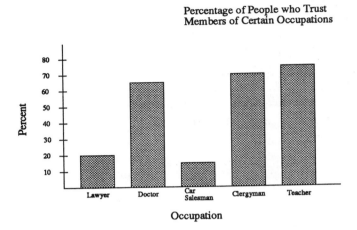

2. See the graph below.

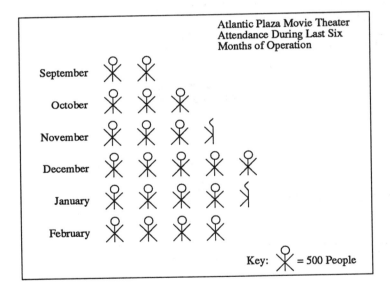

3a. $400,000   3b. $300,000   3c. $800,000   3d. $1,800,000

4. See the graph below.

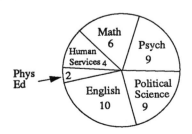

ANSWERS 469

5a and b.

Grade	Frequency
63	4
68	2
73	7
78	3
83	7
88	2
93	1

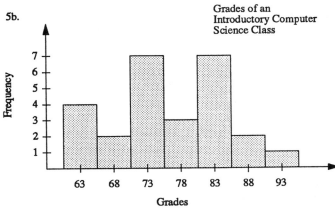

6a and b.

Ages	Frequency
37-45	2
46-54	3
55-63	4
64-72	6
73-81	3
82-90	2
91-99	4

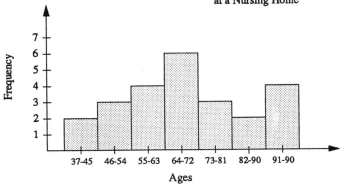

7a and b.

Hours	Frequency
2-6	3
7-11	2
12-16	3
17-21	3
22-26	4
27-31	4
32-36	1

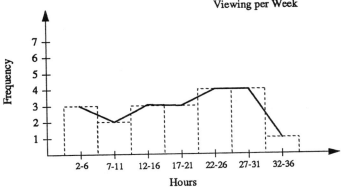

8a and b.

Rent	Frequency
180-260	4
251-321	4
322-392	9
393-463	2
464-534	3
535-605	2

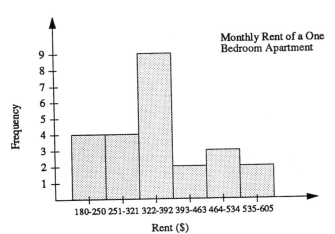

Answers to Supplementary Exercise 4.6

1. Full: 18.4%   Associate: 43.4%   Assistant: 28.5%   Instructor: 9.7%
2a. 28   2b. 26   2c. 572   2d. 818   2e. 2
3a. 8.1%   3b. 91.9%   3c. 5%   3d. 34.5%   3e. 84.9%
4a. 0.115   4b. 0.425   4c. 0.036   4d. 9

Answers to Chapter 4 Test

1a. February   1b. three inches   1c. March and April   1d. approximately 12 inches
2a. 170   2b. 22.5   2c. between Monday and Saturday
3. 51.75   4. 49   5. 49   6. no mode   7. 56
8. 15.9   9. 42   10. 68   11. Milt

12a and b.

Temperature	Frequency
80	2
82	3
84	3
86	6
88	7
90	4

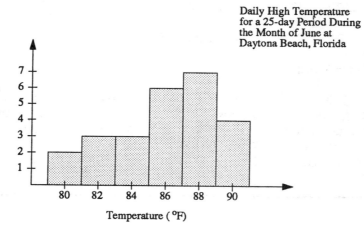

Daily High Temperature for a 25-day Period During the Month of June at Daytona Beach, Florida

13. a. 9.7%   13b. 66.1%   13c. 0.35%   13d. 0.136

CHAPTER 5

Answers to Supplementary Exercise 5.2

1. a   2. d   3. e   4. c   5. b
6. f   7. liter   8. milliliter 9. gram   10. kilogram
11. dekameter 12. centiliter 13. centimeter 14. hectoliter 15. milligram

Answers to Supplementary Exercise 5.3

1. 230 mm   2. 0.7 mm   3. 8600 mm   4. 6,200,000 cm   5. 27 cm   6. 380 cm
7. 20 cm   8. 600,000cm   9. 2.97dm   10. 68.7 dm   11. 213 dm   12. 46,800 dm
13. 0.0563 m   14. 3.87m   15. 0.0296   16. 520 m   17. 3.2 km   18. 2.586 km
19. 3.706 km   20. 12.70309   21. $8 per meter 22. yes; 3.6 mph 23. 2700 cm

Answers to Supplementary Exercise 5.4

1. 60 ml 2. 4200 ml;   3. 620 ml 4. 9200 ml 5. 3800 cl 6. 60 cl
7. 40 cl 8. 380 cl 9. 1.6 l 10. 3820 l 11. 9 l 12. 3.025 l
13. 48,200 l 14. 0.0128 l 15. 25 da$l$ 16. 3.096 hl 17. 8 kl 18. 4.87 kl
19. 9 kl 20. 0.295 kl 21. 350 l 22. $195   23. 7.6 m^3

24. cornstarch: 37.5 ml; sugar: 180 ml; salt: 1.25 ml; orange juice or water: 120 ml; lemon juice: 45 ml; lemon rind: 2.5 ml; butter 15 ml

Answers to Supplementary Exercise 5.5

1. 120 mg   2. 2675 mg   3. 3,620,000 mg   4. 460 mg   5. 80,000 cg   6. 230,000 cg
7. 150 dg   8. 27,000 dg   9. 4000 g   10. 60 g   11. 0.35 g   12. 0.375 g
13. 3900 dag   14. 0.63 dag   15. 100 hg   16. 490 hg   17. 2 kg   18. 3.875 kg
19. 3.25 kg   20. 7.698 kg   21. 72 kg   22. 504 g   23. 18¢

Answers to Supplementary Exercise 5.6

1. 27°C   2. -15°C   3. 6°C   4. 218°C   5. 100°C   6. 38°C   7. -3°C
8. 21°C   9. 50°F   10. 61°F 11. 90°F   12. 158°F 13. 36°F   14. 392°F
15. 84°F   16. 208°F 17. 232°C 18. 37°C   19. 28°C   20. 100°F

21. yes (102°F)   22. 2°C, 1°C   23. *Celsius 233*

Answers to Chapter 5 Test

1. 0.1   2. 0.001   3. 100   4. 1000   5. 10   6. 0.01
7. volume   8. length   9. mass   10. area   11. area   12. volume
13. mass   14. temperature 15. volume   16. length   17. mass   18. length
19. 0.03 m   20. 5000 ml   21. 72,000 g   22. 1.7 km   23. 0.55 g   24. 0.06 l
25. 350 m   26. 14 mm   27. 1,200,000 ml 28. 10,000 mg 29. $31.79   30. Bill

---

CHAPTER 6

Answers to Supplementary Exercise 6.2

1. 6   2. 5   3. 3   4. 9   5. 2   6. 1
7. 11 8. 12 9. 4   10. 8 11. 8 12. 1

13. Inverse property for addition
14. Closure property for multiplication
15. Identity property for multiplication
16. Commutative property for addition

17. True   18. True   19. False   20. True

Answers to Supplementary Exercise 6.3

1. century   2. decade   3. month   4. year   5. week   6. month   7. month
8. week   9. century   10. day   11. yes; 6 12. day   13. does not exist   14. yes

15. no; the elements 2, 3, 4, and 6 do not have an inverse

Answers to Supplementary Exercise 6.4

1. 7   2. 3   3. 4   4. 1   5. 5   6. 1   7. 0   8. 0   9. 7   10. 2   11. 6
12. 7   13. 7   14. 2   15. 4   16. 4   17. 6   18. 5   19. 3   20. 5   21. yes

472 STUDENT STUDY GUIDE

22. no; the elements 0, 2, 4, and 6 do not have an inverse

23. False   24. False   25. True   26. True   27. False   28. False
29. 2   30. 5   31. 0   32. 0   33. 2   34. 9

Answers to Supplementary Exercise 6.5

1. T   2. Y   3. Y   4. T   5. E   6. E   7. Y   8. no answer   9. E

10. No; there is an entry X that is not in the set   11. Y
12. T and P; Y and Y; P and T; E and E   13. No; the system is not closed under #

Answers to Supplementary Exercise 6.6

Axiom III says we must have a house. Axiom I says that this house must have exactly two cats in it. Thus we now know that two cats exist. This satisfies Axiom II. Axiom IV says we must have one other house that corresponds to the house we presently have (which resulted from Axiom III). Moreover, this second house must have two cats that are not in common with the two cats we have from the first house. As a result, we have proven via the axioms the existence of at least four cats.

Answers to Chapter 6 Test

1. A set of elements and one or more operations used to combine the elements
2. Identity property of addition   3. {0,1,2,3,4,5}   4. commutative
5. One example is the false statement $(2 - 3) - 4 = 2 - (3 - 4)$

6. True   7. False   8. True   9. False   10. True   11. 5
12. 8   13. 6   14. 6   15. 10   16. maybe   17. yes
18. yes   19. 4   20. 4   21. 2   22. 2   23. 3
24. 5   25. 2   26. 2   27. %   28. !   29. :
30. :   31. %   32. !   33. no   34. none   35. none

36. no; there is no identity element, each element does not have an inverse, and the system is not closed

---

CHAPTER 7

Answers to Supplementary Exercise 7.2

1. ∩∩||||||||||
2. ∩∩∩||||.||||
3. ∩∩∩∩∩∩|||||
4. e |
5. e e ∩∩∩∩∩∩∩∩||||||
6. e e e ∩|||||
7. e e e e
8. $\overset{\ulcorner D}{\Delta}$ e e e e e e e ∩∩∩∩||
9. $\overset{\ulcorner D}{\Delta}$ $\overset{\ulcorner D}{\Delta}$ ∩∩||
10. | $\overset{\ulcorner D}{\Delta}$ e ∩ |

11. 47 12. 93 13. 111 14. 813 15. 3,543
16. 5,832 17. 12,782 18. 97,000 19. 200,173 20. 354,020

21. e ∩∩∩∩∩||||
22. ∩∩||
23. e e e ∩∩|||
24. |||||||| ⌐△⌐△⌐△⌐△⌐△⌐△⌐△ e ∩∩∩∩
25. e e ∩∩||||

Answers to Supplementary Exercise 7.3

1. ΔΔΔΔIII   四十三

2. ΓΔΔΔΔΓI   九十六

3. HHΔII   二百十二

4. ΓHΓΓ   五百五十五

5. ΓHHHHΔΓII   九百十七

6. XHΔI   一千一百十一 OR 千百十一

7. XXXHHHΓΓ   三千三百五十五

8. ΓXHHΓΔΔIIII   六千二百七十四

9. M X H H ⌐ Δ Γ I I I

Note: 10,000 = 万 OR 万

一万一千二百六十八 OR 万千二百六十八

10. 53 11. 102 12. 568 13. 8,591 14. 26,195 15. 97,502
16. 82 17. 333 18. 916 19. 2,545 20. 6,020

Answers to Supplementary Exercise 7.4

1. ≪≪▽
2. ▽   ▽▽▽▽▽▽▽▽
3. ▽   ≪≪▽▽▽▽▽▽▽
4. ▽▽   ▽
5. ▽▽▽▽▽   ≪▽▽▽▽▽
6. ▽▽▽▽▽▽▽   ≪≪≪≪▽▽▽▽▽▽▽
7. ≪▽▽▽▽▽▽▽▽▽   ≪≪▽▽▽▽
8. ▽   ▽▽▽   ≪   ≪≪▽▽▽▽▽▽

9. 14   10. 52   11. 93   12. 234   13. 438   14. 905   15. 1,000   16. 3,924

17. $(8 \times 10^2) + (2 \times 10^1) + (5 \times 10^0)$
18. $(3 \times 10^3) + (2 \times 10^2) + (8 \times 10^1) + (3 \times 10^0)$
19. $(6 \times 10^3) + (4 \times 10^2) + (8 \times 10^0)$
20. $(4 \times 10^4) + (2 \times 10^3) + (5 \times 10^1) + (4 \times 10^0)$
21. $(4 \times 10^2) + (4 \times 10^0)$
22. $(2 \times 10^4) + (5 \times 10^3) + (8 \times 10^2) + (1 \times 10^1) + (4 \times 10^0)$
23. $(3 \times 10^4) + (7 \times 10^2)$
24. $(1 \times 10^5) + (1 \times 10^3) + (6 \times 10^0)$

25. 216   26. 947,618   27. 700,006,125   28. 6,532   29. 80,307   30. 234,560

Answers to Supplementary Exercise 7.5

1. 12   2. 6   3. 27   4. 84   5. 96   6. 108   7. 353   8. 5,114
9. 105 10. 42 11. 71 12. 129 13. 727 14. 563 15. 1,451 16. 1,847

17. 13_{five} and 8_{twelve} 18. 111_{five} and 27_{twelve}

19. 313_{five} and $6E_{\text{twelve}}$ 20. 434_{five} and $9E_{\text{twelve}}$
21. 2432_{five} and $T7_{\text{twelve}}$ 22. 12203_{five} and 654_{twelve}
23. 31010_{five} and $11E1_{\text{twelve}}$ 24. 330403_{five} and $2E53_{\text{twelve}}$

Answers to Supplementary Exercise 7.6

1. 104_5 2. 122_5 3. 343_5 4. 1103_5 5. 13201_5
6. 3_5 7. 4_5 8. 111_5 9. 33_5 10. 4004_5
11. 1111_5 12. 3223_5 13. 13303_5 14. 32343_5 15. 110303_5
16. 12_5 17. 121_5 R 1_5 18. 2_5 R 20_5 19. 11_5 R 3_5 20. 22_5 R 32_5

Answers to Supplementary Exercise 7.7

1. 9   2. 12   3. 15   4. 14   5. 18
6. 23   7. 29   8. 21   9. 10011_2 10. 10100_2
11. 11000_2 12. 11110_2 13. 101011_2 14. 110011_2 15. 111110_2
16. 1101101_2 17. 1011_2 18. 10011_2 19. 11000_2 20. 11101_2
21. 10_2 22. 11_2 23. 11_2 24. 111_2 25. 11000_2
26. 101010_2 27. 1001011_2 28. 10001111_2 29. 21020_3 30. 1321_9
31. 1303_5 32. 55_8 33. $2BF_{16}$

Answers to Chapter 7 Test

1. False   2. True   3. True   4. False   5. False
6. False   7. True   8. True   9. False   10. False
11. 44   12. 77   13. 101   14. 851   15. 1103_4
16. 19_{12} 17. 2110_3 18. 151_8 19. 1233_6 20. 11101_2
21. 1121_6 22. 2010_3 23. 358_9 24. 1244_8 25. 10010010010_2
26. 1320222_4 27. 14005253_7 28. 13_5 R 4_5 29. 200_5 R 4_5

CHAPTER 8

Answers to Supplementary Exercise 8.2

1. 628 is divisible by 2 and 4
2. 3,917 is not divisible by any of the numbers listed
3. 21,648 is divisible by 2,3,4,8, and 11
4. 109,520 is divisible by 2,4,5,8, and 10
5. 6,314,829 is divisible by 3

6. composite   7. composite   8. composite   9. composite   10. prime
11. composite   12. composite   13. composite   14. composite   15. composite
16. prime   17. composite   18. prime   19. composite   20. composite
21. 2, 2, 2, 3   22. 3, 3, 3   23. 2, 2, 7   24. 3, 13   25. 2, 2, 11
26. 2, 2, 2, 2, 3   27. 2, 2, 2, 3, 5   28. 2, 2, 2, 2, 2, 2, 2,   29. 3, 7, 7   30. 3, 61
31. 3, 3, 23   32. 13, 17   33. 2, 2, 5, 23   34. 2, 2, 5, 5, 5   35. 3, 3, 3, 37

Answers to Supplementary Exercise 8.3

1. 3   2. 4   3. 3   4. 2   5. 1   6. 8
7. 9   8. 10   9. 5   10. 20   11. 5   12. 1
13. 12   14. 56   15. 45   16. 56   17. 90   18. 80
19. 126   20. 150   21. 300   22. 3300   23. 120   24. 990
25. $\frac{1}{3}$ 26. $\frac{1}{3}$ 27. $\frac{1}{4}$ 28. $\frac{3}{5}$ 29. $\frac{7}{8}$ 30. $\frac{10}{17}$
31. $\frac{2}{7}$ 32. $\frac{8}{15}$ 33. $\frac{5}{12}$ 34. $\frac{19}{41}$ 35. $\frac{9}{31}$ 36. $\frac{3}{5}$
37. $\frac{5}{8}$ 38. $\frac{17}{16}$ 39. $\frac{23}{30}$ 40. $\frac{27}{20}$ 41. $\frac{41}{36}$ 42. $\frac{56}{45}$
43. $\frac{1}{6}$ 44. $\frac{1}{2}$ 45. $\frac{11}{40}$ 46. $\frac{1}{12}$ 47. $\frac{13}{30}$ 48. $\frac{7}{36}$

Answers to Supplementary Exercise 8.4

1. 6   2. 3   3. −10   4. −64   5. −1   6. 12   7. −7   8. 4   9. 4
10. −3   11. 0   12. −9   13. −7   14. 21   15. −24   16. −24   17. −28   18. 0
19. 20   20. 1   21. −6   22. −9   23. −53   24. 30   25. −120   26. 350   27. 6
28. −70   29. 2   30. 20   31. −10   32. 10   33. −6   34. 8   35. 18   36. −16
37. <   38. >   39. >   40. <   41. >   42. <

Answers to Supplementary Exercise 8.5

1. $\frac{7}{8}$ 2. $\frac{8}{9}$ 3. $\frac{17}{16}$ 4. $\frac{16}{15}$ 5. $\frac{13}{14}$ 6. $\frac{11}{12}$ 7. $\frac{29}{36}$ 8. $\frac{44}{45}$ 9. $\frac{25}{36}$
10. $\frac{1}{8}$ 11. $\frac{1}{9}$ 12. $\frac{1}{8}$ 13. $\frac{1}{21}$ 14. $\frac{3}{35}$ 15. $\frac{7}{24}$ 16. $\frac{5}{24}$ 17. $\frac{8}{15}$ 18. $\frac{1}{18}$
19. $\frac{15}{28}$ 20. $\frac{8}{15}$ 21. $\frac{21}{32}$ 22. $\frac{25}{33}$ 23. $\frac{1}{15}$ 24. $\frac{2}{9}$ 25. $\frac{3}{2}$ 26. $\frac{21}{4}$ 27. $\frac{9}{4}$
28. $\frac{25}{12}$ 29. $\frac{27}{35}$ 30. $\frac{15}{16}$ 31. $\frac{15}{16}$ 32. $\frac{3}{2}$ 33. $\frac{9}{8}$ 34. $\frac{1}{12}$ 35. $\frac{2}{21}$ 36. 40
37. $\frac{21}{16}$ 38. $\frac{45}{28}$ 39. $\frac{3}{11}$ 40. $\frac{112}{85}$ 41. $\frac{216}{133}$ 42. $\frac{171}{32}$ 43. $\frac{136}{63}$ 44. $\frac{40}{21}$ 45. $\frac{1665}{224}$
46. <   47. >   48. >   49. >   50. <   51. >   52. >   53. <   54. <

Answers to Supplementary Exercise 8.6

1. 0.6   2. 0.95   3. 0.5625   4. 0.25   5. $0.\overline{43}$ 6. $0.2\overline{16}$ 7. $0.7\overline{727}$
8. $0.\overline{809523}$ 9. $\frac{17}{20}$ 10. $\frac{9}{1000}$ 11. $\frac{2}{125}$ 12. $\frac{1}{5}$ 13. $\frac{28}{99}$ 14. $\frac{1}{9}$
15. $\frac{11}{9}$ 16. $\frac{19}{12}$ 17. $\frac{11}{18}$ 18. $\frac{19}{30}$ 19. $\frac{13}{16}$ 20. $\frac{53}{84}$

Answers to Supplementary Exercise 8.7

1. rational 2. rational 3. rational 4. irrational
5. irrational 6. rational 7. rational 8. rational

	Number	Natural Number	Whole Number	Integer	Rational Number	Irrational Number	Real Number
9.	-5			x	x		x
10.	-3.2			x	x		x
11.	0		x	x	x		x
12.	$\sqrt{16}$	x	x	x	x		x
13.	$0.\overline{43}$				x		x
14.	$3\sqrt{2}$					x	x
15.	$\frac{5}{8}$				x		x

Answers to Supplementary Exercise 8.8

1. 3.0×10^9 2. 1.21×10^7 3. 4.03×10^5
4. 2.07001×10^{11} 5. 6.0×10^{-8} 6. 1.02×10^{-1}
7. 2.04×10^{-5} 8. -1.102×10^{-3} 9. 460,000,000
10. 0.00601 11. 99,000 12. $-17,010$
13. 2.94×10^{10} 14. 2.268×10^{-4} 15. 6×10^{-4}
16. 4.8×10^{-3} 17. 4.8×10^{-7} 18. 4.8×10^3
19. 3.35×10^7 20. 3×10^{15} 21. 4.9×10^2

Answers to Chapter 8 Test

1. False 2. False 3. True 4. False 5. True 6. True
7. False 8. True 9. False 10. True 11. Prime 12. Composite
13. Composite 14. Prime 15. 26 16. 18 17. 210 18. 1800
19. $\frac{67}{40}$ 20. $\frac{4}{35}$ 21. $\frac{4}{21}$ 22. $\frac{9}{10}$ 23. -9 24. 7
25. -12 26. 35 27. -25 28. 0.6 29. 0.5625 30. $\frac{5}{33}$
31. $\frac{3140}{999}$ 32. Rational 33. Irrational 34. Rational 35. Rational 36. Rational
37. Irrational 38. 42.32×10^2 39. 0.6×10^{-12} 40. 0.9×10^5

CHAPTER 9

Answers to Supplementary Exercise 9.2

1. -3 2. 6 3. 4 4. 25 5. 19 6. 15 7. 22 8. -1 9. 4 10. 1

478 STUDENT STUDY GUIDE

11. $x > 4$

12. $x \leq -2$

13. $x \leq 11$
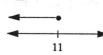

14. $x > 7$
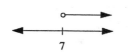

15. $x \leq -1$
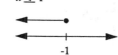

16. $x > 4$
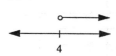

17. $0 \leq x < 2$ (x is an integer)
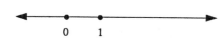

18. $-5 \leq x \leq 3$ (x is a whole number)

19. $-2 < x \leq 5$
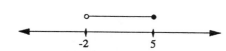

20. $1 \leq x \leq 4$

Answers to Supplementary Exercise 9.3

1. $6x$ 2. $8y$ 3. $4x$ 4. $4x$ 5. $10x$ 6. $9y$
7. x 8. $4y$ 9. $2x$ 10. $11x + 3$ 11. $y + 5$ 12. $3x + 9$
13. $2x - 1$ 14. $3y + 2$ 15. $3x + 3$ 16. $3y - 1$ 17. $2x - 3$ 18. -2
19. $6x - 4$ 20. $11x - 5$ 21. $8x - 4$ 22. $6x - 7$ 23. $10x - 2$ 24. $3x + 3$
25. $17y - 30$ 26. $2y + 18$ 27. $6 - x$ 28. $9x - 7$ 29. $7x + 29$ 30. $11x - 22$

Answers to Supplementary Exercises 9.4

1. -6 2. 24   3. 2   4. -4 5. 28   6. 12   7. 9   8. 2   9. 15
10. 24   11. 10   12. 7   13. 10   14. 6   15. 11   16. -1 17. 0   18. -2
19. 4   20. 5   21. 2   22. 2   23. 9   24. 5   25. -1

Answers to Supplementary Exercise 9.5

1. 4   2. 8   3. 6   4. 11   5. 6 and 24   6. 5 and 13

7. Kelly is 11 and Kristin is 9   8. Cost of shirt = $18, cost of pair of pants = $40
9. Width = 36 feet, length = 78 feet   10. 8 dimes and 24 quarters

Answers to Supplementary Exercise 9.6

1. no   2. yes   3. yes   4. no   5. no   6. yes   7. yes

8. no   9. (6,−3)   10. (−8,10)   11. (2,7)   12. $(5, \frac{9}{2})$ 13. (−3,5)   14. (2,18)

15. (0,−4)   16. $(-3, \frac{13}{3})$ 17. (4,2)   18. $(-1, \frac{-5}{3})$ 19. (2, −1)   20. (−2, 4)

Answers to Supplementary Exercise 9.7

1 through 8, see graph.

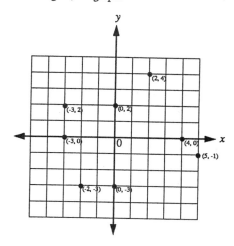

9. $x + y = 3$

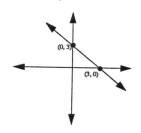

10. $x + y = -2$

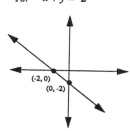

11. $x - y = 4$

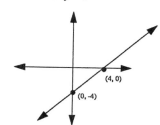

12. $x - y = -5$

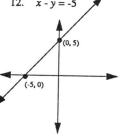

13. $2x - y = 3$

14. $x - 2y = 6$

15. $2x + 3y = 12$

16. $5x - 3y = 15$

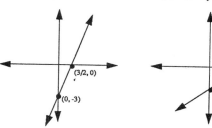
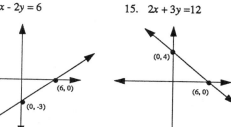
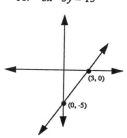

17. $-3x - 2y = 18$

18. $(0, -2)$

19. No point of intersection

20. $(2, 5)$

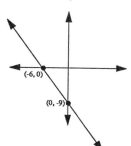
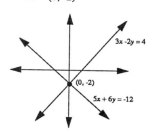
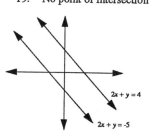
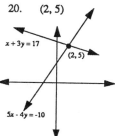

Answers to Supplementary Exercise 9.8

1. 9   2. $\dfrac{13}{3}$ 3. undefined   4. 0   5. -2 6. $\dfrac{-4}{3}$

7. $\dfrac{1}{2}$ 8. -1 9. 1   10. undefined   11. $\dfrac{-3}{2}$ 12. $\dfrac{-1}{4}$

Answers to Supplementary Exercise 9.9

1. $y = \dfrac{3}{4}x + \dfrac{23}{4}$
2. $y = x + 2$
3. $y = \dfrac{-1}{2}x + 9$
4. $y = -3x + 6$
5. $y = \dfrac{4}{9}x - \dfrac{28}{3}$
6. $y = 2x - 1$
7. $y = -4x - 7$
8. $y = \dfrac{x}{2} - \dfrac{3}{2}$
9. $y = \dfrac{-5}{4} + \dfrac{3}{2}$
10. $y = \dfrac{5}{6}x + \dfrac{25}{6}$
11. $y = \dfrac{-2}{7}x - \dfrac{13}{7}$
12. $y = \dfrac{1}{2}x - 4$
13. $y = 2x + 16$
14. $y = -3x + 23$
15. $y = -x + 6$
16. $y = \dfrac{-3}{2}x$
17. $y = \dfrac{1}{2}x - 2$
18. $y = \dfrac{1}{2}x + \dfrac{1}{2}$

19. $m = 2$; y-intercept is $(0, 1)$; See graph below

20. $m = 1$; y-intercept is $(0, 0)$; See graph below

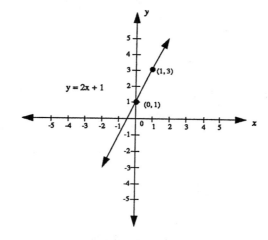
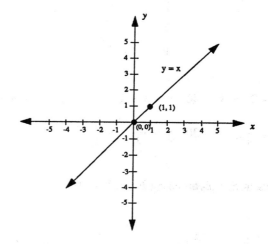

21. $m = \dfrac{-2}{3}$; y-intercept is $(0, 4)$; See graph below

22. $m = 1$; y-intercept is $(0, -3)$; See graph below

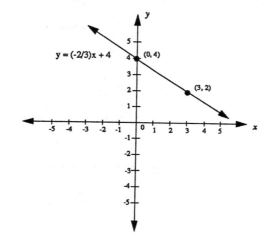
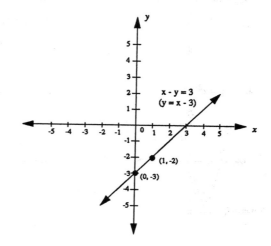

ANSWERS 481

23. $m = \dfrac{-2}{3}$; y-intercept is $(0, \dfrac{8}{3})$; See graph below   24. $m = \dfrac{3}{4}$; y-intercept is $(0, -4)$; See graph below

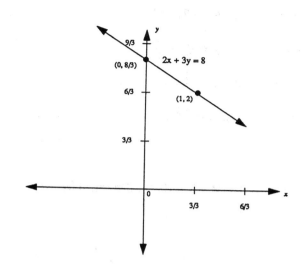 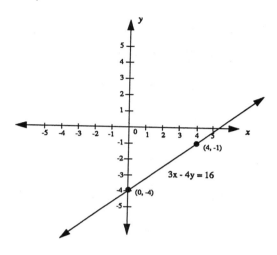

25. $y = 2x + 1$ 26. $y = x - 2$ 27. $y = \dfrac{2}{3}x + 1$ 28. $y = \dfrac{-5}{3}x - 3$

29. $y = \dfrac{1}{3}x$ 30. $y = \dfrac{-1}{4}x + \dfrac{1}{2}$

Answers to Supplementary Exercise 9.10

1. $y = x^2 + 1$ 2. $y = x^2 - 5$ 3. $y = x^2 - 2x$

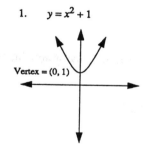 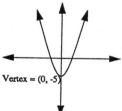 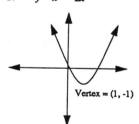

4. $y = x^2 + 6x$ 5. $y = x^2 - 4x - 5$ 6. $y = x^2 - 6x + 5$

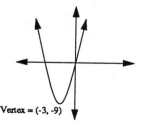 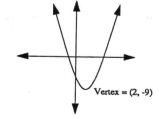 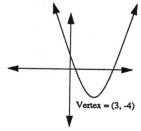

7. $y = -x^2 - 6x - 5$

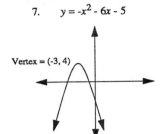

Vertex = (-3, 4)

8. $y = -x^2 + 2x - 3$

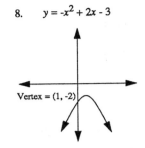

Vertex = (1, -2)

9. $y = x^2 + 2x + 2$

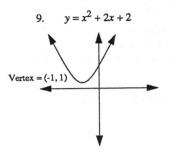

Vertex = (-1, 1)

Answers to Supplementary Exercise 9.11

1. $x + y > 3$

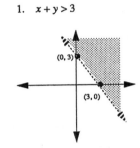

2. $x - y < 2$

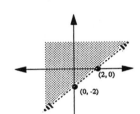

3. $3x - y \geq 0$

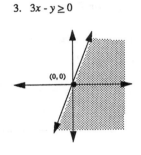

4. $2x + y \leq -5$

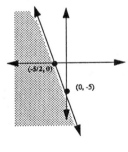

5. $2x - 3y > -12$

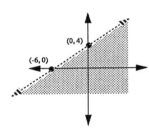

6. $5x - 4y \geq 20$

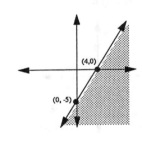

7. $-x + y > 8$

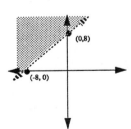

8. $-2x - 3y \geq -6$

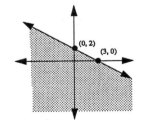

9. $-3x + 4y < 24$

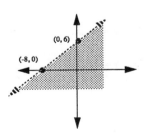

ANSWERS 483

Answers to Supplementary Exercise 9.12

1. Maximum value of P is 50 and occurs at (0,10)
2. Maximum value of P is 36 and occurs at (2,4)
3. Maximum value of P is 90 and occurs at (18,0)

Answers to Supplementary Exercise 9.13

1. $\{2\}$ 2. $\{2, \frac{-3}{2}\}$ 3. $\{-3, 5\}$ 4. $\{-5, 7\}$ 5. $\{-1, \frac{-2}{3}\}$
6. $\{-2, \frac{1}{4}\}$ 7. $\{2, -2\}$ 8. $\{\frac{-1}{2}, \frac{-3}{2}\}$ 9. $\{-1, 3\}$ 10. $\{1, -4\}$
11. $\{1, 4\}$ 12. $\{-2, \frac{-5}{2}\}$ 13. $\{0, 6\}$ 14. $\{3, -7\}$ 15. $\{0, -3\}$

Answers to Chapter 9 Test

1. –6   2. 27   3. –14   4. –9   5. 7   6. 42
7. 9   8. 4   9. 1   10. –3   11. –4   12. 4

13.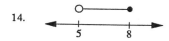
14.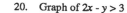
15. Jack: 2 years old; Julia: 6 years old
16. 110 dimes and 80 quarters

17. $3x + 2y = 18$ 18. $2x - 3y = 12$ 19. $y = \frac{2}{3}x - 7$ 20. Graph of $2x - y > 3$

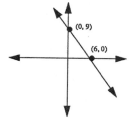 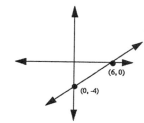 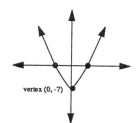 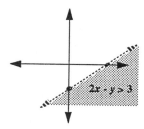

21. -7/8
22. $y = -2x + 22$
23. $y = -3x - 10$

24. Slope = $\frac{-5}{2}$ and the y-intercept = (0, -3)

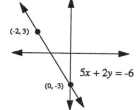

25. $y = 2x + 3$ 26. $y = 1$
27. $\{7, 1\}$ 28. $\{3, 1/3\}$
29. $\{-1/5, -1\}$

CHAPTER 10

Answers to Supplementary Exercise 10.2

1. point D 2. $\angle DCB$ 3. $\emptyset$ 4. point F 5. $\overline{DC}$
6. point C 7. $\angle DAG$ 8. $\overleftrightarrow{AG}$ 9. $\overrightarrow{AF}$ 10. $\angle ABG$ or $\overleftrightarrow{AG}$
11. $\overrightarrow{EA}, \overrightarrow{EB}$ 12. $\overrightarrow{AE}, \overrightarrow{AD}$ 13. $\overleftrightarrow{AD}, \overrightarrow{BD}$ 14. $\angle ACD$ 15. $\overrightarrow{BE}, \overrightarrow{BD}$

Answers to Supplementary Exercise 10.3

1. point A 2. $\angle CAE$ 3. $\overleftrightarrow{BC}$ 4. $\overrightarrow{AB}$ 5. $\overleftrightarrow{AC}$ 6. $\overrightarrow{AE}$
7. $\emptyset$ 8. $\overrightarrow{AD}$ 9. interior $\angle BAC$ 10. exterior $\angle BAD$ 11. false   12. $\angle CAD$

Answers to Supplementary Exercises 10.4

1. $\angle DAC, \angle CAB, \angle BAG, \angle GAF, \angle FAE, \angle EAD, \angle EAC, \angle DAB, \angle CAG, \angle BAF, \angle GAE, \angle FAD$

2. $\angle CAF, \angle BAE, \angle GAD$

3. $\angle CAB$ and $\angle EAF$; $\angle CAE$ and $\angle BAF$

4. $\angle FAG$ 5. $\angle BAG, \angle EAF$ 6. 66° 44′ 37″   7. 111° 31′
8. 50°   9. 45°   10. 45°   11a. $x = 14$
11b. m $\angle COE = 44°$ m $\angle EOF = 22°$ m $\angle FOB = 24°$

12. m $\angle 1 = 157°$; m $\angle 2 = 23°$; m $\angle 3 = 23°$; m $\angle 4 = 157°$; m $\angle 5 = 48°$; m $\angle 6 = 132°$
    m $\angle 7 = 132°$ m $\angle 8 = 48°$ m $\angle 10 = 109°$ m $\angle 12 = 48°$ m $\angle 13 = 109°$ m $\angle 14 = 23°$

Answers to Supplementary Exercise 10.5

1. True   2. False   3. True   4. False   5. True   6. True   7. True
8. False   9. False   10. False   11. 84°   12. 23   13. 17   14. 20 feet

Answers to Supplementary Exercise 10.6

1. 32 sq. in.   2. 3.75 sq. ft.   3. 400 sq. cm   4. 4.5 sq. ft.   5. 900 sq. cm
6. 12 sq. in.   7. 8.2 sq. in.   8. 1.0 sq. ft.   9. 30 sq. ft.   10. 35 sq. ft.
11. 40 sq. in.   12. 6 in.   13. 66 sq. in.   14. 102 sq. in.   15a. 37.7 cm
15b. 113 sq. cm   16a. 9.4 ft.   16b. 7.1 sq. ft   17. 144 sq. in.   18. 104 sq. cm
19. 75 sq. mm   20. 8,826 sq. m   21. 536.9 sq. in.   22. 94.2 in.   23. 152 sq. ft.

Answers to Supplementary Exercise 10.7

1a. $SA = 146$ sq. cm   1b. $V = 90$ cubic cm   2a. $SA = 301.4$ sq. ft.   2b. $V = 301.4$ cubic ft.
3. $V = 17.5$ cubic cm   4. $V = 34$ cubic in.   5a. $SA = 42.1$ sq. in.   5b. $V = 20.7$ cubic in.
6. 314 sq. cm   7. 179.5 cubic in.   8a. $SA = 301.4$ sq. cm   8b. $V = 401.9$ cubic cm
9. 144
10a. 7   10b. 1   10c. 7   10d. 12

Answers to Supplementary Exercise 10.8

1. 13 2. 12 3. 5 4. 65° 5. 65° 6. 90° 7. 5
8. 38° 9a. yes (SAS) 9b. 53 10. 1 11. 4 12. 3 13. 30

Answers to Supplementary Exercise 10.9

1a. 5	2a. 7	3a. 3	4a. 6	5a. 0	6a. 8
1b. 0	2b. 0	3b. 2	4b. 3	5b. 8	6b. 6
1c. yes	2c. yes	3c. yes	4c. no	5c. no	6c. no
1d. any point	2d. any point	3d. point C or D	4d. none	5d. none	6d. none

Answers to Chapter 10 Test

1. f 2. g 3. b 4. j 5. a
6. d 7. h 8. e 9. $\vec{RD}, \vec{RC}$ 10. ∠DRC and ∠ARB
11. ∠ERD 12. interior ∠ERD 13. 11 14. ∠4, ∠5, ∠8 15. 120°
16. 122° 17. 8 18. right 19. equiangular 20. obtuse
21. 18° 22. 59° 23. 26° 24. 33° 25. 59°
26. 121° 27. 60 sq. in. 28. 140 sq. cm 29. 150 sq. cm 30. P = 36 in.; A = 60 sq. in.
31. No 32. m($\overline{YM}$) = 56; m($\overline{HM}$) = 70 33. V = 226.1 cubic ft.
34. A = 108.5 sq. ft. 35. C = 125.6 cm; A = 1,256 sq. cm 36. 2 cubic m
37. yes; start at either point G or D

CHAPTER 11

Answers to Supplementary Exercise 11.2

1. $\frac{1}{3}$ 2. $\frac{1}{5}$ 3. $\frac{21}{5}$ 4. $\frac{13}{22}$ 5. $\frac{4}{1}$ 6. $\frac{36}{55}$
7. $\frac{200 \text{ miles}}{3 \text{ hours}}$ 8. $\frac{2 \text{ children}}{1 \text{ family}}$ 9. $\frac{5 \text{ months}}{18 \text{ months}}$ 10. $\frac{10 \text{ dimes}}{1 \text{ dime}}$ 11. 56 12. 3
13. 36 14. 5.5 15. 16 16. 54 17. 36 18. 104
19. 104 20. $15.35 21. 340 miles 22. 66¢ 23. $82.50 24. 14 hours

Answers to Supplementary Exercise 11.3

1. $\frac{1}{2} = 0.5 = 50\%$ 2. $\frac{3}{10} = 0.3 = 30\%$ 3. $\frac{4}{5} = 0.8 = 80\%$
4. $\frac{1}{8} = 0.125 = 12.5\%$ 5. $\frac{5}{8} = 0.625 = 62.5\%$ 6. $1 = 1.00 = 100\%$
7. $\frac{1}{5} = 0.2 = 20\%$ 8. $\frac{1}{500} = 0.002 = 0.2\%$ 9. $\frac{1}{100} = 0.01 = 1\%$
10. $\frac{751}{250} = 3.004 = 300.4\%$ 11. $\frac{2}{25} = 0.08 = 8\%$ 12. $\frac{4}{125} = 0.032 = 3.2\%$
13. $\frac{11}{16} = 0.6875 = 68.75\%$ 14. $\frac{11}{25} = 0.44 = 44\%$ 15. $\frac{1}{10} = 0.10 = 10\%$

16. $\dfrac{1}{125} = 0.008 = 0.8\%$ 17. $\dfrac{26}{5} = 5.2 = 520\%$ 18. $\dfrac{17}{5} = 3.4 = 340\%$

19. $\dfrac{3}{8} = 0.375 = 37.5\%$ 20. $2\dfrac{7}{8} = 2.875 = 287.5\%$

Answers to Supplementary Exercises 11.4

1. $202.50   2. 13.64%   3. 31.58%   4. Markup: $350
                                        Percent markup: 70%
5. $331.03   6a. $21.46   6b. $140.66
7a. $93.80   7b. $562.80   8a. $136   8b. $561
9a. $26.10   9b. $136.10   10a. $1,020   10b. $5,270
11. Selling price: $61.60   12. Discount: $4.20   13. 17.3%   14. $260
    Markdown: $26.40   Selling price: $23.80   15. $26   16. 12.1%
17. $269 discounted 20%   18. $420   19. 2%

20. No, the markdown is $178. As a result, the sale price should be $889.99 − $178, which is equal to $711, not $719.89.

Answers to Supplementary Exercise 11.5

1. $148,500   2. $740   3. $1,023.75   4. $333   5. $1,434.38
6. $18.67   7. $96   8. $1,560   9. $4,166.67   10. $178,125

Answers to Supplementary Exercise 11.6

1a. $4,722   1b. $4,782   1c. $4,815   2a. $6,413   2b. $6,435   2c. $6,446
3a. $290   3b. $291.50   3c. $292.50   4a. $18,606   5a. $343,575
6a. $155,300   6b. $160,350   7a. $150,900   7b. $152,475   8a. $270,659.24   8b. $276,372.85

9. Howard's interest earned $8,000; Mike's interest earned $9,600; Difference = $1,600

10. $870.50   11. 12 years   12. $164

Answers to Supplementary Exercise 11.7

1. 7.12%   2. 11.46%   3. 8.8%   4. 9.84%   5. 17.23%

6. Bank A: 5.58%; Bank B: 5.38%; Bank A has higher rate

7. No, the correct effective rate is 9.99%

Answers to Supplementary Exercise 11.8

1. $111.98   2. $76.17   3. $465.70   4. $448.81   5. $223.13   6. $684.97
7. $179.01   8. $35,000

9. $32.24 each quarter, or $322.40 for the 10 year period.   10. $61.77

Answers to Supplementary Exercise 11.9

1a. $53.33   1b. 14.55%   2a. $56.83   2b. 23.04%   3. $42.48
4a. 30 payments   4b. ≈ $155.42   5. $384.30   6. $933   7. $224.41   8. $4,392

Answers to Chapter 11 Test

1. 35 2. 5 3. 5 4. 2.25 5. 16 6. 18.9
7. 0.5; 50% 8. 0.8; 80% 9. $\frac{1}{500}$; 0.2% 10. $2\frac{1}{4}$; 225% 11. $\frac{1}{50}$; 0.02 12. $\frac{1}{4}$; 0.25
13. 37.5% 14. $3.85 15. $218.75 16. $541 17. 9.20% 18. $62.61
19a. $30.33 19b. 22.2% 20a. $389.22 20b. $54,412.80

CHAPTER 12

Answers to Supplementary Exercise 12.4

1. 28 2. 10 3. 10 4. 5
5. 58 6. 9 7. 17 8. 12
9. 89 10. 78 11. 2 * (3 + 5) − 8 12. (X + Y) / 5
13. 6 / 2 + 4 * 8 14. 2 * X ↑ 2 + 3 15. 5 ↑ 3 + (3 ↑ 2 + 4) 16. (A − B) / (C + D)
17. 100 − 4 * (6 − 2) ↑ 2 18. B ↑ 2 − 4 * A * C
19. 5 ↑ 3 + (16 + 3) ↑ 2 / 10 ↑ 2
20. A * X ↑ 2 + B * X + C
21. 2 + 2 = 5 22. GOODBYE 23. BEGIN 24. 20.5
                                         -3

Answers to Supplementary Exercise 12.5

1. 5
2. 1 1
   2 3
   3 6
   4 10
   5 15
3. 1 1
   2 4
   3 9
4. 1
   ELEPHANT
   3
   4
5. 10 LET S = 0
   20 FOR X = 1 TO 5
   30 LET S = S + X
   40 PRINT X, S
   50 NEXT X
   60 END

Answers to Chapter 12 Test

1. e 2. j 3. h 4. b 5. i 6. f 7. c 8. g
9. a 10. d 11. 51 12. 34 13. 19 14. (5 − 3) / 2 + 18 / (3 * 2) + 5 15. c

16. 5
    8
    8
    8
    22
17. 80
18. 10 LET A = 85
    20 LET B = 90
    30 LET C = 84
    40 LET D = 76
    50 LET E = 70
    60 LET V = (A + B + C + D + E) / 5
    70 PRINT V
    80 END